Fundamental Principles of Radar

Fundamental Principles of Radar

Habibur Rahman

CRC Press
Taylor & Francis Group
Boca Raton London New York

CRC Press is an imprint of the
Taylor & Francis Group, an **informa** business

CRC Press
Taylor & Francis Group
6000 Broken Sound Parkway NW, Suite 300
Boca Raton, FL 33487-2742

© 2019 by Taylor & Francis Group, LLC
CRC Press is an imprint of Taylor & Francis Group, an Informa business

No claim to original U.S. Government works

Printed on acid-free paper

International Standard Book Number-13: 978-1-138-38779-9 (Hardback)

This book contains information obtained from authentic and highly regarded sources. Reasonable efforts have been made to publish reliable data and information, but the author and publisher cannot assume responsibility for the validity of all materials or the consequences of their use. The authors and publishers have attempted to trace the copyright holders of all material reproduced in this publication and apologize to copyright holders if permission to publish in this form has not been obtained. If any copyright material has not been acknowledged please write and let us know so we may rectify in any future reprint.

Except as permitted under U.S. Copyright Law, no part of this book may be reprinted, reproduced, transmitted, or utilized in any form by any electronic, mechanical, or other means, now known or hereafter invented, including photocopying, microfilming, and recording, or in any information storage or retrieval system, without written permission from the publishers.

For permission to photocopy or use material electronically from this work, please access www.copyright.com (http://www.copyright.com/) or contact the Copyright Clearance Center, Inc. (CCC), 222 Rosewood Drive, Danvers, MA 01923, 978-750-8400. CCC is a not-for-profit organization that provides licenses and registration for a variety of users. For organizations that have been granted a photocopy license by the CCC, a separate system of payment has been arranged.

Trademark Notice: Product or corporate names may be trademarks or registered trademarks, and are used only for identification and explanation without intent to infringe.

Library of Congress Cataloging-in-Publication Data

Names: Rahman, Habib, author.
Title: Fundamental principles of radar / authored by Habib Rahman.
Description: Boca Raton : Taylor & Francis, [2019] | Includes bibliographical references.
Identifiers: LCCN 2019001998| ISBN 9781138387799 (hardback) | ISBN 9780429279478 (ebook)
Subjects: LCSH: Radar.
Classification: LCC TK6575 .R324 2019 | DDC 621.3848—dc23
LC record available at https://lccn.loc.gov/2019001998

Visit the Taylor & Francis Web site at
http://www.taylorandfrancis.com

and the CRC Press Web site at
http://www.crcpress.com

eResource material is available for this title at https://www.crcpress.com/9781138387799.

Dedication

To our grandson ZADE
for the joy he brings to our lives, and

to our son TANVIR
who is a beacon to reconciliation and social justice.

Contents

Preface xiii
Author xvii

Chapter 1 Introduction 1

    1.1 Historical Notes 1
    1.2 Elementary Electromagnetic Waves 3
    1.3 Radar Principles 3
    1.4 Basic Radar Block Diagram and Operation 4
    1.5 Basic Elements of Radar System 5
    1.6 Types of Radar Systems 7
        1.6.1 Primary and Secondary Radar 7
        1.6.2 Monostatic, Bistatic, and MIMO Radars 7
        1.6.3 Search Radars 8
        1.6.4 Tracking Radars 8
        1.6.5 Classification by Frequency Band 8
        1.6.6 Classification by Waveforms and Pulse Rates 8
        1.6.7 Classification by Specific Applications 9
    Notes and References 10

Chapter 2 Radar Fundamentals 11

    2.1 Introduction 11
    2.2 Detection 11
    2.3 Range 11
        2.3.1 Range Ambiguity 12
        2.3.2 Range Resolution 13
    2.4 Velocity Measurement 14
    2.5 Target Location Measurement 18
    2.6 Signature Reflectivity and Imaging 18
    Problems 19

Chapter 3 Radar Equations 21

    3.1 The Radar Equations: An Introduction 21
    3.2 The Pulse Radar Equation 25
        3.2.1 SNR in Low PRF Radars 25
        3.2.2 SNR in High PRF Radars 27
    3.3 The Search Radar Equation 28
    3.4 The Tracking Radar Equation 30
    3.5 The Bistatic Radar Equation 30
    3.6 The Radar Equation with Pulse Compression 32
    3.7 Radar Equation with Jamming 33
        3.7.1 Self-Screening Jamming (SSJ) 34
        3.7.2 Stand-Off Jamming (SOJ) 36
    3.8 The Beacon Radar Equation 37
    Problems 40

vii

viii   Contents

Chapter 4   Targets and Interference 43

    4.1   Introduction 43
    4.2   Radar Cross Section (RCS) 43
        4.2.1   RCS of Sphere 44
        4.2.2   RCS of Cylinder 47
        4.2.3   RCS of Planar Surfaces 47
        4.2.4   RCS of Corner Reflectors 48
        4.2.5   RCS of a Dipole 49
        4.2.6   RCS of Complex Objects 50
    4.3   RCS Fluctuations and Statistical Models 51
    4.4   Radar Clutter 53
        4.4.1   Surface Clutter 54
        4.4.2   Volume Clutter 57
        4.4.3   Point (Discrete) Clutters 61
    4.5   Clutter Statistical Distributions 62
    4.6   Clutter Spectrum 64
    4.7   Radar Receiver Noise 64
        4.7.1   Noise Factor and Effective Temperature of a System 66
        4.7.2   Noise Temperature of Absorptive Network 68
        4.7.3   Overall Effective Temperature of a Composite System 69
    4.8   System Losses 70
        4.8.1   Design Losses 70
        4.8.2   Operational Losses 71
        4.8.3   Propagation Losses 72
    Problems 73
    References 73

Chapter 5   Propagation of Radar Waves 75

    5.1   Introduction 75
    5.2   Reflection Effects due to Multipath Phenomena 75
        5.2.1   Reflection from a Plane Flat Earth 75
        5.2.2   Reflection from a Smooth Spherical Earth 80
    5.3   Refraction of EM Waves 84
        5.3.1   Four-Thirds Earth Model 85
        5.3.2   Anomalous Propagation 86
    5.4   Diffraction of EM Wave 88
    5.5   Attenuation by Atmospheric Gases 89
    5.6   Ionospheric Attenuation 92
    Problems 93
    References 94

Chapter 6   Continuous Wave (CW) Radars 97

    6.1   Introduction 97
    6.2   Unmodulated Continuous Wave (CW) Radar 98
        6.2.1   Homodyne Receiver 98
        6.2.2   Doppler Shift and Range Rate 98
        6.2.3   Superheterodyne Receiver 100
        6.2.4   Sign of the Radial Velocity 101

Contents   ix

6.3	Frequency Modulated CW Radars	102
	6.3.1 Linear Frequency-Modulated CW Radars	102
	6.3.2 Sinusoidal Frequency-Modulated CW Radars	105
6.4	Multiple-Frequency CW Radar	107
6.5	Phase-Modulated CW Radar	108
Problems		110
References		111

Chapter 7   MTI and Pulse Doppler Radars 113

7.1	Introduction	113
7.2	Description of Operation	113
7.3	Single Delay-Line Canceler	117
7.4	Double Delay-Line Canceler	119
7.5	MTI Recursive Filter	120
7.6	MTI Nonrecursive Filter	121
7.7	Staggered Pulse Repetition Frequencies	122
7.8	Pulse Doppler Radar	124
7.9	Range and Doppler Ambiguities	125
	7.9.1 Resolution of Range Ambiguity	125
	7.9.2 Resolution of Doppler Ambiguity	127
Problems		129
References		129

Chapter 8   Pulse Compression Radar 131

8.1	Introduction	131
8.2	The Matched Filter	131
8.3	The Radar Ambiguity Function	135
8.4	Pulse Compression in Radars	138
8.5	Frequency Modulation in Pulse Compression	138
	8.5.1 LFM in Pulse Compression	138
	8.5.2 Frequency Stepping in Pulse Compression	143
	8.5.3 Active Processing in LFM Pulse Compression	144
8.6	Phase-Coded Modulation in Pulse Compression	149
	8.6.1 Barker Coding for Pulse Compression	150
	8.6.2 Frank Coding for Pulse Compression	151
Problems		153
References		153

Chapter 9   Synthetic Aperture Radars 155

9.1	Introduction	155
9.2	SAR History	155
9.3	SAR General Description	157
	9.3.1 Resolution along the LOS Axis	157
	9.3.2 Resolution along the XLOS Axis	159
9.4	SAR Signal Processing	162
9.5	Radar Equation of the SAR System	163
9.6	SAR System Design Consideration	165
Problems		166
References		167

Contents

Chapter 10 Tracking Radars 169

    10.1   Introduction 169
    10.2   Range Tracking 169
    10.3   Angle Tracking 173
           10.3.1 Sequential Lobing 173
           10.3.2 Conical-Scan Tracking 174
           10.3.3 Monopulse Tracking Radar 179
    10.4   Track-While-Scan (TWS) Radar 187
           10.4.1 Target Prediction and Smoothing 188
           10.4.2 The α-β Tracker 189
           10.4.3 The α-β-γ Tracker (Kalman Filtering) 191
    Problems 195
    References 195

Chapter 11 Aperture and Phased Array Antennas 197

    11.1   Introduction 197
    11.2   Fundamental Antenna Parameters 198
           11.2.1 Radiation Pattern 198
           11.2.2 Beamwidths 198
           11.2.3 Directive Gain, Directivity and Effective
                     Aperture Area 200
           11.2.4 Power Gain and Antenna Efficiency 201
           11.2.5 Polarization and Bandwidth 202
    11.3   Aperture Antennas 202
           11.3.1 Uniform Circular Aperture 203
            11.3.2 Tapered Circular Aperture 205
            11.3.3 Parabolic Reflector 206
    11.4   Phased Array Antennas 211
           11.4.1 The Array Factor of Linear Array 211
            11.4.2 N-Element Linear Array 212
           11.4.3 Planar Array 221
            11.4.4 Circular Array 223
            11.4.5 Conformal Array 225
    Problems 226
    References 227

Chapter 12 Radar Height Finder and Altimeter 229

    12.1   Introduction 229
    12.2   Derivation of Radar Heights 229
    12.3   Height-Finding Radars 234
           12.3.1 Nodding Height Finder 235
           12.3.2 V-Beam Radars 236
           12.3.3 Stacked-Beam Radar 237
           12.3.4 Frequency Scanned Radars 238
           12.3.5 Phased Array Height-Finding Radars 238
           12.3.6 Digital Beamforming Height-Finding Radar 239
           12.3.7 Interferometry Height-Finding Radar 239

Contents

xi

12.4	Radar Altimeters	240
	12.4.1 Beam-Limited Altimeter	240
	12.4.2 Pulse-Limited Altimeter	240
	12.4.3 SAR Altimeter	242
References		244

Chapter 13 Radar Electronic Warfare 245

13.1	Introduction	245
13.2	Historical Backgrounds	245
13.3	Electronic Warfare Definitions	246
13.4	Effects of EW Elements	247
13.5	Electronic Support Measures (ESM)	248
13.6	Electronic Countermeasures (ECM)	251
	13.6.1 Active ECM	251
	13.6.2 Passive ECM	253
13.7	Electronic Counter-Countermeasures (ECCM)	255
	13.7.1 Radar Parameter Management	255
	13.7.2 Signal Processing Techniques	260
	13.7.3 Radar Design Philosophy	260
	13.7.4 Operational Doctrines	260
References		261

Chapter 14 Over-the-Horizon Radar 263

14.1	Introduction	263
14.2	Historical Notes	263
14.3	Classification of OTHR Systems	265
	14.3.1 Skywave OTHR System	265
	14.3.2 Surface-Wave OTHR System	265
14.4	Ionospheric Effects on the OTHR System	266
14.5	Ray Path Trajectories	269
	14.5.1 The Thin Layer Model	269
	14.5.2 The Thick Layer Model	270
	14.5.3 The Multiple Layers Model	270
14.6	Principles of OTHR Systems	271
	14.6.1 OTHR Range Equation	271
	14.6.2 OTHR Waveforms	274
	14.6.3 Detection of Targets	275
14.7	Current OTHR Systems in Use	275
References		279

Chapter 15 Secondary Surveillance Radar 281

15.1	Introduction	281
15.2	Principles of SSR	282
15.3	Deficiencies in SSR	284
15.4	Solutions to SSR Deficiencies	284
	15.4.1 Improved Antennas	285
	15.4.2 Monopulse Techniques	285
	15.4.3 Sidelobe Suppressions in SSR	286

xii | Contents

15.5 Range Performance in SSR 287
    15.5.1 The Up-Link Range 287
    15.5.2 The Down-Link Range 288
References 289

Appendixes

Appendix A: Review of Deterministic Signals 291
    A.1 Fourier Series 291
        Complex Exponential Series 292
    A.2 Fourier Transforms 292
Appendix B: Review of Random Signals 293
    B.1 Probability Theory 293
    B.2 Random Variables, Distributions, and Densities 294
    B.3 Statistical Averages 295
        Moments 295
        Characteristic Function 296
    B.4 Random Processes 296
Appendix C: Z-Transforms Table 298
    Table of Common z-Transforms 298
Appendix D: Fourier Transforms Table 300
Appendix E: Probability Density Functions 301
Appendix F: Mathematical Formulas 303
    F.1 Trigonometric Identities 303
    F.2 Integrals 303
    F.3 Infinite Series Expansions 305
References 306

Answers to Selected Problems 307

Bibliography 311

Index 317

Preface

The important and fascinating topics of radar enjoy an extensive audience in industry and government, but they deserve more attention in undergraduate education to better prepare graduating engineers to meet the demands of modern mankind. Radar is not only one of the major applications of electronics and electromagnetic communications but also a mature scientific discipline with significant theoretical and mathematical foundations that may warrant an intellectual and educational challenge specifically in a classroom environment. Although there is a plethora of books on radar systems, all of them are written for the specialist or the advanced graduate students covering the subject of radar in varying degrees of detail and scope. Most of the existing books do not provide end-of-chapter problems to help students exercise the concepts developed, and cover a wider scope of topics usually not suitable for readers seeking an initial exposure to radar.

This book is an attempt to serve as a textbook for a course providing the first exposure to radar principles. It provides a broad concept underlying the basic principle of operations of most of the existing radar systems, and maintains a good balance of mathematical rigor suitable enough to convince readers without causing them to lose interest. The purpose of the book is to provide an extensive exposition of techniques currently being used for radar system design, analysis, and evaluation. It presents a comprehensive set of radar principles—including all features of modern radar applications—with their underlying derivations using the simplest mathematics possible. The coverage is limited basically to the main concepts of radar with an attempt to present them in a systematic and organized fashion. Topics are treated not as abstruse and esoteric to the point of incomprehensibility. Highly specialized topics that cover a wider scope with less detail are not chosen. Every attempt is made to distill the very complex and rich technology of radar into its fundamentals. Emphasis is given on clarity without sacrificing rigor and completeness. Abundant examples and exercises are chosen to reinforce the concepts presented, and to illustrate the radar applications. Hundreds of notes and references are provided leading to further reading and discussion of subtle and interesting concepts and applications. These notes and references provide an unusual degree of completeness for a textbook at this level, with interesting and sometimes thought-provoking content to make the subject even more appealing.

Fundamental Principles of Radar is designed for upper-division undergraduate and junior graduate engineering students, for those seeking their first exposure to radar principles through self-study, and for practicing radar engineers who need an in-house up-to-date reference text. The student using the text is assumed to have completed typical lower-division physics and mathematics. Some exposures to electromagnetic communications and probability theory are also useful. The entire book can be covered leisurely in two semesters. However, most of the material, with the exclusion of some chapters, can be squeezed into one semester. The material covered in the book is broad enough to satisfy a variety of backgrounds and interests, thereby allowing considerable flexibility in making up the course material.

Most chapters are meant to be self-contained and stand alone. The readers may be selective and still benefit. The book is organized into 15 chapters and 6 appendixes. Each chapter includes notes and references that provide suggestions for further readings. Chapters 1 through 5 constitute the first part of the book devoted to introductory material necessary for radar system analysis, design, and evaluation. Chapter 1 presents material to cover historical notes, elementary electromagnetics, and some preliminary descriptions of different radar types. Chapter 2 includes an overview of radar background information such as range, range resolution, Doppler frequency, and extraction of target information. Chapter 3 introduces the concept of radar equation, pointing out the hierarchy of concepts required for overall system evaluation, thus paving the way for detailed presentation of techniques contained in later chapters. It discusses first the general form of the radar equation and then expands and generalizes by filling in missing details, and finally concludes in a set of radar

xiii

equations that apply to different radars such as, low PRF and high PRF radars, radar beacon, bistatic radar, search radar, tracking radar, pulse compression radar, and radar jamming. Chapter 4 explores target scattering, examines various statistical models to describe radar cross section (RCS) fluctuations, and looks at the RCS characteristics of some simple and complex shapes. The necessary system descriptive parameters are also introduced in this chapter for the analysis and prediction of noise effects. In addition to noise figure, it is shown that a more useful measure is a system noise temperature. Some of the mechanisms by which the earth and atmosphere influence radar waves are addressed in Chapter 5, which is concerned with atmospheric effects on the propagation of radar signals. It describes phenomena such as diffraction due to the earth's curvature, atmospheric refraction, multipath reflection, anomalous propagation, and ionospheric attenuation.

The basic theory and modeling approaches developed in the first five chapters are applied to specific types of radar. Chapters 6 through 10 consider specific implementation and applications of radar. Chapter 6 starts with a discussion of continuous wave (CW) radar fundamentals, and then reviews the basis for CW radar detection, especially the Doppler effect. Also presented in this chapter are techniques for predicting CW radar range performance and several modulation techniques, including frequency modulation, phase modulation, and multiple frequency CW radar. Chapter 7 reviews the limitations of radar in moving target detection and then presents an overview of various moving target indication (MTI) and pulsed Doppler radar techniques. Chapter 8 presents the general time-frequency ambiguity function, which provides a composite measure of both resolution and ambiguity capability and limitations of given transmitter waveforms. It discusses key issues of radar resolution, and techniques for coding the transmitted waveform and detecting the reflected signal in a manner that produces signal compression on reception. It also describes in detail the techniques of frequency-modulation pulse compression, frequency stepping and phase modulation, and the application of pseudo-random sequences in signal encoding. It is shown that waveforms based on such sequences can have important properties from the point of view of resolution and ambiguity in certain practical situations. Chapter 9 describes the concepts of synthetic aperture radar (SAR) that allows achieving high-angular and cross-range resolution in long-range airborne search radar. The basic mathematical relationships that allow the design of an SAR and prediction of performance are developed in this chapter. In Chapter 10, there is a detailed presentation of both angle and range tracking configurations, including sequential lobing, conical scan, amplitude and phase comparison monopulse radars, and track-while-scan. Applications, advantages, and disadvantages of each technique are discussed.

A wide variety of antennas have been used in radar systems. The type of antenna selected for a certain application depends on many factors, including specifications, cost, and risk trade-offs. The detailed design of any antenna is typically quite involved. In Chapter 11, only two forms, aperture and array antennas, are presented. Discussions of each are sufficient to grasp the important concepts, many of which are useful in understanding other antennas. It considers reflector and phased-array antennas in detail, which are extremely important and practical devices for use in radar systems. It is shown that the phased-array antenna is particularly attractive for radar applications because of its inherent ability to steer a beam without the necessity of moving a large mechanical structure.

Chapters 12 through 15 present an account of radar topics that have been selected as having present-day relevance and future growth. These topics are chosen to cover materials that are at least vaguely familiar to most radar engineers. The treatment is descriptive and informative, rather than vigorously mathematical. Chapter 12 presents various types of height-finding radars and three-dimensional radar to measure the elevation angle and azimuth angle along with elevation angle, respectively. This chapter also treats many different types of radar altimeters, including beam-limited, pulse-limited, and SAR altimeters. Chapter 13 introduces the complete spectrum of electronic warfare (EW), including electronic countermeasures (ECM), electronic support measures (ESM), and electronic counter-countermeasures (ECCM). Each major element of radar is considered from an ECCM point of view. Only a generalized overview of EW is presented with nomenclature, definitions, and semantics rather than specific technical descriptions. Chapter 14 addresses

Preface

the concept of over-the-horizon radar (OTHR), and describes skywave and surface-wave OTHR systems, with some discussions of ionospheric effects and ray path trajectories, including thin-layer and thick-layer models. Chapter 15 presents the basic principle of secondary surveillance radar (SSR), and discusses systematic problems in SSR and their causes to clarify understanding of the modern solutions.

The book closes with the coverage of some ancillary review material, such as deterministic and random signals (Appendixes A and B), z-transform and Fourier transform tables (Appendixes C and D), probability density functions (Appendix E), and various useful mathematical formulas (Appendix F). Techniques of analyzing both deterministic and random signals are discussed. Also the effects of time variable and nonlinear devices are covered in some important cases. Appendix E provides some insights into probability density functions associated with noise and signal plus noise. It also discusses formalistic expositions of the probability function that arise in radar system theory.

As an aid to the instructor of the course, a detailed solutions manual for all problems at the end of the chapters in the book is available from the publisher.

Much of the work in preparation of this book was performed during a sabbatical leave from the Department of Electrical and Computer Engineering at Saint Louis University. I would like to thank Parks College of Engineering, Aviation and Technology, for providing enough support and services. I am forever grateful for the influence and instruction of my elder brother Professor Zaiz Uddin, one of the best educators of mathematics in the world. I wish to express gratitude to my family members for their unwavering support and encouragement that gave me a sound and healthy outlook on life. Perhaps the deepest appreciation and thanks must go to my wife Pauline for her love, understanding, and encouragement, which far exceeded that normally required in a usual family relationship. Her enthusiasm made it almost impossible to quit, and I appreciate all her patience given the time it took to write this book. I wish to express my profound appreciation and gratitude to my editor, Marc Gutierrez, for his strong support at various stages of writing this book. The help provided by Kari Budyk, editorial assistant and Edward Curtis, project editor at CRC, is gratefully appreciated. Finally, I would also wish to express my appreciation to all the people on the editorial team—particularly Deepti Kapoor, projcet manager at Cenveo Publisher Services, for providing excellent expertise and supports.

Habibur Rahman
Saint Louis, USA

Author

Habibur (Habib) Rahman is a professor of electrical and computer engineering at St. Louis University, where he teaches courses in the areas of electromagnetic fields and waves, radar systems, and satellite communications. He earned a PhD at Syracuse University, New York; MEng at McMaster University, Canada; and BScEng at the Bangladesh University of Engineering and Technology (BUET), all in electrical engineering.

Professor Rahman has extensive experience with education and has been professionally active in research in the general areas of electromagnetics, radar, satellite communications, and engineering education. He has published numerous research papers, mostly as the sole author. His main research focuses on the formulation of a simple and computationally efficient method to treat electromagnetic phenomenon and couplings in conducting cables or wires enclosed in a metallic cavity. He served as the chairman of the Department of Electrical Engineering from 1991 to 1998. He was instrumental in developing the course curricula to establish the electrical engineering program at St. Louis University. He has extensive academic, administrative, and professional service records, including successfully handling assessment and ABET matters.

He is a member of the American Society of Engineering Education (ASEE), a life senior member of the Institute of Electrical and Electronic Engineers (IEEE), and a member of the Electromagnetic Academy.

1 Introduction

"Them bats is smart. They use radar!"

—**David Letterman**

1.1 HISTORICAL NOTES

The principles of radar were formulated and preliminary techniques derived sometime between 1886 and the early 1920s. The earliest roots of radar can be associated with the theoretical work of the Scottish physicist James Maxwell who developed equations governing the behavior of *electromagnetic* (EM) wave propagation in 1864. Experimental work by the German physicist Heinrich Hertz confirmed Maxwell's theory in 1886, and demonstrated that radio waves could be reflected by physical objects known as targets. This fundamental fact forms the basis by which radar can perform the process of detection by sensing the presence of a reflected wave.

Some years later in 1903–1904, a German engineer, Christian Hulsmeyer, proposed the use of radio echoes in a detecting device designed to avoid collisions in marine navigation. In 1922 M. G. Marconi suggested angle-only radar for the avoidance of ship collision. The first successful radio range-finding experiment occurred in 1924, when the British physicist Sir Edward Victor Appleton used radio echoes to determine the height of the ionosphere, an ionized layer of the upper atmosphere that reflects longer radio waves. In 1935 the British physicist Robert Watson-Watt presented a paper entitled "The Detection of Aircraft by Radio Methods" to the Committee for the Scientific Survey of Air Defence, and demonstrated his ideas at Daventry. His idea was based on the bouncing of a radio wave against an object and measuring its travel to provide targeting information called radar detection and ranging.

The first practical radar system, known as *Chain Home* (CH) radar stations operating at 5 megahertz (MHz), was successfully demonstrated in 1937. By 1939 Britain had established a chain of radar stations along its southern and eastern coasts to detect aggressors in the air or on the sea. This system played a critical role in the Battle of Britain, pinpointing the location of German raids rather than having to search for enemy aircraft by patrolling. In the same year, two British scientists were responsible for the most important advance made in the technology of radar during World War II. The development of magnetron for high-power microwave radars was completed by a joint British/US program in 1939. Britain tended to have the best radar system during the early stages of the war. The physicist Henry Boot and the biophysicist John T. Randall invented an electron tube called the resonant-cavity magnetron. This type of tube can generate high-frequency radio pulses with large amounts of power, thus permitting the development of microwave radar, which operates in the very short wavelength band of less than 1 cm by using lasers. Microwave radar, also called *light detection and ranging* (LIDAR), is used in the present day for communications and to measure atmospheric pollution. During the Battle of Britain these stations could detect enemy aircraft at any time of day and in any weather conditions.

Radar was also used by ships and aircraft during the war. Germany was using radar by 1940, and deployed several different types of radars during World War II. Ground-based radars were used for air surveillance and height finding so as to perform *ground control of intercept* (GCI). The work in Japan was very slow but received impetus from disclosures by their German allies in 1940. The development of radar in the Soviet Union was quite similar to others. The United States had

1

a good radar system, which was able to predict the attack on Pearl Harbor an hour before it happened. By 1941 the United States successfully designed and implemented early warning radars that were extensively used during the Korean conflict. During World War II significant improvements and developments in radar applications were accomplished, including the establishment of radar as an indispensable tool for remote sensing of the enemy and the direction of the weapons toward that enemy. Airborne radars proved particularly useful in the war in finding Axis submarines and guiding bomb raids in inclement weather and at night. An unusual early air-to-air system was the *Lichtenstein* radar developed by the Germans for use on the BF-110 fighter.

Since World War II many laboratories were established to develop modern radar technologies, including *pulse compression radar, synthetic aperture radar* (SAR), *tracking radar* techniques, radar *electronic countermeasure* (ECM) and *electronic counter countermeasure* (ECCM), *phased array radar, moving target indication* (MTI) radar, *over-the-horizon radar* (OTHR), and *secondary surveillance radar* (SSR). The developments initiated during World War II were the monopulse tracking radar and the MTI radar, although many more years were required to bring these two radar techniques to full capability. New and better radar systems emerged during the 1950s with the development of the klystron amplifier, which provided a source of stable high power for very long-range radars. SAR and the airborne pulse radar first appeared in the early 1950s. With the introduction of digital signal processing and other state-of-the-art advances, SAR reached the high state of development that we see now. The concept of matched filter theory, developed during the 1950s, was successfully used in pulse compression radars to maximize the signal-to-noise ratio at the radar receiver output for good range resolution and radial velocity measurement.

Significant applications of the Doppler principle to radar began in the 1950s, and became vital in the operation of many radar systems, including continuous wave radars, MTI, and pulse Doppler radars. Police radar guns, SAR imaging radars, and color weather radars also utilized the Doppler principle to identify severe storms and weather conditions. The first large electronically steered phased-array radars and high frequency (HF) over-the-horizon radars were put into operation in the 1960s. In addition, radars for detecting ballistic missiles and satellites were introduced for the first time during this decade.

During the 1970s digital technology underwent a tremendous advance in signal and data processing, which made modern radars possible. Advances in airborne pulse Doppler radar greatly enhanced its ability to detect targets in the presence of ground clutter. The US Air Force's *airborne warning and control system* (AWACS) radar and military airborne-intercept radar were developed based on the pulse Doppler principle. During this decade radar also became feasible in spacecraft for remote sensing.

During the 1980s a series of developments in the production of phased-array radars made possible the air defense (the Patriot and Aegis systems), airborne bomber radar (B-1B aircraft), and ballistic missile detection (PAVE PAWS) systems. Solid-state technology and integrated microwave circuitry permitted new radar capabilities. Advances in remote sensing made it feasible to measure environmental effects such as wind speeds over the sea, ocean roughness, and ice conditions.

During the 1990s advances in computer technology allowed increased information about the nature of targets and the environment to be obtained from radar echoes. The introduction of Doppler weather radar systems utilized computer technology to measure the wind speed as well as the rate of precipitation. *Terminal Doppler weather radars* (TDWR) were installed at or near major airports to facilitate safe takeoff and landing. *High frequency* (HF) over-the-horizon radar systems were operated by several countries, primarily for the detection of aircraft at very long ranges.

The use of tactical ballistic missiles during the Persian Gulf War (1990–1991) brought back the need for radars for defense against such missiles. Russia and Israeli continually enhanced their powerful radar-based air-defense systems to stay engaged in tactical ballistic missiles. Radar systems such as altimeters, scatterometers, and imaging radar systems are now widely recognized as highly

Introduction 3

successful tools for earth observation from aircraft and satellites. More advances in digital technology in the area of signal and data processing led to further improvement and development of radars to meet the demands of mankind

1.2 ELEMENTARY ELECTROMAGNETIC WAVES

Classical electromagnetic theory is an accurate description of electric and magnetic phenomena governed by a compact and elegant set of coupled, partial differential equations. In honor of James Clerk Maxwell (1831–1879) the fundamental equations governing the behavior of electromagnetic fields are called Maxwell's equations,[1] which essentially provide the foundations of all electromagnetic phenomena and their applications. The most important consequence of Maxwell's equations is electromagnetic waves. Time-varying electric and magnetic fields are coupled to each other, resulting in an electromagnetic field. Under certain conditions, time-dependent electromagnetic fields produce electromagnetic waves that radiate from the source. Maxwell's equations are based on experimentally established facts, namely Coulomb's law,[2] Ampere's law,[3] Faraday's law,[4] and the principles of conservation of electric charges. Detailed discussion of experimental bases of Maxwell's equations is available elsewhere.

Electromagnetic wave theory as governed by Maxwell's equations has a vital role in the development of many practical tools of technological society including radio, television, radar, satellite communications, cellular phones, *global positioning system* (GPS), microwave heating, and X-ray imaging.

Electromagnetic waves, also known as radio waves or radar waves, occupy a portion of the electromagnetic spectrum including X-rays as well as very low frequencies. Radar waves travel at the speed of light in free space. Radar is an electromagnetic device used to remote-sense the position, velocity, and identifying characteristics of targets. This is accomplished by transmitting electromagnetic energy into a volume of space and sensing the energy reflected from objects (targets) within the search volume.

1.3 RADAR PRINCIPLES

Almost every species of bat is nocturnal, meaning that these are animals that are active at night. Bats are nocturnal and have to fly around any obstacles to find food. This is a difficult problem to deal with, and bats have come up with a solution by using a system of biological sonar called *echolocation*. Echolocation is a highly technical and interesting tactic. Echolocation—the active use of sonar along with special morphological and physiological adaptations—allows bats to "see" with sound.

Echolocation calls are usually ultrasonic ranging in frequency from 20 to 200 kilohertz (kHz), whereas human hearing normally tops out at around 20 kHz. In general, echolocation calls are characterized by their frequency, their intensity in decibels (dB), and their duration in milliseconds (ms). Most bats produce echolocation sounds by contracting their larynx. Bats are a fascinating group of animals that can use echolocation to navigate at night by making calls and listening to the sounds that bounce back. As such, the bat emits a loud sound that travels out into the environment until it hits something such as an insect, food, and any obstacles. When the sound hits, some is reflected back as an echo. The bat can hear the echo and use that to tell a lot about the obstacle, such as how far away it is and some fine details of its size and shape. This allows bats to capture flying insects by hunting them down like a radar-guided missile. Fortunately the sounds these bats make are generally not within the range of human hearing.

Bats seem to be the masters of using their sonar system to navigate in the presence of interference and to capture prey even in thick foliage. In both cases, the result should be confusion, but bats avoid the problem by producing echolocation calls with both *constant frequencies* (CF calls) and varying frequencies that are *frequency modulated* (FM calls). Most bats produce a complicated sequence of

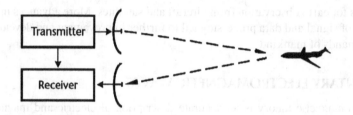

FIGURE 1.1 A radar counterpart of a bat's echolocation process.

calls, combining CF and FM components. Although low-frequency sound travels farther than high-frequency sound, calls at higher frequencies give the bats more detailed information such as size, range, position, speed, and direction of a prey's flight.

The ears and brain cells in bats are especially tuned to the frequencies of the sounds they emit and the echoes that result. The external structure of bats' ears also plays an important role in receiving echoes. The large variation in sizes, shapes, folds, and wrinkles are thought to aid in the reception and funneling of echoes and sounds emitted from prey. A concentration of receptor cells in their inner ear makes bats extremely sensitive to frequency changes. In other words, they appear to log a "mental fingerprint" of each broadcast and corresponding echo in their memories. That allows them to separate signals by slightly altering their frequencies. Professor James Simmons, from Brown University, found in his experimental study that bats change the frequency of their broadcasts by no more than 6 kHz to fly in clutter. Otherwise they would bump into trees and branches.

Figure 1.1 represents a radar counterpart of the bat's echolocation process. This radar transmits unmodulated EM waves and receives the returned echo to detect the presence of targets and to locate their position in space. And this is the first experimental radars developed. This type of radar is useful for detecting targets when the distance between the radar and the target is changing.

1.4 BASIC RADAR BLOCK DIAGRAM AND OPERATION

Radar is an electromagnetic device primarily used for detection and location of objects (targets). The word *radar* is an acronym derived from the phrase **RA**dio **D**etection **A**nd **R**anging. In general, radar operates by transmitting an electromagnetic energy—a pulse-modulated sine wave for example—and detects the returned echo to extract target information such as range, velocity, position, and reflectivity signature.

An elementary form of a typical radar system is described in the block diagram shown in Figure 1.2 that omits many details. It consists of a transmitter routing electromagnetic energy generated by an oscillator of some sort, to the antenna via a *duplexer*, a microwave device that

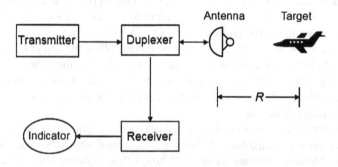

FIGURE 1.2 Simple radar block diagram.

Introduction

enables the same antenna to be used for both transmission and reception. The antenna serves as *transducer* to couple EM energy into free space, where it propagates at the speed of light (approximately 3×10^8 m/s). A portion of the backscattered energy, being reflected by an object, is intercepted by the radar antenna. The time delay, known as the round-trip time, between the transmission by the radar and the reception of the returned echo at a range R is obtained from the simple relationship:

$$t_d = \frac{2R}{c} \tag{1.1}$$

where c is the speed of light. The factor 2 appears in the numerator because of the two-way propagation of the radar signal. The receiver amplifies the weak signal received via the duplexer and translates it into a signal suitable, after required signal processing, for radar display or indicator. The angular position of the target is obtained from the direction of the arrival of the returned EM wavefront. If there is a relative motion between the radar and the target, the shift, known as Doppler shift, in the carrier frequency of the reflected wave is a measure of the relative velocity.

1.5 BASIC ELEMENTS OF RADAR SYSTEM

There are five basic elements in any functional pulse radar: a transmitter, a duplexer, an antenna, a receiver, and an indicator. This configuration is illustrated in Figure 1.3 with some associated elements. The functions of the major elements are described.

Transmitter: The transmitter is one of the basic elements of a radar system, and generates the *radio frequency* (RF) power signal to illuminate the target. The RF signal generated may be *continuous wave* (CW) or pulsed, and its amplitude and frequency are usually designed to meet the specific requirements of the radar system. There are two methods of generating RF power: in the power oscillator approach, the signal is generated at the required level suitable for applying directly to the antenna; and in the master oscillator approach, the oscillator generates a relatively low-power RF signal and then amplified to the appropriate level.

Duplexer: The duplexer enables the same antenna of monostatic[5] radars to be used for both transmission and reception. The duplexer consists of two gas discharge devices, one known as a *transmit-receive* (TR) and the other as an *anti-transmit-receive* (ATR). In this

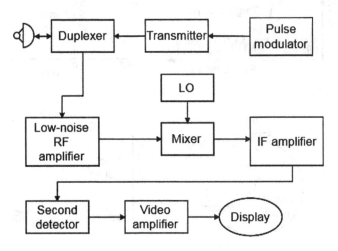

FIGURE 1.3 Pulse radar block diagram.

fast-acting RF electronic switch, the TR protects the receiver from a high-power radar signal during the transmission mode and the ATR directs the echo signal to the receiver during the reception mode. Solid-state ferrite circulators are also used.

Antenna: Antennas used in the radar system are mostly highly directional. A common form of radar antenna is a parabolic dish antenna fed from a feeding antenna at its focus. The beam may be scanned in space by mechanical pointing of the antenna. Phased-array antennas have also been used for radar. In a phased-array antenna the antenna beam is scanned electronically by introducing phases to the phase shifters connected to elements. The functions of the antenna are to concentrate the transmitting signal into a narrow beam in a single preferred direction, intercept the target echo signal from the same direction. Received weak energy is then sent via the transmission line to the receiver.

Receiver: The receiver is basically a superheterodyne type consisting of low-noise radio frequency amplifier, a mixer, an *intermediate frequency* (IF) amplifier, a video amplifier, and a display unit. The front-end RF amplifier is usually a parametric amplifier or a low-noise transistor. The mixer and *local oscillator* (LO) convert the RF signal into an IF signal having a center frequency 30 or 60 MHz and a bandwidth of the order of 1 MHz. The intermediate frequency amplifier is primarily designed as a matched filter[6] and the pulse modulation is extracted by the second detector. The demodulated signal is then amplified by the video amplifier to a level suitable for displaying in an indicator, usually a *cathode-ray tube* (CRT).

Display: Radar displays are used to visually present the information contained in the radar echo signal in a format suitable for operator interpretation and action. The display may usually be connected directly to the video output of the radar receiver. The CRT has been universally used as the radar display. A variety of display formats, some shown in Figure 1.4, are used for presenting target information on CRT displays for surveillance and tracking radars. Some of the commonly used radar displays are defined below according to Institute of Electrical and Electronic Engineers (IEEE) standard definitions:

a. ***A-scope:*** It provides target range and signal strength information. The vertical deflection is proportional to target echo strength and the horizontal coordinate is proportional to range. The A-scope display is used in instrumentation and data collection radars in which the antenna is not scanning.

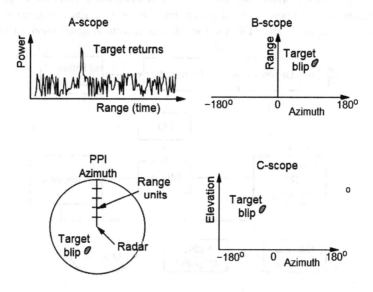

FIGURE 1.4 Some typical radar displays.

Introduction 7

b. ***B-scope:*** It is a rectangular display with azimuth angle indicated by the horizontal coordinate and the range indicated by the vertical coordinate. B-scope displays are often used in air-to-air and fire control applications, and are widely used in airborne radars and in short-range ground surveillance radars.

c. ***C-scope:*** It is a rectangular display with azimuth angle indicated by the horizontal coordinate and the elevation angle by the vertical coordinate.

d. ***E-scope:*** It is a rectangular display with range indicated by the horizontal coordinate and the elevation angle by the vertical coordinate.

e. ***Range-height-indicator (RHI):*** It is a rectangular display with range indicated by the horizontal coordinate and the height (altitude) by the vertical coordinate.

f. ***Plan-position-indicator (PPI):*** It is a circular or polar display with echo signals from reflecting objects indicated by the plan position, and range and azimuth angle indicated by the polar coordinate, forming a map-like display. In surveillance and color weather radar applications the PPI displays a full coverage of 360°.

1.6 TYPES OF RADAR SYSTEMS

There are many different types of radars. The way a particular radar is classified is based on the specific radar characteristics such as mission of the radar, antenna type, frequency band, the specific measurements it is to make, the waveforms it utilizes, the physical environment in which it must operate, and the interferers it is expected to encounter. Some of the numerous classifications are described.

1.6.1 PRIMARY AND SECONDARY RADAR

In primary radars the transmitter transmits the radar signal to illuminate the target, and the receiver receives the returned echo to extract information. The target acts as a passive reflector. The relationship between the echo strength and the range presents a major problem for long-range detection. It is also very difficult to determine the altitude of an aircraft accurately using primary radars. Secondary radars, on the other hand, overcome both problems by triggering an active response from the target, which needs to be cooperative. As a result of integration of the target, the information of both the altitude and identification are obtained. The power requirements of secondary radar transmitters are quite modest because of one-way transmission. The only problem with secondary radars is the expensive installation cost of electronic equipment.

1.6.2 MONOSTATIC, BISTATIC, AND MIMO RADARS

The physical configuration of the transmit and receive antennas also classify radars into monostatic, bistatic, and multistatic radars. Those radars where the same antenna is used for both transmission and reception or where the separate transmit and receive antennas are in essentially the same location are monostatic. On the other hand, in bistatic/multistatic radars the transmit and receive antennas are geographically placed in two/more different locations where the distance(s) of separation between them is (are) significant. It is useful in CW or *frequency-modulated continuous-wave* (FMCW) waveforms to achieve less spillover interference. The functions of the elements in bistatic/multistatic radars are the same as monostatic radars, with the major difference being in the absence of the duplexer. A synchronization link between the transmitter and the receiver is necessary to maximize the receiver knowledge of the transmitted signal. Frequency and phase reference synchronization can also be maintained.

A *Multiple Input Multiple Output* (MIMO) radar system, a subset of multistatic radar, is a system of multiple antennas in which each transmit antenna radiates an arbitrary waveform independently of the other transmitting antennas and each receiving antenna can receive these signals. Due to the different waveforms, the echo signals can be reassigned to the single transmitter.

1.6.3 Search Radars

Search radars continuously scan a volume of space to detect targets, and find their information, including range, angular position, and velocity. Depending on the radar design, different search patterns can be adopted, namely a two-dimensional fan beam, a stacked beam. In the case of a two-dimensional fan beam search radar, the beam width is wide enough to cover the desired search volume along that coordinate while steering in azimuth. In the case of the stacked-beam pattern, the beam is steered in azimuth and elevation.

1.6.4 Tracking Radars

Tracking radar systems are used to measure the position and velocity of a target, and provide data to predict the target path and future position. Tracking radars can track in range, in angle, in velocity, or in any combination thereof by using all or only a part of the available radar data. In general, it is the method by which angle tracking is accomplished that distinguishes the tracking radar from any other radar. It is customary to distinguish between a continuous tracking radar and a track-while-scan radar. The continuous tracking radar provides continuous tracking data on a particular target, while the track-while-scan radar provides sampled data on one or more targets to apply in sophisticated smoothing and prediction filters, namely Kalman[7] filters. Target tracking finds important applications in military radars as well as in most civilian radars, including in fire control, missile guidance, and airport traffic control for incoming and departing airplanes.

The tracking radar utilizes a very narrow pencil beam to find its target before it can track. To accomplish this efficiently a separate search radar is employed to provide rough information—coordinates of the target—necessary to the tracking radar. This search radar used for this purpose is sometimes known as acquisition radar. Then the tracking radar tracks the target by performing a limited search in the area of the designated target coordinates. The surveillance radar that provides tracking is sometimes called a track-while-scan radar, which is primarily used in airport traffic control and missile control.

1.6.5 Classification by Frequency Band

Although radars generally use the radio frequency portion (normally 220–35,000 MHz) of the electromagnetic spectrum, they can also function in any spectral region. The range of radar frequencies outside either end of the limit is shown in Figure 1.5. A few types of radar are found in lower band, except in cases where special propagation and target characteristics dictate their use. At the lower end, over-the-horizon radar uses the band from 6 to 30 MHz to utilize earth's ionosphere for reflecting a radar signal beyond the horizon.

Early in the development of radar, a letter code was employed to designate the radar frequency band. At the upper frequency end of the spectrum, L-band, S-band, C-band, and X-band radars are used where the size of the antenna constitutes a physical limitation. The other higher frequency bands (K_u, K, and K_a) suffer severe weather and atmospheric attenuation. Further information on common usage and applications of the radar frequency bands, as defined by IEEE, are summarized in Table 1.1.

1.6.6 Classification by Waveforms and Pulse Rates

Radars are classed by the type of waveforms they transmit and, in case of pulsed systems, the rate at which they transmit them. Continuous wave radars transmit and receive electromagnetic energy continuously using separate transmit and receive antennas. CW is a high-energy waveform, and is capable of very long-range applications. Unmodulated CW radars can accurately measure target radial velocity and angular position. Utilizing some form of time-variant modulation, the range information can be extracted.

Introduction

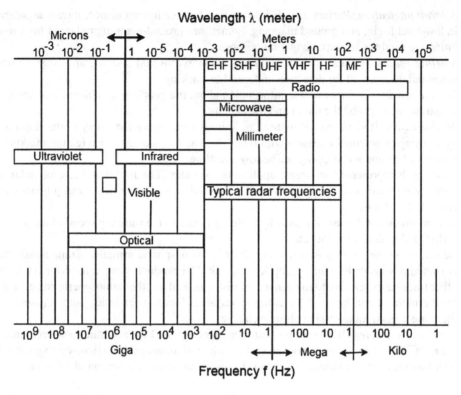

FIGURE 1.5 Radar frequencies and the electromagnetic spectrum.

1.6.7 Classification by Specific Applications

Radar systems may also be classified according to specific mission and application. A number of these are already mentioned. There remains a long list, some of them are briefly described here:

a. *Air traffic control (ATC) radars:* Radars are used by civilian and military aviation to maintain adequate separation between aircrafts, and to guide them safely into and out of the airport. High-resolution radars are widely used to monitor aircraft and large vehicular traffic at large airports.

TABLE 1.1
Radar Frequency Bands and Usages

Band	Frequency (GHz)	Usage
HF	0.003–0.03	OTH surveillance
VHF	0.03–0.3	Very long-range surveillance
UHF	0.3–1.0	Very long-range surveillance
L	1.0–2.0	Long-range military and air traffic control search
S	2.0–4.0	Moderate-range ground-based and shipboard search
C	4.0–8.0	Search and fire control radars, weather detection
X	8.0–12.5	Short-range tracking, missile guidance, marine radar
K_u	12.5–18.0	High-resolution mapping, satellite altimetry
K	18.0–26.5	Police speed-measuring, airport surface detection
K_a	26.0–40.0	Very high-resolution mapping, airport surveillance
MM-Wave	40.0–300.0	Laser range finders and optical targeting systems

b. *Aircraft navigation:* Radars are used on aircraft for weather avoidance, terrain avoidance in low-level flight, and ground mapping. Sometimes ground-mapping radars of high resolution are used for aircraft navigation purposes.

c. *Satellite radars:* Radars used by satellite for rendezvous and docking and large ground-based radars are used for detection and satellite tracking.

d. *Remote sensing:* Radars provide information about the geophysical objects, and are used in astronomy to probe the moon and the planets.

e. *Marine safety:* Radars are widely used by ships for enhancing the safety of the ship travel by warning of potential collision with other ships and for detecting navigation markers in poor weather, and for mapping the nearby coastline.

f. *Military:* It represents the largest application of radar. The traditional role of radar for military application has been for navigation, surveillance, and for the proper guidance and control of weapons.

g. *Law enforcement:* Radars are used by highway police to measure the speed of automobile traffic, and to detect the intruders.

h. *Global ozone monitoring experiment (GOME):* Atmospheric available ozone monitoring is sometimes needed for many applications. GOME products can be used for retrieving other trace gases relevant to the ozone chemistry as well as other atmospheric constituents. Furthermore, it can be used for climatic variable clouds, solar index, and aerosols. All these are crucial for assessing climate change.

i. *Wind scatterometer (WSC):* Wind scatterometers are used for accurate measurements of the radar backscatter from the ocean surface when illuminated by a microwave signal with a narrow spectral bandwidth to derive information on ocean surface wind velocity.

NOTES AND REFERENCES

1. J. C. Maxwell, *A Treatise in Electricity and Magnetism,* vol. 2 (Oxford: Clarendon Press, 1892), 247–262.
2. Coulomb's law states that electric charges attract or repel one another in a manner inversely proportional to the square of the distance between them.
3. Ampere's law states that current-carrying conductors create magnetic fields and exert forces on another, with the amplitude of the magnetic field depending on the inverse square of the distance.
4. Faraday's law states that magnetic fields that vary with time induce electromotive force or time-varying electric field.
5. Monostatic radar is the system where the same antenna is used for both transmission and reception, or where separate transmit and receive antennas are physically very close to each other (same location).
6. The matched filter was developed by Dwight North and Lamont Blake. The unique characteristic of the matched filter is that it achieves the maximum instantaneous signal-to-noise (SNR) at its output when the useful signal and the white noise are present at the input. This filter passes a large fraction of the signal energy while restricting the amount of noise energy passed.
7. The Kalman filter, known for some time, is inherently useful for application for the dynamical and maneuvering target. It can, in principle, utilize a wide variety of models for measurement of assumed white noise and trajectory disturbance. The Kalman filter is sophisticated and accurate, but is more costly to implement than the several other methods commonly used for smoothing and prediction of tracking data.

2 Radar Fundamentals

"It was a pity that there was no radar to guide one across the trackless seas of life. Every man had to find his own way, steered by some secret compass of the soul. And sometimes, late or early, the compass lost its power and spun aimlessly on its bearings."

—**Arthur C. Clarke**

2.1 INTRODUCTION

The target signal received by the radar is usually embedded in and corrupted by random extraneous unwanted signals such as thermal noise, electromagnetic interference, atmospheric noise, and electronic countermeasures. If the desired signal, after being processed by the receiver, is greater than the interfering signals, the information about the targets can be extracted. In simple forms the delay of the returned echo yields the information on the range. The frequency shift, also known as the Doppler frequency, provides information on the radial velocity. The antenna pointing direction yielding maximum strength of the returned signal provides the information on the azimuth and elevation of the target relative to the radar. This chapter discusses some accounts on this information derived from the radar signal.

2.2 DETECTION

The primary function of radar is to detect whether the target is present or not. The noise energy limits the detection of the weak signal received by the radar. In practice the receiver signal is corrupted by thermal noise and interference. Noise is mostly apparent in regions with a low signal level, such as the weak echo signal corrupting the desired signal in a radar receiver. The weakest signal the radar receiver can detect is called the *minimum detectable signal*, which is statistical in nature. Detection is based on establishing a threshold level above which a useful signal is assumed to be present, and is called threshold detection. The signal threshold level must be set at an appropriate level so that the envelope will not generally exceed the threshold if the noise alone is present or would exceed if a strong signal is present. Consider the output of a typical radar receiver as a function of time, as shown in Figure 2.1. A target is said to be detected, without any difficulties, if the receiver output exceeds the threshold such as at A and B. The noise voltage accompanying the signal at C is large enough to just exceed the threshold, but the target is not actually present resulting in what is called a *false alarm*. If the signal is at D, noise is not as large, and the resultant signal plus the noise does not exceed the threshold. In some cases the noise corrupting the signal will sometimes enhance the detection of weak signals. Thus the selection of the threshold level is a compromise between the probability of a miss and the probability of a false alarm.

2.3 RANGE

The measurement of range is one of radar's most important functions. The most common radar waveform used in range measurement is a train of rectangular pulse with narrow pulse width. This electromagnetic energy is transmitted by the pulsed radar, as shown Figure 2.2, toward the target, a portion of which returns to the radar receiver after being reflected from the target.

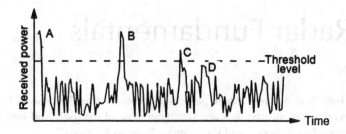

FIGURE 2.1 Typical envelope of the radar receiver.

The rectangular pulse represents transmission, and the shorter triangular pulse represents the signal echo from the target.

The range R of the target is obtained by measuring the time delay t_d taken by the pulse to travel a two-way path between the radar and the target. Since the electromagnetic energy propagates at the speed of light (3×10^8 m/s) in free space, the range, R, is

$$R = \frac{ct_d}{2}. \quad (2.1)$$

In general, a pulsed radar transmits and receives a train of pulses, as illustrated by Figure 2.3, where T is the *time period* and τ is the *pulse width*. The time period, T, is often referred to as *pulse repetition interval* (PRI). The inverse of the PRI is the PRF, denoted by f_r, and is given by $f_r = 1/T$. The duty factor, denoted by d_t, is expressed as $d_t = \tau/T$. The average transmitted power is then

$$P_{av} = P_t d_t = \frac{P_t \tau}{T} = P_t \tau f_r \quad (2.2)$$

where P_t is the peak power transmitted by radar. The pulse energy is

$$E_p = P_t \tau = P_{av} T = \frac{P_{av}}{f_r}. \quad (2.3)$$

2.3.1 Range Ambiguity

Once the transmitted pulse is emitted by the radar, a sufficient length of time must elapse to allow any echo signals to return and be detected before the next pulse may be transmitted. Therefore, the rate at which the pulses may be transmitted is determined by the longest range at which the targets are expected. If the pulse repetition frequency is too high, echo signals from some targets might arrive after the second pulse is transmitted, which results in an ambiguity in the measurement of range. Echoes that arrive after the next pulse is transmitted are called *second-time-around* echoes.

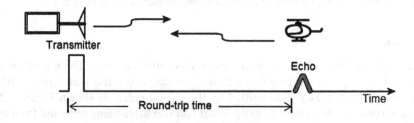

FIGURE 2.2 Radar range concept.

Radar Fundamentals

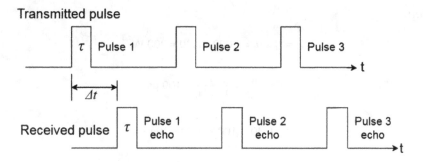

FIGURE 2.3 Transmitted and received pulse.

Such an echo would appear to be at a much shorter range than the actual, and could be misleading. The range beyond which targets appear as second-time-around echoes is called the *maximum unambiguous range*, denoted by R_u, which is given by

$$R_u = \frac{c}{2f_r}. \qquad (2.4)$$

2.3.2 Range Resolution

Range resolution is the radar metric that describes its ability to identify separate multiple targets at the same angular position, but at different ranges. Targets are resolved in four dimensions: range, horizontal cross-range, vertical cross-range, and Doppler shift. Range resolution depends largely on the radar's signal bandwidth. The wider bandwidth results in a better resolution. It can be easily demonstrated that the targets to be resolved must be separated by at least the range equivalent of the width of the processed echo pulse. Accordingly,

$$\Delta R = \frac{c\tau}{2} \qquad (2.5)$$

where
 ΔR = the range resolution, m
 c = velocity of propagation, m/s
 τ = width of the processed echo pulse, s.

The effective bandwidth of any pulse radar is approximately the reciprocal of the processed pulse width. The range resolution can also be expressed in terms of bandwidth as

$$\Delta R = \frac{c}{2B} \qquad (2.6)$$

where B is the transmitted matched filter bandwidth expressed in hertz (Hz).

Example 2.1

A low PRF radar with a pulse repetition frequency of 10 kHz radiates 10 kW of peak power. The duty cycle is 20%.

  a. Calculate the average transmitted power, pulse repletion interval, width of the pulse, and the pulse energy radiated in the first 20 ms.
  b. What is the maximum unambiguous range?
  c. Calculate the corresponding range resolution and the required bandwidth.

Solution:

a.
$$P_{av} = P_t \times d_t = 1{,}000 \times 0.20 = 200 \text{ W}$$

$$T = \frac{1}{PRF} = \frac{1}{10{,}000} = 100 \text{ } \mu s$$

$$\tau = 0.20 \times (100 \times 10^{-6}) = 20 \text{ } \mu s$$

$$E_p = P_t \tau = 10{,}000 \times (20 \times 10^{-6}) = 0.2 \text{ Joules}$$

b.
$$R_u = \frac{cT}{2} = \frac{(3 \times 10^8)(100 \times 10^{-6})}{2} = 15 \text{ km}$$

c.
$$\Delta R = \frac{c\tau}{2} = \frac{(3 \times 10^8)(20 \times 10^{-6})}{2} = 3000 \text{ m}$$

$$BW, \ B = \frac{1}{\tau} = \frac{1}{20 \times 10^{-6}} = 50 \text{ kHz}$$

2.4 VELOCITY MEASUREMENT

It is well-known in the fields of optics and acoustics that if the source of oscillation or the observer of oscillation is in motion, an apparent shift in frequency will result. This is the Doppler effect. When the radar signal reflects from the target, which is moving toward or away from the radar, its frequency is shifted causing the received echo to be at a slightly different frequency from the transmitted signal. This shift in frequency is known as Doppler shift. Radars use Doppler frequency shift to extract the radial velocity of targets and to distinguish targets occurring at the same time but moving at a different velocity. The latter method is used to discriminate moving targets from stationary targets or objects such as clutter. The Doppler shift is positive when caused by targets approaching the radar, and is negative when caused by targets receding (moving away) from the radar. Mathematically,

$$f_d = f_0' - f_0, \text{ Hz} \tag{2.7}$$

where
f_d = Doppler shift, Hz
f_0' = frequency of the returned echo, Hz
f_0 = frequency of the transmitted signal, Hz.

Consider a radar that transmits electromagnetic energy with a frequency f_0 and time period T_0, and a target that is moving at a constant radial velocity v_r toward the radar, as shown in Figure 2.4. Radial velocity is the component of the target velocity directly toward or away from the radar. The velocity v_r will be positive if the target is approaching (closing) the radar, and negative if the target is receding (opening). Assume, for this derivation, that the target is approaching. For the sake of simplicity, assume that $v = v_r$. Assume that the first crest of the wave (point A in Figure 2.4) is transmitted at time $t = t_0$ corresponding to where the target is located at $R = R_0$. Also assume that the second crest of the wave (point B in Figure 2.4) is transmitted at time $t = t_0 + T_0$ corresponding to where the

Radar Fundamentals

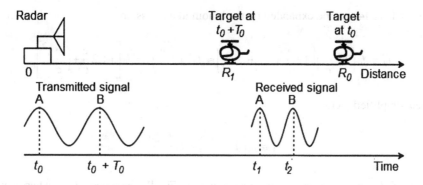

FIGURE 2.4 Transmitted and received waveforms for Doppler shift.

target is located at $R = R_1$, where T_0 is the period of the transmitted wave. The time Δt necessary for point A on the wave to arrive at the target is

$$\Delta t = \frac{R_0 - v\Delta t}{c} \tag{2.8}$$

where c is the velocity of propagation of the wave. Equation (2.8) can now be simplified to

$$\Delta t = \frac{R_0}{c+v}. \tag{2.9}$$

The total round-trip time for point A is, therefore, $2\,\Delta t$. Point A returns to the radar at time

$$t_1 = t_0 + \frac{2R_0}{c+v}. \tag{2.10}$$

Similarly, point B returns radar at time

$$t_2 = t_0 + T_0 + \frac{2R_1}{c+v}. \tag{2.11}$$

Thus the period of the receive wave is $(t_2 - t_1)$, which is written as

$$T_0' = T_0 - \frac{2(R_0 - R_1)}{c+v}. \tag{2.12}$$

Now, since $vT_0 = R_0 - R_1$, (2.12) can be simplified to

$$T_0' = T_0 \left(\frac{c-v}{c+v}\right) = T_0 \left(\frac{1-v/c}{1+v/c}\right) \tag{2.13}$$

The quantity $\left(\frac{c-v}{c+v}\right)$ is referred to as the *dilation factor*. Thus the received frequency of the echo signal is the reciprocal of the corresponding time period given by

$$f_0' = f_0 \left(\frac{1+v/c}{1-v/c}\right). \tag{2.14}$$

Since $v/c \ll 1$, (2.14) can be expanded by the binomial expression

$$f'_0 = f_0\left(1+\frac{v}{c}\right)\left(1+\frac{v}{c}+\frac{v^2}{c^2}+\ldots\ldots\right) \approx f_0\left(1+\frac{v}{c}\right)^2 = f_0\left[1+2\frac{v}{c}+\left(\frac{v}{c}\right)^2\right] \quad (2.15)$$

which when simplified yields

$$f'_0 = f_0\left(1+\frac{2v}{c}\right) = f_0 + \frac{2v}{\lambda} \quad (2.16)$$

where $\lambda = c/f_0$ has been applied to the above equation. Finally the received wave has been shifted from the transmitted wave by the Doppler shift, and is expressed as

$$f_d = \frac{2v}{\lambda}. \quad (2.17)$$

The value of f_d will be negative, if the target is receding from the radar. Reverting back to $v = v_r$ the Doppler shift is then

$$f_d = \frac{2v_r}{\lambda} = \frac{2v_r f_0}{c}. \quad (2.18)$$

If f_d is in hertz, v_r in km/hour, and λ in meters, we get

$$f_d = \frac{0.536 v_r}{\lambda}. \quad (2.19)$$

A plot of Doppler frequency given by (2.19) is shown in Figure 2.5. Thus the Doppler shift is directly proportional to the radial velocity of the target, and is a measure of the target velocity knowing the actual direction of the target.

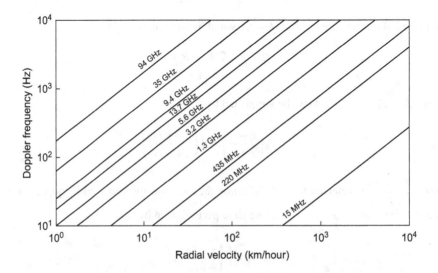

FIGURE 2.5 A plot of Doppler frequency as a function of radial velocity of the target for various frequencies.

Radar Fundamentals

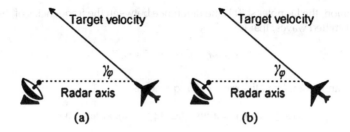

FIGURE 2.6 (a) The vertical and (b) the horizontal angles between the radar axis and the target velocity.

Accounting for the difference between the actual and radial velocities of the target, (2.18) must be modified to

$$f_d = \frac{2v}{\lambda}\cos\gamma \qquad (2.20)$$

where v is the velocity difference between the target and the radar, and γ is the total angle that the direction of the target makes with the radar line of sight (LOS). The general expression for f_d that accounts for the difference between target and radial velocities can be written as

$$f_d = \frac{2v}{\lambda}\cos\gamma_\theta\cos\gamma_\varphi \qquad (2.21)$$

where γ_θ and γ_φ are the vertical and the horizontal angles, respectively, between the radar axis and the target velocity, as illustrated in Figures 2.6.

Example 2.2

A K-band radar operating at 24.5 GHz is used by the police. A car is moving toward the stationary police radar at 126 km/hour. The car's velocity vector makes an angle of 30° with the axis of the radar.

a. Calculate Doppler shift.
b. Repeat part (a) if the car is moving away from the radar with the car's velocity vector at an angle of 45°.
c. What are the frequencies of the received echo signals?

Solution:

$$v = 126 \text{ km/hour} = 35 \text{ m/s}$$

a. Using (2.20) we get

$$f_d = \frac{2v}{\lambda}\cos\gamma$$

$$\text{where } \lambda = \frac{3\times10^8}{24.5\times10^9} = 0.01224 \text{ m}$$

$$\text{Thus } f_d = \frac{2(35)\cos 30°}{0.0124} = 4888.85 \text{ Hz}$$

b. Following the step in part (a) and assuming v as negative, we get

$$f_d = \frac{2(-35)\cos 45°}{0.0124} = -3991.73 \text{ Hz}.$$

c. By definition, the Doppler shift is the difference between the frequencies of the received and transmitted waves, that is

$$f_d = f_r - f_t \Rightarrow f_r = f_t + f_d.$$

Thus, for an inbound car, the received frequency is

$$f_r = 24.5 \times 10^9 + 4888.85 = 24.50000488885 \text{ GHz}$$

and, for an outbound car, the received frequency is

$$f_r = 24.5 \times 10^9 - 3991.73 = 24.49999600927 \text{ GHz}.$$

2.5 TARGET LOCATION MEASUREMENT

The position of a target is described in spherical coordinates (R, θ, ϕ), as shown in Figure 2.7. It is assumed that the radar and the target are located at $(0, 0, 0)$ and (R, θ, ϕ), respectively. In this figure R represents the range of the target, θ is the angle that R vector from the origin to the radar makes in reference to z-axis, and ϕ is the angle that the projection of R vector on the horizontal x-y plane makes in reference to x-axis, measured counterclockwise. The elevation angle is represented by $\gamma = 90° - \theta$. The process of measuring the distance of the target from the radar is a direct function of the round-trip time for the radar signal to return after being reflected from the target, and has already been discussed. Elevation angle is the angle measured between the antenna beam position and the local vertical, which is a line perpendicular to the local horizontal usually through the center of radiation of the antenna.

Azimuth angle is the antenna beam angle that it makes on the local horizontal plane—passing through the antenna's center of radiation and perpendicular to the earth's radius passing also through the same point—measured counterclockwise from the reference x-axis. Azimuth angle reference is based on various radar platforms, namely land-based, shipboard, and airborne.

2.6 SIGNATURE REFLECTIVITY AND IMAGING

Radar signatures relate to the determination of target physical characteristics, including *radar cross section* (RCS), Doppler spectrum, and imaging. RCS is basically the target cross section as seen by the radar. This is a quantitative measure of the ratio of the power density in the vector signal

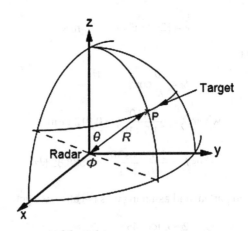

FIGURE 2.7 Target position in spherical coordinates.

Radar Fundamentals

scattered in the direction of the receiver to the power density of the radar signal incident upon the target. The returned energy, hence the RCS, is dependent on many parameters such as wavelength of the transmitted electromagnetic energy, target geometry, orientation, polarization combination, and reflectivity. Thus, in most cases of practical concern, RCS may vary over rather wide limits because of the changes in any of these parameters. RCS is usually extracted from the knowledge, a priori, of the radar design parameters, and by measuring the strength of the received echo and the target range.

Target identification can be achieved through appropriate filtering or correlation methods. Filtering identification uses the knowledge of Doppler spectral density of the returned signal from targets under consideration. Cross-correlation identification can also be used as an alternative to filtering. The signals are passed through a bank of correlators, which correlate data with various target signatures.

It may be possible to construct a radar imagery of the terrain if sufficient resolution is achieved. Sufficient resolution for imaging is found in a radar that uses a form of synthetic aperture that is much longer than the real antenna. In SAR high resolution is achieved by taking advantage of the motion of the vehicle carrying the radar to synthesize the effect of a large antenna aperture. The imaging of the terrain on the earth's surface by SAR to provide an imagery display can be applied to many remote sensing applications such as military reconnaissance, geological and mineral explorations, and the measurement of sea and ocean wave conditions.

PROBLEMS

2.1 The distance of the moon from the radar transmitter located on the surface of the earth is 3.84×10^8 m. Calculate the elapsed round-trip time of a radar signal transmitted from the radar antenna.

2.2 Consider a low PRF pulsed radar with a PRF of 1,500 pps and a bandwidth of 0.5 MHz. Calculate the maximum unambiguous range, pulse width, range resolution, and the duty factor.

2.3 A C-band radar transmits a peak power of 1 MW at a frequency of 5.5 GHz with the pulse length of 1μs and the PRF of 200 Hz.
   (a) Find the average transmitted power.
   (b) Find the bandwidth and the range resolution of the radar

2.4 A pulsed radar has a PRF of 1,500 pps and transmit rectangular pulse train of duration 15 μs.
   (a) What maximum range can a target have if no part of its first time around returned echo is to overlap any part of the transmitted pulse?
   (b) What is the minimum distance of separation so that targets can be identified?

2.5 The speed of a missile toward a radar is 300 m/s. Assume an X-band radar operating at a frequency of 12 GHz.
   (a) Calculate the exact Doppler frequency at the receiver.
   (b) Calculate the receiver Doppler frequency assuming $v_r \ll c$.

2.6 Assuming that the target is receding (opening), derive the expression for the Doppler shift.

2.7 For an approaching (closing) target whose radial velocity is 300 m/s, find the Doppler shift and the unambiguous range when the PRF is 8,000 pps and the transmitting frequency is 15 GHz.

2.8 Consider an S-band pulsed radar transmitting 250 kW of peak power with a pulse width of 1.5 μs and a PRF of 500 pps. The radar is transmitting at a frequency of 3,000 MHz.
   (a) Calculate the maximum unambiguous range of this radar, range resolution, and duty factor.
   (b) Calculate the average transmitted power and the energy radiated in the first 10 ms.
   (c) Calculate the Doppler shift for a target approaching the radar with a radial velocity of 30 m/s.

Fundamental Principles of Radar

2.9 Find an expression for the Doppler shift when the radial velocity of the approaching target is in km. Plot the Doppler frequency as a function of radial velocity of the target for various transmitted frequencies.

2.10 An L-band radar capable of transmitting a peak power of 500 W at 1,000 MHz is designed to provide an unambiguous range of 100 km and range resolution of at least 100 m.

    (a) Find the maximum required pulse width and the PRF.

    (b) Find the average transmitted power.

2.11 An L-band radar operates at a frequency of 1,500 MHz. Find the Doppler shift associated with an outbound target moving at the velocity of 100 m/s when the target velocity vector makes an angle of 45°, and 90° with the radar line of sight. In each case, calculate the time dilation factor.

3 Radar Equations

"Faith is like radar that sees through the fog—the reality of things at a distance that the human eye cannot see."

—**Corrie ten Boom**

3.1 THE RADAR EQUATIONS: AN INTRODUCTION

The radar range equation, or simply the radar equation, is the most descriptive mathematical relationship that accounts for not only the effects of each major parameter of the radar system but also those of the target, target background, propagation path, and the medium, It is useful to determine the maximum distance of the target from the radar. It can also serve both as a tool for understanding radar operation and as a basis for radar system analysis and design. In practice, however, knowledge of the radar and target parameters and propagation factor is not perfect; and many loss factors are included to make the results align with actual performance. The calculation or estimation of these loss terms is an important step in the evaluation of practical radar. Despite all these estimations, this section provides a simple form of radar equation—pertaining to the calculation of the amount of signal returned to the radar as a function of the known parameters and the range.

Assume that a transmitter generates P_t watts of power and delivers it to an isotopic antenna, which radiates the EM energy uniformly in all directions. This results in a constant power density p_t' at a distance R from the radar in a lossless propagation medium, equal to the transmitter power divided by the surface area $4\pi R^2$ of an imaginary sphere of radius R. This is expressed as

$$\hat{P}_t' = \frac{P_t}{4\pi R^2}. \tag{3.1}$$

Radar systems use directional antennas instead of isotropic antennas to concentrate the power density in a certain desired direction. The gain of an antenna is a measure of the increased power radiated in the desired direction as compared with the power that would have been radiated from an isotropic antenna. By definition, the gain of a directional antenna is the ratio of the intensity in the direction of maximum radiation to the radiation intensity from a lossless, isotropic antenna, both measured at a constant range with the same power input. Directional antennas used are characterized by the antenna gain G_t and antenna effective aperture A_t. The relationship between these parameters is expressed by

$$G_t = \frac{4\pi A_t}{\lambda^2} \tag{3.2}$$

where λ is the wavelength of the transmitted signal, and G_t and A_t are referred to the transmitting antenna. The effective aperture area A_t of an antenna is related to its actual area A as

$$A_t = \rho_a A, \\ 0 \le \rho_a \le 1 \tag{3.3}$$

where ρ_a is the antenna efficiency. In this derivation an antenna efficiency $\rho_a = 1$ is assumed for the sake of simplicity. The power density at the target at a distance R from the radar using a directive antenna with a transmitting gain G_t is then expressed as

$$\hat{p}_t = \frac{P_t G_t}{4\pi R^2}.$$ (3.4)

The radiated EM wave is intercepted by the target, and as a result, the incident energy will be scattered in various directions. The amount of energy scattered by the target is the function of many unknown parameters such as target size, orientation, physical shape, and material, which are all lumped together in one target-specific parameter known as RCS, denoted by σ. It is a characteristic of a particular target, and is a measure of its size as seen by the radar. The RCS describes the ability of the target to scatter the incident energy, and is defined as 4π times the ratio of the power per unit solid angle reflected by the target in the direction of the radar to the power density of the incident wave at the target. Mathematically,

$$\sigma = 4\pi \frac{P_b / 4\pi}{P_t / 4\pi R^2} = 4\pi R^2 \frac{P_b}{P_t}$$ (3.5)

where $P_b / 4\pi$ is the power per unit solid angle reflected back in the direction of the radar. Thus the total reflected power by the target is

$$P_\sigma = \frac{P_t G_t \sigma}{4\pi R^2}.$$ (3.6)

Considering the reflected power as a uniform source of power located at the target, the power density $\hat{p}_\sigma$ of the energy at the radar can be expressed as

$$\hat{p}_\sigma = \frac{P_t G_t \sigma}{4\pi R^2} \left(\frac{1}{4\pi R^2} \right) = \frac{P_t G_t \sigma}{\left(4\pi R^2 \right)^2}.$$ (3.7)

The radar antenna captures a portion of the returned echo. Assuming that the area capturing this energy is equal to the effective area of the receiving antenna A_r, the total power received by the radar is then given by

$$P_r = \frac{P_t G_t \sigma}{\left(4\pi R^2 \right)^2} A_r.$$ (3.8)

For the monostatic radar, the same antenna is used for both transmission and reception. Assuming $A_t = A_r = A_e$ and $G_t = G_r = G$ for the sake of simplicity, (3.8) is simplified to

$$P_r = \frac{P_t G \sigma}{\left(4\pi R^2 \right)^2} A_e.$$ (3.9)

Solving (3.2) for A_e and substituting this value in (3.9), the expression for P_r becomes

$$P_r = \frac{P_t G^2 \lambda^2 \sigma}{\left(4\pi \right)^3 R^4}.$$ (3.10)

This is a simple form of radar equation, and is useful for rough calculations. It does not predict the range performance of actual radar systems to a satisfactory degree of accuracy. This is because of the failure to account for the various losses occurring throughout the system, and the statistical or

Radar Equations

23

unpredictable nature of some of the parameters. For more realistic and accurate estimation of the radar range, one must consider many factors such as radar system losses including thermal and signal processing losses, propagation medium, and atmospheric noise.

Example 3.1

A spaceship in lunar orbit (orbit distance is 3.8×10^8 m) employs an L-band radar to transmits an average power of 3 MW isotopically with an antenna operating at 3000 MHz.

a. Find the power density at the surface of the earth.
b. Find the time it takes for the signal to travel from the spaceship to the earth.
c. Find the power recei ved by a receiver on the earth's surface with a 50-inch diameter dish antenna. Assume a lossless transmission.

Solution:

a. The power density is

$$\hat{p} = \frac{P_{av}}{4\pi R^2} = \frac{3 \times 10^6}{4\pi \left(3.8 \times 10^8\right)^2} \approx 1.653 \times 10^{-12} \text{ W/m}^2$$

b. The time from spaceship to the earth is

$$t = \frac{R}{c} = \frac{3.8 \times 10^8}{3.0 \times 10^8} = 1.27 \text{ s}$$

c. The power received by the receiver is obtained as follows:

$$\lambda = \frac{3 \times 10^8}{3 \times 10^9} = 0.1 \text{ m}$$

$$A_r = \pi \left(\frac{D}{2}\right)^2 = (3.1416)\left(\frac{50 \times 0.0254}{2}\right)^2 = 1.27 \text{ m}^2$$

$$G = \frac{4\pi A_r}{\lambda^2} = \frac{(4)(3.1416)(1.27)}{(0.1)^2} = 1595.93$$

$$P_r = \hat{p} A_r G_r = (1.653 \times 10^{-12})(1.27)(1595.93) = 3.35 \times 10^{-9} \text{ W}$$

For a realistic operational performance of the radar, the returned signal received by the radar receiver will be corrupted with receiver thermal noise as well as signal losses. Noise is the main factor limiting receiver sensitivity. Noise is an unwanted EM energy, which interferes with the ability of the receiver to detect the desired signal. The thermal noise power at the receiver input terminals is random in nature. This is generated by the random motion of electrons in ohmic portions of the receiver input stages. This is directly proportional to the temperature of ohmic portions of the circuit at the antenna output terminal (also called effective temperature) and the receiver bandwidth, and is given by

$$N_i = KT_e B \tag{3.11}$$

where $K = 1.38 \times 10^{-23}$ Joules/degree Kelvin is a Boltzmann's constant, and B is the receiver noise bandwidth at radar receiver input temperature T_e. If the temperature T_e is the standard room temperature T_0 (usually 290° Kelvin as the reference), the factor KT_0 becomes equal to 4×10^{-21}. The standard temperature T_0 is taken to be 290° Kelvin, according to the IEEE definition. The noise in a practical receiver is greater than the thermal noise. Thus the total noise at the output of the receiver may be considered to be equal to the thermal noise power from the ideal receiver multiplied by a factor called noise factor. The noise factor, denoted by F, is defined by the equation:

$$F = \frac{N_0}{KT_0BG_a} = \frac{\text{output noise of practical receiver}}{\text{output noise of ideal receiver at std. temp } T_0} \tag{3.12}$$

where G_a is the power gain of the receiver. The gain G_a is the ratio of the output signal S_0 to the input signal S_i, and the input noise $N_i = KT_0B$. equation (3.12) can be written as

$$F = \frac{S_i / N_i}{S_o / N_o} = \frac{(SNR)_i}{(SNR)_o} \tag{3.13}$$

where $(SNR)_i$ and $(SNR)_o$ represent the SNRs at the receiver input and output, respectively. S and N indicate the signal and noise power levels, and the subscripts i and o indicate input and output, respectively. Thus the input signal S_i corresponding to P_r used in (3.10) is written as

$$S_i = KT_0BF(S_o / N_o). \tag{3.14}$$

Elimination of P_r in (3.10), using (3.13), and proper rearrangement of terms yield

$$\left(\frac{S_o}{N_o} \right) = \frac{P_t G^2 \lambda^2 \sigma}{(4\pi)^3 R^4 FKT_0 B}. \tag{3.15}$$

Equation (3.15) was derived without incorporating signal losses. Three kinds of losses exist in radar systems: losses within the system, denoted by L_S, losses in propagation medium and path, denoted by L_P, and losses due to multiple signal paths from ground reflections, denoted L_R. All these losses are incorporated in one term, called L,—the product of L_S, L_P and L_R—resulting in a reduction of the output SNR. Accordingly, (3.15) must be modified to

$$\left(\frac{S_o}{N_o} \right) = \frac{P_t G^2 \lambda^2 \sigma}{(4\pi)^3 R^4 FKT_0 BL} \tag{3.16}$$

where L is the total signal losses consisting of all losses external and internal to the radar system. This expression for the output SNR is the most useful form, and is widely known as the radar equation. An alternative form of this equation to explicitly calculate the range can be written as

$$R = \left[\frac{P_t G^2 \lambda^2 \sigma}{(4\pi)^3 FKT_0 BL(S_o / N_o)} \right]^{1/4}. \tag{3.17}$$

The maximum detection range—distance beyond which the target cannot be detected—with some specified RCS is an important radar performance measure. It occurs when the received echo signal P_t (hence S_i) just equals the minimum detectable signal, the specification of which is sometimes difficult because of its statistical nature. The maximum radar range R_{max}, corresponding to the minimum value of the output SNR $(S_0 / N_0)_{min}$, is obtained from (3.16) as

$$R_{max} = \left[\frac{P_t G^2 \lambda^2 \sigma}{(4\pi)^3 FKT_0 BL(S_o / N_o)_{min}} \right]^{1/4}. \tag{3.18}$$

The dependence of R_{max} on minimum output SNR is shown in Figure 3.1.

Radar Equations

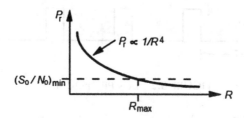

FIGURE 3.1 Received signal versus range.

It is customary to radar engineers to assume that maximum detection range corresponds to unity output SNR. The range for unity output SNR is denoted by R_0, and is written directly from (3.17) as

$$R_o^4 = \frac{P_t G^2 \lambda^2 \sigma}{(4\pi)^3 FKT_0 BL}. \tag{3.19}$$

In (3.19) all the terms, except for σ, are constants describing the radar system. Once a design is established for a given σ, R_0 can be easily determined by using (3.19) in (13.16). It follows then

$$\frac{S_o}{N_o} = \left(\frac{R_o}{R}\right)^4. \tag{3.20}$$

This indicates that the output SNR is inversely proportional to the fourth power of range R.

The calculated value of the SNR as found from (3.16) is just a mere prediction, and cannot be checked experimentally with any degree of accuracy. Precise values of some parameters are very difficult to determine in the radar equation. Nevertheless, to achieve the safest range performance, the design must be carried out conservatively with a reasonable safety factor.

3.2 THE PULSE RADAR EQUATION

Pulse radars are classified based on the pulse repletion frequency (PRF) being used in radar operations: low PRF and high PRF, also known as pulse Doppler. In radars where a single antenna is used for transmitting and receiving by utilizing a duplexer in the system, the receiver is off when the transmitter is on, and vice versa, thus generating a train of pulse-like waveform. The SNR ratio in low PRF radar is usually calculated on a per pulse basis, and this value is then multiplied by the number of pulses integrated to obtain the SNR for a given duration of target illumination. On the other hand, in high PRF pulse Doppler radars, the SNR is calculated by using the centerline power spectral density of the transmitter. The SNR calculations for both types of pulse radars will be demonstrated.

3.2.1 SNR in Low PRF Radars

In general, a pulse radar transmits and receives a train of pulses with pulse width τ and pulse period T, as illustrated in Figure 3.2, which actually depicts a low PRF pulse transmission. The transmit and receive duty factors d_t and d_r can be written as

$$d_t = \frac{\tau}{T} = \tau f_r \tag{3.21}$$

$$d_r = \frac{T-\tau}{T} = 1 - \tau f_r \cong 1.0 \tag{3.22}$$

FIGURE 3.2 Low PRF pulse train.

where f_r represents the pulse repetition frequency. Then the single pulse radar equation expressing the output SNR is given by

$$\left(\frac{S_o}{N_o}\right)_1 = \frac{P_t G^2 \lambda^2 \sigma}{(4\pi)^3 R^4 FKT_0 BL}. \tag{3.23}$$

If a number of pulses that strike the target are coherently integrated, the total SNR will be the product of the per-pulse SNR and the number of integrated pulses. Thus

$$\left(\frac{S_o}{N_o}\right)_n = \frac{P_t G^2 \lambda^2 \sigma n}{(4\pi)^3 R^4 FKT_0 BL} \tag{3.24}$$

where n is the number of pulses. If the target illumination time is denoted by T_i, the number n can be expressed as

$$n = T_i f_r. \tag{3.25}$$

Using $B = 1/\tau$ for a receiver matched filter in (3.23), we can modify (3.23) for a low PRF radar as

$$\left(\frac{S_o}{N_o}\right)_n = \frac{P_t G^2 \lambda^2 \sigma T_i f_r \tau}{(4\pi)^3 R^4 FKT_0 L} = \frac{(P_t \tau f_r) G^2 \lambda^2 \sigma T_i}{(4\pi)^3 R^4 FKT_0 L}. \tag{3.26}$$

Or

$$\left(\frac{S_o}{N_o}\right)_n = \frac{P_{av} G^2 \lambda^2 \sigma T_i}{(4\pi)^3 R^4 FKT_0 L} \tag{3.27}$$

where $P_{av} = P_t \tau f_r$ has been is applied.

Example 3.2

A low PRF C-band radar operating at 6.0 GHz and having a dish antenna with a 12-foot diameter transmits a pulse train with a peak power of 1 MW, the pulse width of 2 µs, and the PRF of 250 Hz. Assuming the equivalent noise temperature of 600 K, the total signal losses of 15 dB, and the target RCS of 10 m², calculate

    a. the maximum unambiguous range,
    b. the range R_0 at unity S/N, and
    c. the S/N of the target at half the maximum unambiguous range R_u.

Solution: We have

$$f_0 = 6.0 \text{ GHz} \Rightarrow \lambda_0 = \frac{3 \times 10^8}{6.0 \times 10^9} = 0.05 \text{ m}$$

$$L = 20 \text{ dB} = 100, \ B = \frac{1}{\tau} = \frac{1}{2 \times 10^{-6}} = 0.5 \text{ MHz}$$

Radar Equations

a. The maximum unambiguous range is

$$R_u = \frac{c}{2f_r} = \frac{3 \times 10^8}{(2)(250)} = 600 \text{ km}$$

b. The range at unity SNR is obtained as follows:

$$D = 12 \text{ feet} = 1.89 \text{ m}$$
$$A = \pi(D/2)^2 = \pi(1.89/2)^2 = 2.80 \text{ m}^2$$
$$G = \frac{4\pi A}{\lambda_0^2} = \frac{4\pi(2.80)}{(0.05)^2} = 14.10 \times 10^3 = 41.49 \text{ dB}$$

$$R_0^4 = \frac{P_t G^2 \lambda^2 \sigma}{(4\pi)^3 FKT_0 BL} = \frac{(10^6)(14.10 \times 10^3)^2(0.05)^2(10)}{(4\pi)^3(1)(1.38 \times 10^{-23})(600)(5 \times 10^5)(100)}$$
$$\Rightarrow R_0 \approx 278.596 \text{ km}$$

c. Using (3.20) gives

$$\left(\frac{S}{N}\right) = \left(\frac{R_0}{R_u/2}\right)^4 = \left(\frac{278.596 \times 10^3}{6 \times 10^5/2}\right)^4 = 0.7437 = -1.286 \text{ dB}$$

3.2.2 SNR in High PRF Radars

The transmitted signal, in case of high PRF radars, is also a periodic pulse train where the pulse width τ and the time period T are indicated in Figure 3.2. The value of the central line in the power spectral density is the dominant one, which is equal to the square of the transmit duty factor given by $(\tau/T)^2$. The SNR for a high PRF radar can now be written in terms of central line power spectral density and the transmit duty factor as follows:

$$\left(\frac{S_o}{N_o}\right) = \frac{P_t G^2 \lambda^2 \sigma d_t^2}{(4\pi)^3 R^4 FKT_0 BLd_r}. \tag{3.28}$$

In this case the receive duty factor can no longer be ignored of its significant magnitude compared to the transmit duty factor. For all practical purpose it can be assumed that $d_r \approx d_t = \tau f_r$. Additionally, the operating bandwidth is matched to the radar integration time T_i, and is given by $B = 1/T_i$. Incorporating all these in (3.28), the SNR of a high PRF radar is written as

$$\left(\frac{S_o}{N_o}\right) = \frac{P_t G^2 \lambda^2 \sigma T_i \tau f_r}{(4\pi)^3 R^4 FKT_0 L}. \tag{3.29}$$

It is noted that the average power $P_{av} = P_t \tau f_r$. By grouping terms for P_{av} in the numerator, (3.29) can be modified to

$$\left(\frac{S_o}{N_o}\right) = \frac{P_{av} T_i G^2 \lambda^2 \sigma}{(4\pi)^3 R^4 FKT_0 L}. \tag{3.30}$$

The energy product $P_{av} T_i$ indicates that the high PRF radar can enhance the detection performance by using relatively low power and longer target illumination time.

Example 3.3

A multimode high PRF airborne radar operating at 10.5 GHz transmits a peak power of 10 kW to detect a target with RCS of 2.0 m² at a distance of 100 km. The radar has the following specifications: pulse width $\tau = 1.2$ μs, PRF $f_r = 250,000$ pps, antenna gain $G = 35$ dB, system losses $L = 10$ dB, noise figure $F = 5$ dB, dwell interval $T_i = 2$ s. Find the single pulse S/N.

Solution: We have

$$P_{av} = P_t \tau f_r = (10^4)(1.2 \times 10^{-6})(2.5 \times 10^5) = 3 \times 10^3 \; W, \quad \lambda = 3 \times 10^8 / 10.5 \times 10^9 = 0.02857$$
$$G = 35 \; dB = 3162.28, \quad L = 10 \; dB = 10, \quad F = 5 \; dB = 3.162$$

Using (3.30) gives

$$S/N \frac{P_{av} T_i G^2 \lambda^2 \sigma}{(4\pi)^3 R^4 FKT_0 L}$$

$$S/N = \frac{(3 \times 10^3)(2)(3162.28)^2(0.0286)^2(2)}{(4\pi)^3(100 \times 10^3)^4(3.162)(1.38 \times 10^{-23})(290)(10)} \approx 3.90 \times 10^3 = 35.9 \; dB$$

3.3   THE SEARCH RADAR EQUATION

A search radar is used primarily for the detection of targets in a particular volume in space, while a surveillance radar provides for the maintenance of track files on the selected traffic. Thus the search radar is the more general term for the detection of targets within a volume scanned by the radar antenna.

Search radars continuously scan a selected volume in space searching for targets, and extract target information such as range, position, and velocity. Search radars must scan rapidly to cover their assigned scan volume in as short a time as possible, but also must scan slowly enough to obtain the required number of hits within the antenna beamwidth. The search radar equation is unique primarily because of the need to use the antenna scan rate, beamwidth, and the PRF to account for the required integration number. Assuming a uniform search, without overlap, of an assigned solid angle Ω the number of antenna beam positions n_p within the scan coverage is

$$n_p = \frac{\Omega}{\theta_a \theta_e} = \frac{\Omega}{\theta^2} \tag{3.31}$$

where θ_a and θ_e are the azimuth and elevation 3-dB beamwidths, respectively, and $\theta = \theta_a = \theta_e$ for circular aperture of symmetrical antennas with diameter D. Using the relation $\theta = \lambda / D$ in (3.31) gives

$$n_p = \frac{D^2}{\lambda^2} \Omega. \tag{3.32}$$

For a scan duration or radar frame time t_s the time on target can be expressed as

$$T_i = \frac{t_s}{n_p} = \frac{\lambda^2 t_s}{D^2 \Omega}. \tag{3.33}$$

Starting now with (3.30), and using (3.33), we get

$$\left(\frac{S_o}{N_o}\right) = \frac{P_{av} G^2 \lambda^2 \sigma}{(4\pi)^3 R^4 FKT_0 L} \frac{\lambda^2 t_s}{D^2 \Omega} \tag{3.34}$$

Radar Equations

and by using (3.2) for A in (3.34), the SNR for a search radar can be expressed as

$$\left(\frac{S_o}{N_o}\right) = \frac{P_{av}A\sigma}{16R^4 FKT_0 L}\frac{t_s}{\Omega}. \tag{3.35}$$

Equation (3.35) indicates the SNR for a search radar with a scan coverage of Ω and radar time on target t_s, and that it is independent of radar transmit frequency. The quantity $P_{av}A$ in (3.35) is known as the power aperture product, which is widely used to assess the radar ability to meet its search mission. Thus for a given power aperture product, (3.35) can be solved for the range R in terms radar parameters as follows:

$$R = \left[\frac{P_{av}A\sigma}{16FKT_0 L(S_o / N_o)}\frac{t_s}{\Omega}\right]^{1/4}. \tag{3.36}$$

Example 3.4

A millimeter wave (MMW) radar operating at 80 GHz transmits a peak power of 7.5 kW with a pulse width of 30 μs and a PRF of 10 kHz has the following parameters: noise factor $F = 5$ dB, antenna gain $G = 60$ dB, circular antenna with diameter $D = 0.3$ m, RCS $\sigma = 2$ m^2, radar scan rate $t_s = 2$ s, search volume $\Omega = 3.053$ steradian, system losses $L = 10$ dB. Calculate: (a) angular coverage and antenna beamwidth, (b) range resolution, (c) the unambiguous range, (d) dwell time on target (e) the number of antenna beam positions needed to cover the search volume, (f) the S / N as a function of the range, and (g) the range corresponding to $S / N = 10$ dB.

Solution: We have

$$F = 5 \text{ dB} = 3.162, \quad G = 60 \text{ dB} = 10^6, \quad L = 10 \text{ dB} = 10$$
$$\lambda = (3 \times 10^8 / 75 \times 10^9) = 0.004 \text{ m}$$

a. The angular and antenna beamwidth are obtained as follows:

$$\Omega = \frac{\theta^2}{(57.23)^2} = 3.053 \Rightarrow \theta = 100°$$

$$\theta_{3dB} = \frac{\lambda}{D} = \frac{0.004}{0.3} = 0.0133 \ rad = 0.764 \text{ degree}$$

b. The range resolution is

$$\Delta R = \frac{c\tau}{2} = \frac{(3 \times 10^8)(30 \times 10^{-6})}{2} = 4.5 \text{ km}$$

c. The unambiguous range is

$$R_u = \frac{c}{2f_r} = \frac{3 \times 10^8}{(2)(10^4)} = 15 \text{ km}$$

d. Using (3.33), we can find the dwell time is

$$T_i = \frac{t_s\lambda}{D^2\Omega} = \frac{(2)(.004)}{(0.3)^2(3.053)} \approx 30 \text{ ms}$$

e. Antenna beam positions needed:

$$n_p = \frac{\Omega}{\theta_{3dB}^2} = \frac{3.053}{(0.01333)^2} \approx 17,182$$

30   Fundamental Principles of Radar

f. The S/N as a function of the range by using (3.34) as follows:

$$S/N = \frac{P_{av} A \sigma t_s}{16 FKTL\Omega},$$

where $P_{av} = P_t \tau f_r = (7.5 \times 10^3)(30 \times 10^{-6})(10^4) = 2250$ W

$$A = \frac{G\lambda^2}{4\pi} = \frac{(10^6)(.004)^2}{4\pi} = 1.27 \text{ m}^2,$$

$$S/N = \frac{(2250)(1.27)(2)(2)}{R^4(16)(3.162)(1.38 \times 10^{-23})(290)(10)(3.053)} = \frac{1.85 \times 10^{21}}{R^4}$$

g.

$$\frac{S}{N} = \frac{1.85 \times 10^{21}}{R^4} \Rightarrow R^4 = \frac{1.85 \times 10^{21}}{10} \Rightarrow R = 116.62 \text{ km}$$

3.4   THE TRACKING RADAR EQUATION

The integration time, in tracking radars, is essentially infinite and irrelevant, since the antenna is always pointing at the target. If the integration is performed by the track filters or algorithms, the integration number or the total number of beams n_p, at least for the tracking channels, is given by

$$n_p = T_i f_r. \tag{3.37}$$

When the beams are coherently integrated with n_p, we get from (3.29) corresponding to a high PRF radar as

$$\left(\frac{S_o}{N_o}\right) = \frac{P_{av} G^2 \lambda^2 \sigma T_i}{(4\pi)^3 R^4 FKT_0 L}. \tag{3.38}$$

Because of the tracking filters, the time on target T_i is now given by the reciprocal of the tracking *servo bandwidth* B_s. It follows that

$$\left(\frac{S_o}{N_o}\right) = \frac{P_{av} G^2 \lambda^2 \sigma}{(4\pi)^3 R^4 FKT_0 LB_s}. \tag{3.39}$$

Equation (3.39) gives the SNR for a tracking radar with a tracking servo bandwidth of B_s Hz, and can be solved for the range R in terms of tracking radar parameters as follows

$$R = \left[\frac{P_{av} A \sigma}{16 FKT_0 L(S_o/N_o)B_s}\right]^{1/4} \tag{3.40}$$

where the relation $A = G\lambda^2/(4\pi)$ has been used.

3.5   THE BISTATIC RADAR EQUATION

A bistatic radar is one in which the transmit and receive antennas are separated by a considerable distance as shown in Figure 3.3, and the separation must be comparable with the target distance. The bistatic RCS must now be stated as a bistatic value σ_b, which measures the ability of the real target to scatter energy incident from the direction of the transmitter into the direction of the receiver. The bistatic angle, denoted by β, largely determines the size of the bistatic RCS: for a small angle

Radar Equations

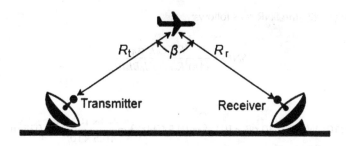

FIGURE 3.3 Geometry of a bistatic radar.

bistatic RCS is similar to monostatic RCS, and for a large bistatic angle approaching 180° the bistatic RCS is determined by

$$\sigma_b = \frac{4\pi A_e^2}{\lambda^2}. \tag{3.41}$$

The bistatic radar equation can be easily derived in a similar fashion as followed in the case of the monostatic radar. Thus the monostatic radar equation (3.16) for the SNR is modified to calculate the SNR of the bistatic radar

$$\left(\frac{S_o}{N_o}\right) = \frac{P_t G_t G_r \lambda^2 \sigma_b}{(4\pi)^3 R_t^2 R_r^2 F K T_0 B L_t L_r L_p} \tag{3.42}$$

where most of the symbols have their usual meanings with the exception of

- G_t = gain of the transmit antenna in the direction of the target
- G_r = gain of the receive antenna in the direction of the target
- R_t = range from the radar transmitter to the target
- R_r = range from the radar receiver to the target
- σ_b = bistatic radar cross section
- L_t = radar transmitter loss
- L_r = radar receiver loss
- L_p = medium propagation loss including transmit and receive paths

Example 3.5

An X-band aircraft radar and air-to-air semi-active missile receiver act as a bistatic radar system. The transmitter illuminates the target continuously. The bistatic radar has the following specifications:

Radar transmit power	5 kW
Radar antenna gain	30 dB
Radar Operating frequency	10.5 GHz
Radar system losses	0 dB
Missile effective aperture of antenna	0.02 m²
Missile bandwidth	600 Hz
Missile noise figure	10 dB
Missile overall losses	3 dB
Total medium propagation losses	0 dB

Find the SNR at the missile if a target with bistatic RCS $\sigma = 3$ m² is 40 km from the transmitter and 24 km from the missile.

Solution: Using (3.42), the SNR is as follows:

$$(SNR) = \frac{P_t G_t G_r \lambda^2 \sigma_b}{(4\pi)^3 R_t^2 R_r^2 FKT_0 BL_t L_r L_p}$$

where

$$P_t = 5 \times 10^3 \text{ W, } G_t = 30 \text{ dB} = 10^3, \, f_t = 10.5 \text{ GHz, } \lambda = \frac{3 \times 10^8}{10.5 \times 10^9} = 0.028 \text{ m}$$

$$G_r = \frac{4\pi(.02)}{(0.028)^2} = 320.57, \, B_r = 600 \text{ Hz, } F_r = 10 \text{ dB} = 10, \, L_t = 0 \text{ dB} = 1, \, L_r = 3 \text{ dB} = 2,$$

$$L_p = 0 \text{ dB} = 1, \, \sigma_b = 3 \text{ m}^2, \, R_t = 40 \text{ km, } R_r = 24 \text{ km}$$

$$(SNR) = \frac{(5 \times 10^3)(10^3)(320.57)(0.028)^2(3)}{(4\pi)^3 (40 \times 10^3)^2 (24 \times 10^3)^2 (10)(1.38 \times 10^{-23})(290)(600)(1)(2)(1)} = 16.33 \text{ dB}$$

This is not sufficient for the proper operation of the missile. However, by changing the location of the missile such that the distance of the missile from the target is decreased, results in the improvement of the SNR at the missile to a desired value for proper operation.

3.6 THE RADAR EQUATION WITH PULSE COMPRESSION

The average transmitted power of a given radar can be raised by increasing the pulse length of the transmitted energy. However, the increased pulse length results in an undesirable effect of reducing the range resolution capability of the radar. But the technique of pulse compression allows a radar to use a long pulse to achieve high radiated energy and simultaneously to obtain the range resolution of a short pulse. This is accomplished by employing the frequency or phase modulation to widen the signal bandwidth. The received signal is processed in a matched filter that compresses the long pulse to a duration that is inversely proportional to the transmitted bandwidth. Pulse compression is useful when the peak power required of a short-pulse radar cannot be achieved with practical transmitters. Pulse compression is a method for achieving most of the benefits of a short pulse while keeping the practical constraints of the peak power limitation. With pulse compression the classic detection/range resolution paradox is no longer prevailing.

The process of pulse compression correlates the received pulse with the delayed version of the transmitted one. The received signal can be considered as consisting of a group of subpulses, delayed in time, each having the width of the compressed pulse. The SNR, with pulse uncompressed, for a total of n number of pulses striking the target is given by (3.24), which is repeated here for convenience as

$$\left(\frac{S_o}{N_o}\right)_{pu} = \frac{P_t G^2 \lambda^2 \sigma n}{(4\pi)^3 R^4 FKT_0 BL} \tag{3.43}$$

where the symbols have their usual meanings. It must be mentioned that the pulses are integrated coherently. It is instructive and useful to consider the relationship between the compressed pulse width τ_c and bandwidth B of the matched filter, which is reciprocal of one another. Mathematically,

$$B \approx \frac{1}{\tau_c}. \tag{3.44}$$

Radar Equations

Eliminating B from (3.43) by (3.44), we get the expression for the SNR for each compressed pulse segment width τ_c in a pulse compression echo given by

$$\left(\frac{S_o}{N_o}\right)'_{pc} = \frac{P_t \tau_c G^2 \lambda^2 \sigma}{(4\pi)^3 R^4 F K T_0 L}. \tag{3.45}$$

It is noted that the SNR for the entire pulse results from the coherent summation of the number of compressed pulse widths in the received wave, and is then written directly from (3.45) as follows:

$$\left(\frac{S_o}{N_o}\right)_{pc} = \frac{P_t (CR) \tau_c G^2 \lambda^2 \sigma n}{(4\pi)^3 R^4 F K T_0 L} \tag{3.46}$$

where CR is the pulse compression ratio, which is a measure of the degree to which the pulse is compressed. It is defined as the ratio of the uncompressed pulse width to the compressed pulse width, that is, $CR = \tau / \tau_c$. Substitution of the expression for CR in (3.46) simplifies to

$$\left(\frac{S_o}{N_o}\right)_{pc} = \frac{P_t \tau G^2 \lambda^2 \sigma n}{(4\pi)^3 R^4 F K T_0 L}. \tag{3.47}$$

Recognizing that the bandwidth of the transmitter and receiver filters are fairly equal, and that $B = 1 / \tau_c$, (3.47) becomes

$$\left(\frac{S_o}{N_o}\right)_{pc} = \frac{P_t G^2 \lambda^2 \sigma n}{(4\pi)^3 R^4 F K T_0 (B / CR) L}. \tag{3.48}$$

Equations (3.43) and (3.47) indicate that, with pulse compression, detection can be maintained and range resolution improved by keeping the same transmitted pulse width but increasing the bandwidth. On the other hand, the range resolution can remain the same and the detection can be improved by increasing the transmitted pulse width and maintaining the same bandwidth.

3.7 RADAR EQUATION WITH JAMMING

All radars must be able to operate in a hostile electromagnetic environment where they may be subject to deliberate interference designed to degrade their performance. The various methods of interfering electronically with radars are usually referred to as ECM, sometimes called *jamming*. ECM defines actions taken to prevent or reduce the effectiveness of hostile electromagnetic radiation. It includes jamming, disrupting, and deceiving the sought-after information. Noise is a fundamental limitation to radar performance, and can be an effective countermeasure. Jamming using noise can be divided into general categories of *barrage jamming* and *repeater jamming*.

A jammer that radiates over a wide range of frequencies is called a barrage jammer. A barrage jammer attempts to increase the noise level across the entire radar operating bandwidth, and lowers the receiver SNR making it difficult to detect the desired targets. Increasing the noise level by external means, as with barrage jamming, further degrades sensitivity of the radar receiver. It becomes difficult, however, to keep the noise out when the jammer is located in the radar main beam. If the noise is illuminated by the antenna sidelobes, the entire display can be obliterated masking the target information. In general, radars operating at a higher frequency are less vulnerable to jamming, since it causes the jammer to spread its available power to a large bandwidth. A higher frequency also lowers the sidelobe levels making it even more difficult for sidelobe jamming.

34   Fundamental Principles of Radar

A jamming that retransmits the same signal a target would reradiate is called *repeater jamming*. A target can generate false target-like echoes by delaying the received radar signals and retransmitting at a slightly later time in order to confuse the operation of the radar. Delaying the retransmission causes the repeated signals to appear at a location different from that of the jammer. A repeater jammer can be very effective against unprepared radar systems. In general, the repeater jamming is easier to counter than a noise jamming. It is really difficult to design a repeater jammer that will mimic the exact radar signal.

Although mathematical analyses can be developed for a number of cases, only noise jamming methods such as *self-protection or self-screening jamming* (SSJ) and *stand-off jamming* (SOJ) will be considered.

3.7.1 Self-Screening Jamming (SSJ)

Self-screening jamming (SSJ) is the ECM that is carried by the vehicle the jammer is protecting. The relation derived here also holds approximately for escort jamming, where the vehicle carrying the jammer remains near the target it is protecting. It is assumed that the target to the radar is approximately the same as the range of the jammer from the radar. It is assumed that the power out of the jammer is constant. The power from the single hit on a target is repeated here as (3.10) with the receiver loss L included:

$$S = \frac{P_t G^2 \lambda^2 \sigma}{(4\pi)^3 R^4 L} \tag{3.49}$$

where the symbols have their usual meanings. The power received by the radar from the jammer is the result of one-way travel of the electromagnetic energy, and is expressed as

$$S_{ssj} = \left(\frac{P_j G_j}{4\pi R^2} \right) \left(\frac{\lambda^2 G}{4\pi} \right) \left(\frac{B_r}{B_j L_j} \right) \tag{3.50}$$

where P_j, G_j, B_j, L_j are, respectively, the peak power, antenna gain, effective bandwidth, and system losses, all related to the jammer. And B_r represents the radar's receive filters. The bandwidth ratio (B_r / B_j) is never greater than unity, because jammers are normally designed to operate against a wide variety of radar systems with different bandwidths. The signal-to-jamming ratio can now be computed using (3.49) and (3.50) as

$$\frac{S}{S_{ssj}} = \frac{P_t G \sigma B_j L_j}{4\pi P_j G_j R^2 B_r L}. \tag{3.51}$$

As observed in (3.51) there will be a certain range that exists where the signal and jamming have the same power. This particular range is called *cross-over* or *burn-through* range, denoted by R_{co}. The cross-over range is found by setting the left side of (3.51) equal to unity, and solving for range yields

$$(R_{co})_{ssj} = \left[\frac{P_t G \sigma B_j L_j}{4\pi P_j G_j B_r L} \right]^{1/2}. \tag{3.52}$$

The range at which the (S/S_{ssj}) ratio is sufficient for target detection in the presence of the interference is called the *detection range R_d*. This is obtained by solving (3.51), given by

$$(R_d)_{ssj} = \left[\frac{P_t G \sigma B_j L_j}{4\pi P_j G_j B_r L (S / S_{ssj})_{min}} \right]^{1/2}. \tag{3.53}$$

Radar Equations

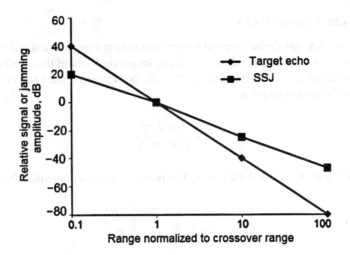

FIGURE 3.4 Target echo signal and self-screening jamming.

Dividing (3.53) by (3.52) yields

$$(R_d)_{ssj} = \frac{(R_{co})_{ssj}}{\sqrt{(S/S_{ssj})_{\min}}}. \tag{3.54}$$

The relationship between target echo signal, self-screening jamming power, and normalized range is plotted in Figure 3.4.

Example 3.6

A radar transmits a peak power of 60 kW with an antenna having a gain of 50 dB and pulse width of 2 μs is subject to interference by a self-screening jammer. Assume the following specifications: radar losses $L = 5$ dB, jammer power $P_j = 200$ W, and jammer antenna gain in the direction radar $G_j = 10$ dB, jammer bandwidth $B_j = 50$ MHz, and jammer loss $L_j = 0$ dB. Find (a) the crossover range for a 10-m² target and (b) the detection range if the required SNR for detection is 20 dB.

Solution: We have

$$G = 50 \text{ dB} = 10^5, \ L = 5 \text{ dB} = 3.162, \ G_j = 10 \text{ dB} = 10,$$

$$L_j = 0 \text{ dB} = 1, \ (S/S_{ssj})_{\min} = 20 \text{ dB} = 100, \ B = \frac{1}{2 \times 10^{-6}} = 5 \times 10^5 \text{ Hz}$$

a. Using (3.52), we can calculate the crossover range as

$$(R_{co})_{ssj} = \left[\frac{P_t G \sigma B_j L_j}{4\pi P_j G_j BL}\right]^{1/2} = \left[\frac{(60 \times 10^3)(10^5)(10)(50 \times 10^6)(1)}{(4\pi)(200)(10)(5 \times 10^5)(3.162)}\right]^{1/2}$$

$$\Rightarrow (R_{co})_{ssj} = 8.689 \text{ km}$$

b. Using (3.54), we can find the detection range as

$$(R_d)_{ssj} = \frac{(R_{co})_{ssj}}{\sqrt{(S/S_{ssj})_{\min}}} = \frac{8,689}{\sqrt{100}} = 868.9 \text{ m}$$

3.7.2 Stand-Off Jamming (SOJ)

A jammer that operates outside the range of normal defenses is known as a stand-off jammer. Thus in SOJ, the ECM is emitted from a vehicle at a range longer than that of the target being protected. The relationships developed here also assume that the power out of the jammer is constant. The power from a single hit on a target is

$$S = \frac{P_t G^2 \lambda^2 \sigma}{(4\pi)^3 R^4 L} \tag{3.55}$$

where the symbols have their usual meanings. The power received by the radar from the jammer is

$$S_{soj} = \left(\frac{P_j G_j}{4\pi R_j^2}\right)\left(\frac{\lambda^2 G'}{4\pi}\right)\left(\frac{B_r}{B_j L_j}\right) \tag{3.56}$$

where symbols have their usual meanings except for G'. The term G' represents the radar antenna gain in the direction of the jammer, and is normally considered to be the side lobe gain. Dividing (3.55) by (3.56) provides the signal-to-standoff jammer ratio as

$$\frac{S}{S_{soj}} = \frac{P_t G^2 R_j^2 \sigma B_j L_j}{4\pi P_j G_j G' R^4 B_r L}. \tag{3.57}$$

Again the detection occurs when the signal-to-jamming at the radar is sufficient for detection of the target in the presence of the interference. Accordingly, it follows that

$$(R_{co})_{soj} = \left[\frac{P_t G^2 R_j^2 \sigma B_j L_j}{4\pi P_j G_j G' B_r L}\right]^{1/4}, \tag{3.58}$$

and the detection range from (3.57) is

$$(R_d)_{soj} = \left[\frac{P_t G^2 R_j^2 \sigma B_j L_j}{4\pi P_j G_j G' B_r L (S / S_{soj})_{min}}\right]^{1/4}. \tag{3.59}$$

Dividing (3.59) by (3.58) yields

$$(R_d)_{soj} = \frac{(R_{co})_{soj}}{\sqrt[4]{(S / S_{soj})_{min}}}. \tag{3.60}$$

The relationship between target echo signal, stand-off jamming (SOJ) power, and normalized range is plotted in Figure 3.5.

Example 3.7

The radar in Example 3.6 with gain 30 dB is subject to interference by a stand-off jammer, and is transmitting 7.5 kW with a jammer antenna gain in the direction of radar of 20 dB from a range of 20 km. Assume that the radar views the jammer through the main lobe only. The other jammer specifications remain the same. Find (a) the crossover range for a 10-m² target, (b) the detection range if the required SNR for detection is 20 dB, and (c) repeat parts (a) and (b) if the radar views the jammer through 10 dB sidelobes.

Radar Equations

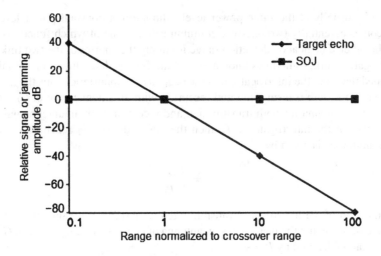

FIGURE 3.5 Target echo signal and stand-off jamming.

Solution: We have

$G = 30$ dB $= 10^3$, $L = 5$ dB $= 3.162$, $G_j = 20$ dB $= 100$, $B_j = 50 \times 10^6$, $G' = 0$ dB $= 1$

$L_j = 0$ dB $= 1$, $(S/S_{ssj})_{min} = 20$ dB $= 100$, $B = \dfrac{1}{2 \times 10^{-6}} = 5 \times 10^5$ Hz

a. Using (3.58), we can find the crossover range as

$$(R_{co})_{soj} = \left[\frac{P_t G^2 R_j^2 \sigma B_j L_j}{4\pi P_j G_j G' B L}\right]^{1/4} = \left[\frac{(7.5 \times 10^3)(10^3)^2 (20 \times 10^3)^2 (50 \times 10^6)(1)}{(4\pi)(200)(10^2)(1)(5 \times 10^5)(3.162)}\right]^{1/4}$$

$$\Rightarrow (R_{co})_{soj} = 4.42 \text{ km}$$

b. Using (3.60), we can find the detection range as

$$(R_d)_{soj} = \frac{(R_{co})_{soj}}{\sqrt[4]{(S/S_{soj})_{min}}} = \frac{4{,}407.88}{\sqrt[4]{100}} = \frac{4{,}407.88}{3.162} \approx 1.4 \text{ km}$$

c. In this case, $G' = 10$ dB $= 10$ leading to

$$(R_{co})_{soj} = \left[\frac{P_t G^2 R_j^2 \sigma B_j L_j}{4\pi P_j G_j G' B L}\right]^{1/4} = \left[\frac{(7.5 \times 10^3)(10^3)^2 (20 \times 10^3)^2 (50 \times 10^6)(1)}{(4\pi)(200)(10^2)(10)(5 \times 10^5)(3.162)}\right]^{1/4}$$

$$\Rightarrow (R_{co})_{soj} = 2.48 \text{ km}$$

$$(R_d)_{soj} = \frac{(R_{co})_{soj}}{\sqrt[4]{(S/S_{soj})_{min}}} = \frac{2{,}478.73}{\sqrt[4]{100}} = 783.84 \text{ m}$$

3.8 THE BEACON RADAR EQUATION

SSR, also known as beacon radar, has a number of similarities with primary surveillance radar. In SSR the interrogator send out a signal to transponder and receives its reply. Unlike primary radar, it is essentially a cooperative system. When an interrogation is detected, a transponder in the beacon always transmits usually at the same power level. Maintaining constant power level, the beacon equation becomes essentially two one-way communications, one of which functions over a longer range than the other. The beacon detection range is simply the shorter of the two links.

The interrogation link, sometimes known as up-link, is considered first. For the sake of simplicity it is assumed that both the interrogator and the beacon are monostatic. Thus the interrogator uses the same antenna for both transmission and reception with an antenna gain of G_i, and the beacon also uses the same antenna for both transmission and reception with an antenna gain of G_b. If the transmitter power of the interrogator is P_i, then the power density $\tilde{p}_b$ reaching the beacon at the center of the main lobe is given by

$$\tilde{p}_b = \frac{P_i G_i}{4\pi R_i^2} \tag{3.61}$$

where G_i is the gain of the interrogator transmit antenna, and R_i is the range from the radar to the beacon. Now, assuming that the beacon has an antenna of aperture area A_b and gain G_b, the received power by the beacon denoted by P_{rb}, is

$$P_{rb} = \frac{P_i G_i A_b}{4\pi R_i^2} = \frac{P_i G_i G_b \lambda_i^2}{16\pi^2 R_i^2} \tag{3.62}$$

where λ_i is the interrogation wavelength. If losses are all combined together by a single term L, (3.62) becomes

$$P_{rb} = \frac{P_i G_i G_b \lambda_i^2}{(4\pi)^2 R_i^2 L}. \tag{3.63}$$

The SNR of the interrogator in the beacon is obtained from (3.63) and the beacon receiver noise $N_b = F_b k T_0 B_i$ where the symbols have their usual meanings except for B_i representing the matched filter bandwidth of the interrogator signal, and F_b representing the noise factor of the beacon receiver. It, therefore, follows that the SNR of the interrogation at the beacon in up-link is

$$\frac{S_i}{N_b} = \frac{P_{rb}}{N_b} = \frac{P_i G_i G_b \lambda_i^2}{(4\pi)^2 R_i^2 F_b k T_0 B_i L}. \tag{3.64}$$

In (3.64) the beacon processing gain to noise has been assumed equal to unity. Solving (3.64) for range, the maximum range at which the beacon will be successfully interrogated is

$$(R_i)_{\max} = \left[\frac{P_i G_i G_b \lambda_i^2}{(4\pi)^2 F_b k T_0 B_i L (S_i / N_b)_{\min}} \right]^{1/2} \tag{3.65}$$

where $(R_i)_{\max}$ is the maximum range at which the beacon can be interrogated and $(S_i / N_b)_{\min}$ is the minimum SNR required for interrogation.

Now turning to the down-link, the power received by the interrogator receiver from the beacon is expressed as

$$P_{ri} = \frac{P_b G_b G_i \lambda_b^2}{(4\pi)^2 R_b^2 L} \tag{3.66}$$

where the symbols have their usual meanings except that λ_b is the beacon reply wavelength, R_b is the range from the beacon to the radar, and P_b is the beacon transmitter power. Following the same steps as in up-link, the SNR of the beacon at the interrogator in down-link is

$$\frac{S_b}{N_i} = \frac{P_{ri}}{N_i} = \frac{P_b G_i G_b \lambda_b^2}{(4\pi)^2 R_b^2 F_i k T_0 B_b L}. \tag{3.67}$$

Radar Equations   39

In (3.67) the interrogator processing gain to noise has been assumed equal to unity. Solving (3.67) for the maximum range $(R_b)_{max}$, corresponding to $(S_b / N_i)_{min}$ for reply at which the beacon reply will be successfully detected by the interrogator, gives

$$(R_b)_{max} = \left[\frac{P_b G_i G_b \lambda_b^2}{(4\pi)^2 F_i K T_0 B_b L (S_b / N_i)_{min}} \right]^{1/2}.$$

(3.68)

The maximum range at which the interrogator produces a useful reply is, therefore, the shorter of the two ranges found from (3.65) and (3.68).

Example 3.8

Consider a C-band beacon radar system with the following parameters:

Transmitter peak power	1.5 MW
Transmitter frequency	5.6 GHz
Antenna gain	40 dB
Transmit pulse width	0.2 µs
Receiver bandwidth	1.2 MHz
Receiver noise figure	3 dB
Receiver system loss	2 dB
Minimum S/N for interrogation $(S_i / N_b)_{min}$	20 dB
Beacon peak power	2 kW
Beacon frequency	5.4 GHz
Beacon pulse width	1 µs
Beacon antenna gain	10 dB
Beacon system loss	1 dB
Beacon receiver bandwidth	3.0 MHz
Beacon noise figure	10 dB
Minimum S/N for operation $(S_b / N_i)_{min}$	25 dB

Assume that the beacon and interrogator processing gains to noise are both unity, and propagation path loss each way and the ground plane loss are 2 dB and 0 dB, respectively. Find (a) the maximum interrogation range and maximum reply range and (b) the maximum range for which this beacon system can be used.

Solution:

$G_i = 40 \text{ dB} = 10^4$, $G_b = 10 \text{ dB} = 10$, $F_i = 3 \text{ dB} = 2$, $F_b = 10 \text{ dB} = 10$, $L = 2 \text{ dB} = 1.58$,

$(S_i / N_b)_{min} = 23 \text{ dB} = 316.23$, $(S_b / N_i)_{min} = 20 \text{ dB} = 100$, $KT_0 = (1.38 \times 10^{-23})(290) = 4 \times 10^{-21}$,

$$\lambda_i = \frac{3 \times 10^8}{5.6 \times 10^9} = 0.0536 \text{ m}, \quad \lambda_b = \frac{3 \times 10^8}{5.4 \times 10^9} = 0.0556 \text{ m}$$

a. Using (4.65), we can find the maximum interrogation range as

$$(R_i)_{max} = \left[\frac{P_i G_i G_b \lambda_i^2}{(4\pi)^2 F_b (KT_0) B_i L (S_i / N_b)_{min}} \right]^{1/2}$$

$$= \left[\frac{(1.5 \times 10^6)(10^4)(10)(0.0536)^2}{(4\pi)^2 (10)(4 \times 10^{-21})(1.2 \times 10^6)(1.58)(316.23)} \right]^{1/2}$$

$$\Rightarrow (R_i)_{max} = 337,326 \text{ km}$$

40 Fundamental Principles of Radar

Using (4.67), we can find the maximum reply range as

$$(R_b)_{max} = \left[\frac{P_b G_i G_b \lambda_b^2}{(4\pi)^2 F_i (KT_0) B_b L (S_b / N_i)_{min}} \right]^{1/2}$$

$$= \left[\frac{(2 \times 10^3)(10^4)(10)(0.0556)^2}{(4\pi)^2 (2)(4 \times 10^{-21})(4 \times 10^6)(1.58)(100)} \right]^{1/2}$$

$$\Rightarrow (R_i)_{max} = 27,827.6 \text{ km}$$

b. The maximum range for which this beacon system can be used is 27,827 km.

PROBLEMS

3.1 Calculate the maximum gain of an X-band antenna operating at 8 GHz and having a diameter of 1 m. Repeat this problem with the diameter changed to 1.5, 2.0 m. Assume $A_e = \rho_a A$ with $\rho_a = 1$ in each case.

3.2 Calculate the maximum gain of 2 m radius antenna operating in the L-, S-, and C-bands. Assume $A_e = \rho_a A$ with $\rho_a = 1$ in each case.

3.3 Find the size of a circular aperture antenna of X-band radar operating at $f_0 = 10$ GHz to attain $G = 30, 40, 50$ dBs. Assume $A_e = \rho_a A$ with $\rho_a = 0.7$ in each case.

3.4 An L-band radar operates at highest gain of 30 dB. The radar duty factor is 0.2 and the average power transmitted is 30 kW. Find the size of the antenna and the power density at a range of 55 km. Assume $\rho = 1$.

3.5 An L-band radar operating at frequency 1.5 MHz with an antenna of gain 36 dB is designed to obtain a single pulse minimum SNR of 20 dB. Assume the receiver bandwidth of 4 MHz, RCS of 10^2, noise figure of 10 dB, and the maximum range of 120 km. Find the minimum detectable signal, the peak power, and the pulse width for this radar.

3.6 A C-band radar operating at a frequency of 6 GHz with an antenna having a gain of 50 dB transmits a peak power of 1.5 MW. Assume the receiver bandwidth of 5 MHz, the minimum output SNR $(SNR)_{min}$ of 20 dB, and the radar cross section of 0.2 m² for this radar system. Find the maximum range for the receiver noise figure of 5 dB and overall radar loss of 0 dB.

3.7 Consider a C-band radar operating at a frequency of 4.6 GHz that must provide a minimum received signal power of 10^{-12} W. Assume that $P_t = 10$ kW, the antenna aperture area is 2.0 m², aperture efficiency is $\rho_a = 0.80$, radar cross section is $\sigma = 2$ m², and overall loss is $L = 5$ dB. Calculate the maximum range.

3.8 A C-band radar operating at a frequency of 4 GHz with an antenna having a gain of 45 dB transmits a peak power of 50 kW. Assume a total system loss of 2 dB. For a target located at a range of 100 km, find the minimum radar cross section to produce an available received signal power of $P_r = (2 \times 10^{-12})$ W.

3.9 An X-band radar employs the same circular aperture antenna for both transmission and reception at 8 GHz with its diameter of 3 m, antenna efficiency of 0.8. The radar is designed to produce an average received power of (3×10^{-14}) W when the radar cross section is 1 m² at a maximum range of 100 km. If the total system loss is 3 dB, what transmitter peak power is required?

3.10 An MMW radar uses a single antenna at 35 GHz to transmit a peak power of 650 W. The diameter of the antenna is 1.2 m, and antenna efficiency is 0.6. If a target at a range of 50 km has a radar cross section of 10 m², calculate the available received power. Assume that the overall system loss is 0 dB.

Radar Equations 41

3.11 A C-band monostatic radar operating at a frequency of 5.4 GHz transmits a peak power of 1 MW and has the following parameters: total system loss $L = 3$ dB, $R = 120$ km, $\sigma = 1.2$ m^2, and the antenna has a circular aperture with aperture efficiency $\rho_a = 0.6$. Find the diameter of the antenna in order to produce an available receiver power $P_r = 2.0 \times 10^{-14}$ W.

3.12 A high PRF airborne radar operating at a frequency of 10.5 GHz transmits a peak power of 10 kW and has the following parameters: pulse width $\tau = 1.2$ μs, PRF = 250 kHz, antenna gain $G = 35$ dB, radar cross section of the target $\sigma = 10$ m^2, receiver noise figure $F = 3$ dB, and the overall system loss including the propagation path loss $L = 5$ dB.
  (a) Find the maximum range at which the radar can detect the target if the minimum SNR for detection is 15 dB.
  (b) Repeat part (a) for 0 dB SNR.

3.13 A Doppler radar with a 1.3 m diameter antenna transmits 1.2 kW of power at a frequency of 3 GHz. The equivalent noise bandwidth is 1 kHz and the noise figure is 4.4 dB, and the overall loss factor is 10 dB. Assume a radar cross section of 10 m^2.
  (a) Find the SNR at target ranges of 32 and 160 km.
  (b) Find the target range at unity SNR.

3.14 A high PRF radar operating at 5.4 GHz transmits a peak power of 10 kW and has the following parameters: antenna gain $G = 20$ dB, overall loss $L = 10$ dB, noise figure $F = 3$ dB, time on target $T_i = 2.5$ s, duty factor $d_t = 0.25$, radar cross section $\sigma = 0.02$ m^2. For target range $R = 45$ km, find the single pulse SNR.

3.15 Consider an X-band radar operating at 10 GHz with the following parameters: antenna gain $G = 50$, $\theta_a = \theta_e = 3°$, scan time $T_s = 3.0$ s, overall system loss $L = 5$ dB, noise figure $F = 4.41$ dB, radar cross section $\sigma = 0.1$ m^2, SNR = 12 dB, and the range $R = 275$ km.
  (a) Find the power aperture product.
  (b) Find the transmitted power corresponding to $d_t = 0.3$.

3.16 An MMW search radar has the following specifications: $P_t = 5$ W, PRF = 12 KHz, pulse width $\tau = 6 \times 10^{-8}$ s, overall system loss $L = 10$ dB, circular aperture antenna with diameter $D = 0.3048$ m, target RCS $\sigma = 25$ m^2, noise figure $F = 6.17$ dB, azimuth scan $\theta_a = \pm 25°$, elevation scan $\theta_e = 3°$, and $t_s = 3.5$ s.
  (a) Find the power aperture product.
  (b) Find the SNR to detect a target at a range of 10 km.

3.17 A typical MMW search radar operating at a frequency of 94 GHz is used in a sector defined by $\pm 30°$ azimuth and $4°$ elevation scan, and has the following specifications:

Antenna gain	40 dB
Antenna diameter	0.25 m
Radar cross section	25 m^2
System losses	10 dB
Noise figure	3 dB
Transmit peak power	5 W
Pulse width	40 ns
Pulse repetition frequency	10 kHz

  (a) Find the detection range for an SNR of 10 dB.
  (b) Find the antenna coverage rate and the time on-target (dwell time) if the coverage is obtained in a radar frame time of 6 seconds.
  (c) Find the number of integrated pulses.
  (d) Find the detection range when an integration loss of 3 dB is included.
  (e) Justify that it is below the maximum unambiguous range.

Fundamental Principles of Radar

3.18 A radar is subject to interference by a self-protection jammer. The radar and jammer specifications are:

Radar transmit power	60 kW
Radar antenna gain	50 dB
Radar pulse width	2.5 µs
Radar losses	10 dB
Jammer power	180 W
Jammer antenna gain	10 dB
Jammer bandwidth	45 MHz
Jammer losses	0 dB

   (a) Find the cross over range for a target of RCS $\sigma = 5$ m².
   (b) Find the detection range if the required SNR for detection is 10 dB.

3.19 The radar in problem 3.17 is now subject to SOJ with the following parameters: $P_j = 200$ W, $G_j = 20$ dB, $L_j = 3$ dB, $G' = 10$ dB, and $R_j = 20$ km.
   (a) Find the cross over range for a target of RCS $\sigma = 5$ m².
   (b) Find the detection range if the required SNR for detection is 10 dB.

3.20 Work Example 3.7 to find the SNR at the missile of the bistatic system with the atmospheric attenuation of 0.08 dB/km during the propagation.

3.21 In a bistatic radar the two stations use identical antennas at 40 GHz with a gain of 30 dB. In this system $P_t = 55$ kW, $F = 1.66$ dB, $B = 10$ MHz, total loss $L = L_t L_r L_p = 3$ dB, and $(SNR)_r = 13.01$ dB, and target RCS $\sigma = 5$ m². What are the target ranges R_t and R_r if it is found that $R_t = 1.65\ R_r$.

4 Targets and Interference

"To affirm a person is to see the good in them that they cannot see in themselves and to repeat it in spite of appearances to the contrary. Please, this is not some Pollyanna optimism that is blind to the reality of evil, but rather like a fine radar system that is tuned in to the true, the good, and the beautiful."

—Brennan Manning

4.1 INTRODUCTION

The practical, real-world aspects of radar system performance in detection and measurement can be analyzed and predicted with confidence only when the target characteristics agree with nonideal environments. Unfortunately, radars must encounter a wide variety of targets and interfering signals backscattered from objects other than targets. The nonideal environment includes undesirable electrical signals called noise, which may include signals that the radar is not intended to detect—known as *clutter*. This chapter discusses the real-world aspects of radar operation and performance prediction in the presence of many factors, which include radar cross-section fluctuations, radar clutter, and receiver noise and system losses.

4.2 RADAR CROSS SECTION (RCS)

The radar range equation expresses the range at which a target may be detected with a given probability by a radar having a given set of parameters. This equation includes the *radar cross section* (RCS) of an equivalent isotropic radiator, an important parameter that defines the scattering efficiency of a target. Knowledge of the RCS is essential in the calculations of radar range and the SNR. The target's RCS may be viewed as a comparison of the strength of the reflected signal from a target to the reflected signal from a perfectly smooth sphere of cross-sectional area of 1 m². In fact, the RCS is defined as an equivalent cross section of the target producing the same amount of energy returned to the radar as would be produced by an isotropic radiator. The RCS and the actual area of the target are, therefore, not directly related through some simple equation.

The RCS of a target is a measure of its ability to reflect electromagnetic energy in the direction of the radar receiver, and its value is expressed as an area. This reflected energy is dependent on a multitude of parameters such as transmitted wavelength, target geometry, orientation, and reflectivity. The RCS, denoted by the symbol σ, is the area intercepting that amount of power which, when scattered equally in all directions, produces an echo signal at the radar equal to that from the target. Simply stated, the RCS is proportional to the far-field ratio of reflected to incident power density, that is

$$\sigma = \frac{\text{power reflected back to receiver per unit solid angle}}{(\text{incident power density intercepted by the target})/4\pi}. \tag{4.1}$$

Mathematically,

$$\sigma = \lim_{R \to \infty} 4\pi R^2 \left| \frac{E_r}{E_i} \right|^2 \tag{4.2}$$

where R = distance between the radar and target
E_r = reflected electric field strength at radar
E_i = incident electric field strength at target

TABLE 4.1

Maximum RCS for Typical Calibration Targets

Objects	Maximum RCS	Advantages	Disadvantages
Sphere: a = radius	$\sigma = \pi a^2$	Nonspecular	Lowest RCS for size
Cylinder: a = radius, and L = length	$\sigma = \dfrac{2\pi a L^2}{\lambda}$	Nonspecular along the radial axis	Low RCS for size
Flat plate: A = area	$\sigma = \dfrac{4\pi A^2}{\lambda^2}$	High RCS, nonspecular	RCS drops off sharply with angle of incidence
Dipole	$\sigma = 0.88\lambda^2$	High RCS in broadside	Very low RCS along dipole axis
Infinite cone with half angle α	$\sigma = \dfrac{\lambda^2 \tan^4 \alpha}{16\pi}$	High RCS along the axis	Low RCS other than axial axis
Dihedral: a,b = sizes, α = angle between two sides	$\sigma = \dfrac{16\pi a^2 b^2}{\lambda^2}\sin^2\alpha$	Nonspecular	Cannot be used for cross-polarized measurements
Square trihedral: a = size of all sides	$\sigma = \dfrac{12\pi a^4}{\lambda^4}$	Large RCS for size, nonspecular	Cannot be used for cross-polarized measurements
Triangular trihedral: a = size of all sides	$\sigma = \dfrac{4\pi a^4}{3\lambda^2}$	Nonspecular	Cannot be used for cross-polarized measurements

Under free-space conditions assumed here, the ratio $|E_r / E_i|^2$ is the same as the power densities of reflected and incident waves.

The RCS can be determined by solving Maxwell's equations with the proper boundary conditions applied. The determination of the RCS with Maxell's equations is not easy. Thus numerical solutions are often implemented using high-speed computers. However, Maxwell's equations can be solved to determine the RCS for targets of simple shapes such as sphere, cone, paraboloid, cylinder, large flat plate, triangular corner reflector, and dipole. The RCS of these regular, simple conducting objects are shown in Table 4.1. It may be noticed from this table that the maximum radar cross section of a large flat plate is directly proportional to the square of its area and is inversely proportional to the square of the transmitted wavelength implying that the higher transmitted frequency will present larger RCS. Table 4.1 is a summary of some simple, regular conducting shapes, which are useful in visualizing the RCS of more complex targets.

In all cases except for the perfectly conducting sphere, only optical region approximations are considered. The perfectly conducting sphere is considered as the simplest target for RCS calculations. Even in this case, the complexity of the exact solution, when compared to the optical region approximation, is overwhelming. Most formulas presented are approximations for the backscattered RCS measured by a far-field radar in the direction (θ, φ) in the spherical coordinate system as shown in Figure 4.1.

4.2.1 RCS OF SPHERE

A sphere is the simplest object susceptible to exact calculation and is isotropic and independent of viewing aspect. For a sphere, the variation of the RCS as a function of its circumference measured in transmitted wavelength is shown in Figure 4.2 as calculated by Blake (1972).

There are three distinct regions of behavior of the RCS of a sphere of radius a. The region for which $(2\pi a / \lambda) \ll 1$ is called *Rayleigh region*; and the RCS in this region is inversely proportional to the fourth power of the wavelength. The region defined by $1 < (2\pi a / \lambda) < 10$, the *Mie* or *resonance region,* occurs where the RCS oscillates as shown in Figure 4.2. In the high-frequency limit $(2\pi a / \lambda) > 10$, known as the *optical region*, the RCS settles down to the optical cross section $\sigma = \pi a^2$.

Targets and Interference 45

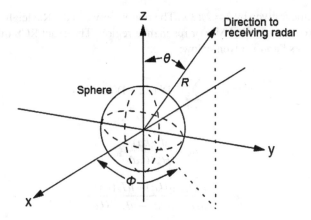

FIGURE 4.1 RCS measurement by far-field radars.

In the optical region (large sphere: $a \gg \lambda$) the asymmetric expression for the RCS of a sphere is given by

$$\sigma = \pi a_1 a_2 \tag{4.3}$$

where a_1 and a_2 are the principal radii of curvature at the surface normal to incident wave. Setting $a_1 = a_2 = a$ for a regular sphere in (4.3) yields

$$\sigma = \pi a^2. \tag{4.4}$$

In the optical region the RCS behavior with wavelength is monotonic although the RCS does not necessarily converge to a constant value. In the Rayleigh region (small sphere: $a \ll \lambda$) the asymmetric expression for the RCS of a sphere is given by

$$\sigma = 9\pi a^2 \left(\frac{2\pi a}{\lambda}\right)^4 = 1.4027(10^4)\left(\frac{a}{\lambda}\right)^4, \quad 0 < (a/\lambda) \leq 0.1 \tag{4.5}$$

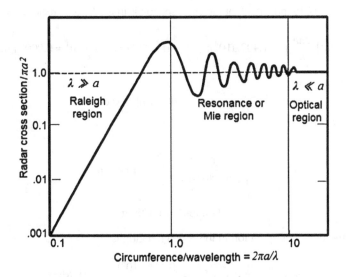

FIGURE 4.2 Radar cross section of a sphere: a = radius, λ = wavelength.

where k is the wavenumber given by $k = 2\pi / \lambda$. The region between the Rayleigh and optical regions is oscillatory in nature, and is called Mie or resonance region. The exact RCS of a sphere has been solved by several authors,[1] and is given below:

$$\sigma / \pi a^2 = \sum_{n=1}^{\infty} (-1)^n (2n+1)(b_n - a_n) \tag{4.6}$$

where

$$a_n = \frac{J_n(ka)}{H_n^{(1)}(ka)} \tag{4.7}$$

$$b_n = \frac{ka J_{(n-1)}(ka) - n J_n(ka)}{ka H_{(n-1)}(ka) - n H_n^{(1)}(ka)} \tag{4.8}$$

and a is the radius of the sphere, $j = \sqrt{-1}$, J_n is the spherical Bessel function of the first kind of order n, $H_n^{(1)}$ is the Hankel function of order n, which is given by

$$H_n^{(1)}(ka) = J_n(ka) + j Y_n(ka) \tag{4.9}$$

where Y_n is the spherical Bessel function of the second kind of order n. The plot of the normalized RCS of a sphere as a function of its circumference in wavelength is shown in Figure 4.2.

In the optical region where the circumference is much greater than the wavelength, the right-hand side of (4.6) becomes unity, consistent with (4.3). In the Rayleigh region, where the circumference is much smaller than the wavelength, the normalized RCS becomes proportional to a^6, consistent with (4.5).

Example 4.1

Find the monostatic cross section of a conducting sphere that occurs at approximately $a / \lambda = 0.015$. Calculate the corresponding diameter of the sphere and the maximum frequency of operation in this region.

Solution:

The criterion $a / \lambda = 0.015$ indicates that the cross section is in the Raleigh region. Then from (4.5):

$$\sigma = 9\pi a^2 \left(\frac{2\pi a}{\lambda} \right)^4 = 1.4027(10^4) \left(\frac{a}{\lambda} \right)^4 = (1.4027)(10^4)(0.015)^4 = 0.00071 \text{ m}^2$$

$$\Rightarrow 9\pi a^2 \left(\frac{2\pi a}{\lambda} \right)^4 = 9\pi a^2 [2\pi(0.015)]^4 = 0.00071$$

Thus

$$a = \left[\frac{0.00071}{9\pi(2\pi(0.015))^4} \right]^{1/2} = 56.41 \text{ cm}$$

$$D = 2a = (2)(.5641) = 1.128 \text{ m}$$

The maximum frequency of operation in this Raleigh region is

$$\frac{a}{\lambda_{min}} = \frac{a f_{max}}{c} = 0.1 \Rightarrow f_{max} = \frac{(0.1)(3 \times 10^8)}{0.5641} = 53.18 \text{ MHz}$$

Targets and Interference

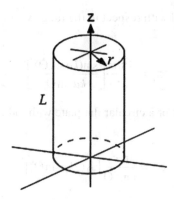

FIGURE 4.3 Geometry of the circular cylinder.

4.2.2 RCS of Cylinder

A target as a regular conducting cylinder has the maximum gain in the direction normal to the surface, and is proportional to the wavelength of the radar signal as long as the dimensions of the cylinder are large compared to the wavelength. The geometry of the conducting circular cylinder is shown in Figure 4.3. The gain of the cylinder is

$$G_c = \pi L / \lambda \qquad (4.10)$$

and the projected area of the cylinder as viewed from the radar is

$$A_c = 2rL \qquad (4.11)$$

where r is the radius and L is the length of the cylinder. The RCS σ_c of the cylinder can be written as the product of A_c, G_c and Γ_c. More precisely,

$$\sigma_c = A_c G_c \Gamma_c. \qquad (4.12)$$

Substituting (4.10) and (4.1) into (4.12), and assuming $\Gamma_c = 1$, we get

$$\sigma_c = \frac{2\pi r L^2}{\lambda}. \qquad (4.13)$$

The RCS of a cylinder is highly aspect-dependent.

4.2.3 RCS of Planar Surfaces

A large, smooth surface reflects most of the EM energy, assuming normal incidence from the radar, back in the perpendicular direction with a gain related to the aperture A as follows:

$$G_p = \frac{4\pi A}{\lambda^2}. \qquad (4.14)$$

Thus, the RCS of a large, smooth plane surface in the normal direction is

$$\sigma_p = A G_p = \frac{4\pi A^2}{\lambda^2} \qquad (4.15)$$

At oblique incidence Θ measured with respect to the range vector, the RCS of a square flat plate is given by[2]

$$\sigma_p = \frac{4\pi a^2}{\lambda^2} \left[\frac{\sin(ka\sin\Theta)}{ka\sin\Theta} \right]^2 \tag{4.16}$$

where a is the length of the side. For a circular flat plate with radius a, the expression for the RCS is modified to

$$\sigma_p = \frac{\pi a^2}{\tan^2\Theta} \left[J_1(2ka\sin\Theta) \right]^2 \tag{4.17}$$

where the symbols have their usual meanings. It should be observed that (4.16) simplifies to (4.15) at $\Theta = 0$, but drops off rapidly following the sinc function pattern. Also, the large reflection of a flat plate resembles the pattern that becomes narrower for a plate of larger size. It is possible to obtain a larger RCS at other than normal incidence by utilizing a corner reflector.

Example 4.2

Consider a calibration target of a right circular cylinder such that its broadside (cylinder) and end (flat plate) RCS are both 1.0 m^2 at 10 GHz. Find the length and diameter of the cylinder.

Solution:

From (4.15):

$$1.0 = \frac{4\pi A^2}{\lambda^2} = \frac{4\pi A^2}{c^2} f^2 = \frac{(4\pi)(10^{10})^2 A^2}{(3\times10^8)^2} \Rightarrow A = \left[\frac{(1.0)(3\times10^8)^2}{(4\pi)(10^{10})^2} \right]^{1/2} = 0.00846 \text{ m}^2$$

Then the diameter of a circle with this area is

$$D = \left[\frac{(4)(0.00846)}{\pi} \right]^{1/2} = 0.104 \text{ m}$$

And the length L is obtained from (4.13):

$$\sigma_c = \frac{2\pi r L^2}{\lambda} \Rightarrow L = \left[\frac{\sigma_c(c/f)}{2\pi r} \right]^{1/2} = \left[\frac{(1.0)(3\times10^8)}{\pi(0.104)(10^{10})} \right]^{1/2} = 0.303 \text{ m}$$

4.2.4 RCS OF CORNER REFLECTORS

Corner reflectors can provide a large cross section that reflects electromagnetic energy over a wide range of aspect angles. The large echoes from these targets arise from multiple reflections between two or three mutually orthogonal flat surfaces forming the corner reflectors. Corner reflectors behave very much like flat plates with similar RCS and frequency response. The difference is that, for a given size, the peak RCS of the corner reflector is much wider than the plate. When utilized deliberately they are target augmenters. When they occur accidentally as at the intersection of a wing and fuselage of an aircraft, they contribute to higher RCS. There are two main types of corner reflectors, dihedral and trihedral, as shown in Figure 4.4. The dihedral has two surfaces that are orthogonal planes, the trihedral has three. Dihedral and trihedral corner reflectors are used for

Targets and Interference

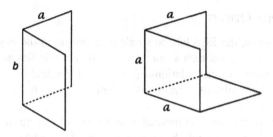

FIGURE 4.4 Corner reflectors: dihedral and trihedral.

external radar calibration. The approximate maximum cross section available from a dihedral corner reflector is expressed as

$$\sigma_{dc} = \frac{16\pi a^2 b^2}{\lambda^2}\sin^2\alpha \qquad (4.18)$$

where a and b are indicated in Figure 4.4. In (4.18) it is assumed that the angle of incidence of the electromagnetic energy is α in the range defined by $0 \le \alpha \le 45°$. Setting $\alpha = 45°$ in (4.18) yields the maximum RCS available from a dihedral corner reflector as expressed by

$$\sigma_{dc} = \frac{8\pi(ab)^2}{\lambda^2}. \qquad (4.19)$$

The square trihedral, with all sides being equal to a, is the one which provides a larger maximum RCS but narrower angular coverage. The approximate expression of the RCS of such a trihedral corner reflector is

$$\sigma_{tr} = \frac{12\pi a^4}{\lambda^2}. \qquad (4.20)$$

4.2.5 RCS of a Dipole

A dipole antenna of finite length is basically a linear antenna consisting of a large number of very short conductors in a connected series. The backscattering from such antenna is strongly dependent on the load at the antenna terminals. The transmitted field may be constructively or destructively added to the portion of the field directly reflected from the dipole. Dipole backscattering and the effect of different loading at its center were studied by Harrington and Mautz.[3] They showed that the RCS of a short-circuited small dipole of length L and wire radius a, for $L < 0.4\lambda$, is given by

$$\sigma_d = \frac{0.176\lambda^2}{1+(12)^2(kL)^{-6}[3\ln(2L/a)-7]^2}. \qquad (4.21)$$

For a small open-circuited dipole for $L < 0.8\lambda$, the RCS is given by

$$\sigma_d = \frac{\lambda^2(kL)^6}{(96)^2[\ln(L/a)-2]^2}. \qquad (4.22)$$

Both dipoles can reach resonance at which the RCS is modified to

$$\sigma_d = 0.818\lambda^2. \qquad (4.23)$$

It should be noted that the resonance occurs at $L \approx 0.45\lambda$ for the short-circuited dipole and at $L \approx 0.87\lambda$ for the open-circuited dipole.

4.2.6 RCS of Complex Objects

In the case of simple objects, the RCS has no single relationship to the physical dimension of the target. The RCS of complex targets such as aircrafts, ships, and missiles are very complicated and difficult, so that the attempts to write relationships based on physical dimensions are seldom successful. In such cases the best radar RCS prediction techniques based on elaborate computer simulation[4] can lead to useful results.

The RCS of complex targets can be computed in many cases by using the RCS of simple shapes. The most common procedure is to break the target into component parts, each of which is assumed to lie within the optical region, and to combine them according to

$$\sigma_m = \left| \sum_{n=1}^{M} \sqrt{\sigma_n} \exp\left(\frac{j4\pi R_n}{\lambda}\right) \right|^2 \qquad (4.24)$$

where σ_n is the RCS of the nth scatterer, and R_n is the distance between the nth scatterer and the receiver. Note that round trio distance is accounted for in the exponent by having a 4π rather than a 2π factor. The model in (4.24) ignores shadowing and multiple reflections between the individual scatterers. This formula has an important application in estimating the RCS of an array of multiple scatterers. This approach is usually used to determine the average RCS over an aspect angle change of several degrees. Since the precise target aspect is generally unknown and time variant, the RCS is best described in statistical terms. The statistical distribution of the RCS is of great importance in predicting detectability. As an illustration of the RCS of complex objects, Figure 4.5[5] shows

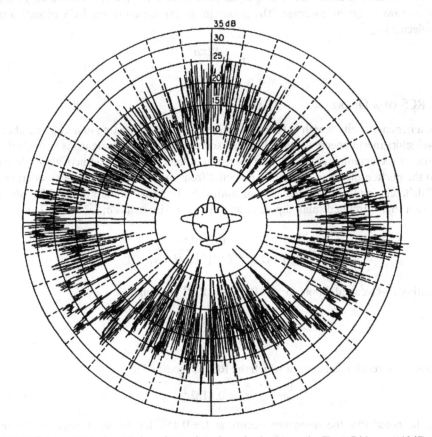

FIGURE 4.5 RCS of B-26 aircraft as a function of azimuth angle (From Ridenour, 1947).

Targets and Interference   51

the RCS as a function of aspect angle for an aircraft. This is a classic radar cross section of a B-26 bomber plotted on polar axes as a function of azimuth angle. Note that the RCS exceeds 35 dBsm (3100 m^2) from certain aspect angles. In contrast the RCS of the B-2 stealth bomber is about 40 dBsm.

4.3 RCS FLUCTUATIONS AND STATISTICAL MODELS

The performance of a radar system in detection and measurement are usually predicted with confidence when the target characteristics agree with their predicted values. Knowledge of the radar cross section of the target is essential in the calculation of SNR to determine the radar system performance. In general, the RCS is found to vary considerably for each target as the aspect angle changes, as internal motions of the target change its shape, and as radar frequency is varied. These changes warrant the use of statistical methods to describe the radar target. The echo signal from such a target may consist of many components of energy scattered from points distributed over the surface of the target. The amplitude and phase of each component vary as a function of time, aspect angle, and radar frequency.

Scintillation is a fluctuation in the amplitude of a target on a radar display. It is closely related to target *glint* or *wander*, an apparent displacement of the target from its mean position. This effect can be caused by a shift of the effective reflection point on the targets, but has other causes as well. The fluctuations can be slow (scan-to-scan) or rapid (pulse-to-pulse). Scintillation and glint are actually two manifestations of the same phenomenon, and are most properly linked to one another in target modeling.

Because typical targets may be directive and frequency sensitive, it is easy to postulate how changes in target aspect angle or frequency and polarization can cause the RCS to fluctuate. An efficient method to assess the effects of a fluctuating cross section is to postulate a reasonable model for the fluctuations and to analyze it mathematically. Marcum[6] and Swerling[7] described the theory, methods, and procedures for predicting the detection performance of nonfluctuating and fluctuation targets, respectively. Swerling extended Marcum's works to incorporate what has become known as the four *Swerling* models. He calculated the detection probabilities of four different Swerling models of fluctuating cross section of targets. In two of the four cases it is assumed that the fluctuations are completely correlated during a particular scan, but are completely uncorrelated from scan to scan (slow fluctuation). In the other two cases the fluctuations are assumed to be more rapid, but the measurements are pulse to pulse uncorrelated. The targets that do not have any fluctuations are normally referred to as *Swerling 0* or *Swerling V*. When the detection probability is large, all four Swerling cases in which the target cross section is not constant require greater signal-to-noise than the constant cross section of Swerling case V. The concept of Swerling targets is illustrated in Figure 4.6. The four fluctuation models for radar cross section are as follows:

Case I. In Swerling I, the echo pulses received by the radar from a target on any scan are of constant amplitude throughout the entire scan, but are uncorrelated from scan to scan. This is sometimes referred to as scan-to-scan fluctuation. The RCS in this case can be described by a chi-square probability density function with a single duo-degree of freedom. The probability density function of the RCS σ corresponding to Swerling I targets is given by

$$p(\sigma) = \frac{1}{\sigma_{av}} \exp\left(-\frac{\sigma}{\sigma_{av}}\right), \quad \sigma \geq 0 \tag{4.25}$$

where σ_{av} is the average RCS over all target fluctuations.

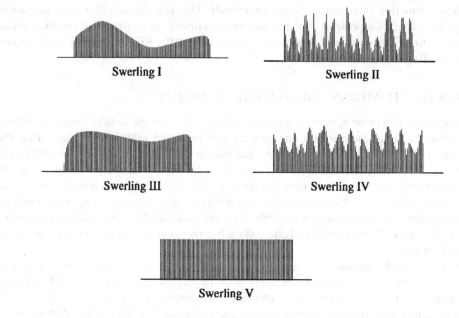

FIGURE 4.6 Radar returns from targets with different Swerling fluctuations (From Mahafza, 1998).

Case II. In Swerling II, target fluctuations are more rapid than Swerling I, but the measurements are taken pulse to pulse uncorrelated instead of scan to scan. The probability density function for the RCS is also given by (4.25).

Case III. In Swerling III, target fluctuations are independent from scan to scan as in Swerling I, bur are correlated pulse to pulse. The RCS in this case can be described by a chi-square probability density function with two duo-degrees of freedom (actually gamma distribution resulting from normalization of the chi-square). The probability density function of the RCS is, therefore, given by

$$p(\sigma) = \frac{4\sigma}{\sigma_{av}} \exp\left(-\frac{2\sigma}{\sigma_{av}}\right), \quad \sigma \geq 0. \tag{4.26}$$

Case IV. In Swerling IV, target fluctuations are rapid and independent from pulse to pulse instead of scan to scan as in Swerling II. The probability density function for the RCS is also given by (4.26).

The typical plot of the probability density function for the Swerling models is shown in Figure 4.7. The probability density function, in cases of Swerling I and II, applies to complex targets consisting of many independent reflectors of equal size. But in the cases of Swerling II and IV, it applies to targets consisting of one large reflector together with other small reflectors. In all the previous cases, the value of RCS to be substituted in the radar equation is the average σ_{av}.

It is observed that RCS fluctuations generally reduce radar range performance. This is especially true for Swerling case I targets giving strong echoes, hence a high nominal probability of detection. The use of case I statistics in radar calculations, therefore, tend to produce substantially more conservative range estimates than are given by nonfluctuating targets.

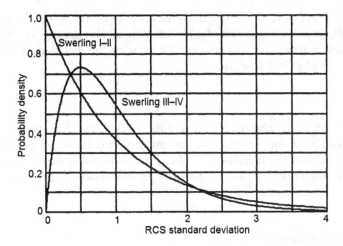

FIGURE 4.7 Probability density functions for Swerling targets (From Mahafza, 1998).

4.4 RADAR CLUTTER

Radar clutter is defined as any unwanted radar echo that can clutter the radar output and make the detection of a wanted target difficult. In fact, one radar's clutter is another radar's target. For example, to an engineer developing a missile to detect and track a tank, the return from vegetation and other natural objects would be considered to be "clutter." However, a remote sensing scientist would consider the return from natural vegetation as the primary target. Clutter is thus defined as the return echo from a physical object or a group of objects that is undesired for a specific application. Examples of clutter include unwanted reflections from land, sea, rain or other precipitation, chaff, birds, insects, air turbulence, aurora, and meteor trails.

The presence of clutter almost always causes a degradation of radar performance as the clutter returns compete with target returns in the radar detection process. When clutter echoes are sufficiently intense, they can impose serious limitations in the performance of the radar operations. Clutter may be divided into sources distributed over a surface, within a volume, or concentrated at discrete points.

Ground or sea returns are typical surface clutter. Returns from geographical land masses are generally stationary; however, the effect of wind on trees, etc., means that the target can introduce a Doppler shift to the radar return. This Doppler shift is an important method of removing unwanted signals in the signal processing part of a radar system. Clutter returned from the sea generally also has movement associated with the waves.

Weather or *chaff* are typical volume clutter. In the air, the most significant problem is weather clutter, which can be produced from rain or snow and can have a significant Doppler content. Birds, windmills, and individual tall buildings are typical point clutter and are not extended in nature. Moving point clutter is sometimes described as angels. Birds and insects produce clutter, which can be very difficult to remove because the characteristics are very much like aircraft.

Clutter can be *fluctuating* or *nonfluctuating*. Ground clutter is generally nonfluctuating in nature because the physical features are normally static. On the other hand, weather clutter is mobile under the influence of wind and is generally considered fluctuating in nature. Clutter can be defined as *homogeneous* if the density of all the returns is uniform. Most types of surface and volume clutter are analyzed on this basis, however, in practice this simplification does not hold true in all cases. *Nonhomogeneous* clutter is nonuniform clutter where the amplitude of the clutter varies significantly from cell to cell. Typically nonhomogeneous clutter is generated by tall buildings in built-up areas.

Because of the distributive nature of clutter, the measure of the backscattering echo from such clutter is generally expressed in terms of a clutter *cross-section density*. For the case of surface clutter, the backscatter parameter commonly called the *normalized clutter reflectivity* per unit area illuminated is defined by Goldstein[8] as:

$$\sigma° = \sigma_s / A_c \quad (4.27)$$

where σ_s is the RCS of the clutter area A_c, and the symbol $\sigma°$ represents the clutter scattering parameter, or normalized clutter reflectivity, per unit area illuminated. Clearly $\sigma°$ is a dimensionless quantity and is commonly specified in decibel. For a volume distributed clutter, the backscatter parameter per unit volume called *clutter reflectivity per unit volume*, is defined as

$$\eta = \sigma_v / V_c \quad (4.28)$$

where σ_v is the RCS of the clutter volume V_c.

4.4.1 Surface Clutter

Surface clutter is a signal echo from approximately planner surfaces whose area exceeds the radar resolution cell on the clutter surface. The radar cross section of surface clutter depends on the clutter characteristics and the area of clutter within a radar resolution cell. Three distinct regions of clutter behavior are recognized separated by grazing angles: the *low angle grazing region*, the *plateau region*, and the *near vertical incidence region* as discussed by Long[9] and shown in Figure 4.8. Within each of the regions the dependence of $\sigma°$ on grazing angle and wavelength can be described in a general way, but the borders of the three regions change with wavelength, surface irregularities, and polarization.

In the low grazing region, $\sigma°$ increases rapidly with increasing incidence angle up to a critical angle below which a surface is smooth by Raleigh's distribution and above which it is rough. By definition a surface is smooth if

$$\Delta h \sin \theta \leq \lambda / 8 \quad (4.29)$$

where

Δh = root mean square (rms) height of the surface irregularities
ψ_g = grazing angle between the incident ray and the average irregularities
λ = transmitted wavelength

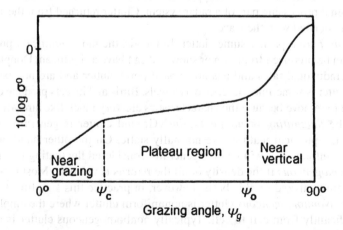

FIGURE 4.8 Dependence of $\sigma°$ on the grazing angle.

Targets and Interference

In the plateau region a clutter surface appears rough, and the scattering is almost incoherent. The grazing angle dependence of $\sigma°$ is much less than in the low grazing region. Finally, for a large incidence region, specular reflection becomes dominant so that $\sigma°$ increases rapidly with a grazing angle up to a maximum incidence 90°. The magnitude of $\sigma°$ at 90° incidence depends on the root mean square (rms) surface roughness and the dielectric properties of the clutter. The angular dependence of $\sigma°$ has been discussed by Hayes and Dyer[10] as

$$\sigma° = C \sin^n \theta \tag{4.30}$$

where C is an arbitrary constant, θ is the incidence angle, and n is roughly +1.

In the determination of surface reflectivity, the computation of the illuminated area depends on the geometry, as shown in Figure 4.9. Two different geometries can be specified as determined by the elevation beamwidth θ_{el}, azimuth 3-dB beamwidth θ_{az}, slant range R, grazing angle ψ_g, and the radar pulse width. The intersection of the antenna beamwidth with the ground defines an elliptically shaped footprint. The area of the largest resolution cell in this case is approximately rectangular and can be obtained by multiplying the projected range resolution length ($\rho = c\tau \sec \psi_g / 2$), by the width of the illuminated ellipse, that is, $2R \tan(\theta_{az}/2)$. In the pulse length limited case, the surface illuminated clutter area A_c is given by

$$A_c = 2R(c\tau/2)\tan(\theta_{az}/2)\sec\psi_g \tag{4.31}$$

where

$$\tan \psi_g < \frac{2R\tan(\theta_{el}/2)}{c\tau/2}.$$

By applying small angle approximation to (4.31) we get

$$A_c = R(c\tau/2)\theta_{az}\sec\psi_g. \tag{4.32}$$

The returned signal strength S_r resulting from the interception of the entire main beam with the ground will be computed using the radar equation (3.10) as

$$S_r = \frac{P_t G^2 \lambda^2 \sigma_t}{(4\pi)^3 R^4} \tag{4.33}$$

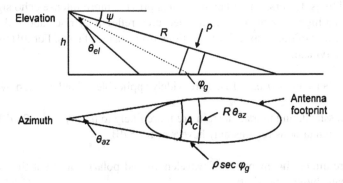

FIGURE 4.9 Geometry of radar clutter showing the antenna footprint and clutter patch.

where the symbols have their usual meanings. Similarly, the received power from the clutter S_c, with clutter area A_c, is given by

$$S_c = \frac{P_t G^2 \lambda^2 \sigma_c}{(4\pi)^3 R^4} \tag{4.34}$$

where

$$\sigma_c = \sigma_0 A_c = \sigma_0 R(c\tau/2)\theta_{az} \sec \psi_g. \tag{4.35}$$

In (4.35) the relation $\rho = c\tau/2$ has been used. Finally, the signal-to-clutter ratio (SCR) can be obtained using (4.33) and (4.34), and eliminating σ_c in (4.35) as

$$(SCR)_{A_c} = \frac{2\sigma_t}{Rc\tau\sigma_0\theta_{az} \sec \psi_z}. \tag{4.36}$$

Land Clutter

The characterization of *land clutter* has been of primary importance for many years, and is a very difficult problem because there are many varieties of natural and man-made structures. Because of the statistical nature of clutter, the mean reflectivity is most often quoted. A convenient mathematical way to describe this mean value for surface clutter is the constant γ model in which the surface reflectivity is modeled as

$$\gamma = \frac{\sigma^\circ}{\sin \psi_g} \tag{4.37}$$

where γ is the parameter describing the scattering effectiveness of clutter. For the beam-filled scenario, γ is equivalent to the ground cross-sectional area intercepted by the beam as mapped onto the plane perpendicular to the line of sight between the radar and the ground. γ is thus approximately independent of the grazing angle ψ_g for small angles. Although knowledge of sea clutter is far from complete, it is better understood than is the knowledge of land clutter. The reflectivity of land clutter is much more difficult to characterize than that of sea. The constant γ model is still followed for land clutter at higher grazing angles. The average values of land clutter depend on clutter type, grazing angle, frequency, and polarization. Barton (1988) discussed the effect of grazing angle on clutter reflectivity for different clutter types, including flatlands, farmlands, rolling hills, wooded hills, and mountains. Furthermore, the radar echo will depend on the terrain roughness as well as the moisture content of the surface scatterers, snow cover, and the stage of growth of any vegetation. Buildings, towers, and other structures provide more intense echo signals than forests and vegetation. At high grazing angles, the measured reflectivity rises above the value predicted by the model because of quasi-specular reflections from surface facets. For different surface types, the following are typical:

a. Values of γ between -10 and -15 dB are widely applicable to land covered by crops, bushes, and trees.
b. Desert, grassland, and marshy terrain are more likely to have γ near -20 dB
c. Urban or mountainous regions may have γ near -5 dB

These values are almost independent of wavelength and polarization, but they only apply to the modeling of mean clutter reflectivity.

TABLE 4.2
The Beaufort Scale, K_B

Beaufort No.	Description	Wind Speed (kts) 1 kts = 1.852 km/h
0	Calm	<1
1	Light air	1–3
2	Light breeze	4–6
3	Gentle breeze	7–10
4	Moderate breeze	11–16
5	Fresh breeze	17–21
6	Strong breeze	22–27
7	Near gale	28–33
8	Gale	34–40
9	Strong gale	41–47
10	Storm	48–55
11	Violent storm	56–63
12	Hurricane	>64

At low grazing angles, propagation considerations become dominant. In fact, the calculation grazing angle becomes quite indefinite for ground-based radar looking out over typical terrain.

Sea Clutter

Sea clutter suffers temporal and spatial variations, and is largely dependent on wave height. Several quantities have been used to describe wave height including sea state, average wave height, peak wave height, wind speed, frequency, polarization, radar look direction relative to wave direction, and grazing angle. An excellent discussion of sea backscatter reflectivity dependencies is given by Long (1983). The dependencies of sea clutter reflectivity changes as the frequency is increased from the microwave region to the millimeter region.

As in land clutter, average values for sea clutter are usually γ model as applied to sea clutter, averaging over all wind directions and wave heights. It is found that γ depends on the Beaufort wind scale K_B as shown Table 4.2, and radar wavelength λ according to the following empirical relationship:

$$10\log\gamma = 6K_B - 10\log\lambda - 64.\tag{4.38}$$

A composite average of sea clutter data[11] is shown in Figure 4.10 displaying a plot of the mean cross section per unit area ($\sigma°$) as a function of grazing angle for various frequencies and polarization.

4.4.2 VOLUME CLUTTER

Volume clutters include echoes from chaffs, rain, or weather atmospheric phenomena. In general, radars at lower frequencies are not affected by weather conditions, but at higher frequencies weather echoes may be quite strong and mark the desired target signals just as any other unwanted clutter signal. For the measurement of foliage or rain where some penetration occurs, the volume clutter formulation is appropriate. In this formulation an analogous normalization parameter to $\sigma°$ called η exists for volume scatterers. The resolution cell for volume clutter is approximately a cylinder with

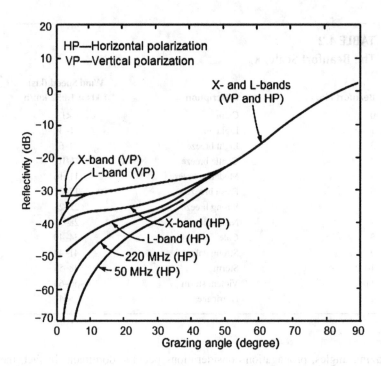

FIGURE 4.10 Effect of grazing angle on sea clutter reflectivity (From Skolnik, 1980).

an elliptical cross section, as shown in Figure 4.11. The cell volume V_c is calculates by multiplying the elliptical cross-sectional area $\pi/4(R\theta_{el})(R\theta_{az})$ by the range resolution $c\tau/2$ as follows:

$$V_c = \frac{\pi}{4} R^2 \theta_{el} \theta_{az} (c\tau/2). \tag{4.39}$$

The volume of V_c computed from (4.37) is often reduced by a factor of 2 to account for the two-way illumination of the clutter volume of the resolution cell. The radar cross section is then the product of the volume reflectivity η and the volume V_c as follows:

$$\sigma_v = \eta V_c. \tag{4.40}$$

This model assumes that there is minimal attenuation of the radar signal over the length of the cell. This may be a valid assumption at lower frequencies where foliage and rain penetration is good, but it is not in the millimeter-wave band where the measured two-way attenuation in dry foliage exceeds 4dB/m and is even higher in dense green foliage.

In this case a more complex formulation to take this attenuation into account is required as given by

$$\sigma_v = \eta V_c \exp(-c\tau\alpha) \tag{4.41}$$

where α represents the beam shape factor of 1.33 for a Gaussian shaped beam.

Rain

Rain is generally the result of some form of precipitation, which contributes significant scattered energy. Austin[12] provides information on atmospheric scattering from rain and other sources. The reflectivity properties of rain depend on the transmitted frequency, polarization, and number and

Targets and Interference

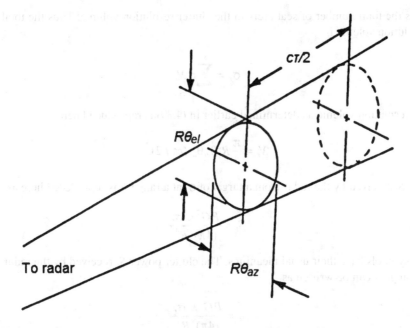

FIGURE 4.11 Geometry of volume clutter resolution cell.

size of rain drops. The volume reflectivity η, as discussed by Gunn and East,[13] varies strongly with the rain rate occurring for 9 and 35 GHz. For 70 and 95 GHz, however, the dependence is weaker. Since the raindrops are very small compared to typical volumetric resolutions of search radars, there are usually a large number of scatterers within a radar resolution volume resulting in rain reflectivity, which is a very dynamic process.

This data is determined using the relationship between the reflected and incident power on small spherical targets as discussed earlier in the section on the RCS of a sphere. Though a given rainfall rate does not imply a specific drop-size distribution, the trend that the drops get bigger as the rainfall rate increases, generally holds true. In the Rayleigh region ($\pi D / \lambda < 1$) the RCS is given by the following formula:

$$\sigma_v = 4\pi R^2 \frac{S_{ref}}{S_{inc}} = \pi^5 |K|^2 \frac{D^6}{\lambda^4} \qquad (4.42)$$

where

$$K = \frac{\varepsilon - 1}{\varepsilon + 1} \qquad (4.43)$$

and ε is the relative dielectric constant, D is the droplet diameter, and S_{ref} and S_{inc} are the power reflected and power incident, respectively. When $\pi D / 4 > 10$, the equation for RCS reduces to the geometric optics form

$$\sigma = \frac{\pi D^2}{4}. \qquad (4.44)$$

Defining η as RCS per unit resolution volume, it is calculated as the sum of all individual scatterers RCS within the volume

$$\eta = \sum_{i=1}^{N} \sigma_i \qquad (4.45)$$

where N is the total number of scatterers in the clutter resolution volume. Thus the total RCS of a single resolution volume is

$$\sigma_r = \sum_{i=1}^{N} \sigma_i V_r \tag{4.46}$$

where the resolution volume as determined earlier in (4.39) is reproduced here

$$V_r = \frac{\pi}{4} R^2 \theta_{el} \theta_{az} (c\tau / 2). \tag{4.47}$$

The power S_t received by the radar from a target of σ_t at a range R is reproduced here as

$$S_t = \frac{P_t G^2 \lambda^2 \sigma_t}{(4\pi)^3 R^4} \tag{4.48}$$

where the symbols have their usual meanings. The clutter power S_r received by the radar from rain of σ_r at a range R can be written as

$$S_r = \frac{P_t G^2 \lambda^2 \sigma_r}{(4\pi)^3 R^4}. \tag{4.49}$$

Using (4.44) and (4.45) into (4.47) yields

$$S_r = \frac{P_t G^2 \lambda^2}{(4\pi)^3 R^4} \left(\frac{\pi}{8} R^2 \theta_{el} \theta_{az} c\tau \right) \sum_{i=1}^{N} \sigma_i. \tag{4.50}$$

The SCR for rain clutter, denoted by $(SCR)_r$, is then

$$(SCR)_r = \frac{S_t}{S_r} = \frac{8\sigma_t}{\pi \theta_{el} \theta_z c\tau R^2 \displaystyle\sum_{i=1}^{N} \sigma_i}. \tag{4.51}$$

Dust

The volume of *dust* that can be supported in the atmosphere is extremely small; and so the reflectivity can often be neglected for EM radiation with wavelengths of 3 mm or more. However, under certain circumstances, if the dust density is very high (such as in rock crushers) or if the propagation path through dust is very long (in dust storms), then it can be useful to determine the reflectivity and the total attenuation.

Chaff

One of the earliest forms of countermeasures used against radar performance and efficient detection was *chaff*. Chaff consists of very small passive reflectors or absorbers suspended in the atmosphere, and may also consist of a large number of dipole reflectors, usually in the form of metallic foil strips packaged in bundles or even aerosols. Historically, chaff is made of aluminum foil; however, in recent years most chaff is made of fiberglass with conducting coating. A large bundle of chaffs when dispensed from an aircraft are scattered by the wind to form a highly reflecting cloud, while a relatively small bundle of chaff can form a cloud with a radar RCS comparable to a large aircraft.

The maximum RCS of an individual dipole of chaff occurs when dipole length L is one half wavelength of the radar signal. When viewed broadside, the RCS of an individual dipole is

$$\sigma_{chaff} \approx 0.88\lambda^2 \tag{4.52}$$

Targets and Interference

and when averaged over all aspect angles, this drops to

$$\sigma_{chaff} \approx 0.88\lambda^2. \tag{4.53}$$

The RCS of chaff can also be expressed as a function of its weight W in kg,

$$\sigma_{chaff} = 6600(W / f) \tag{4.54}$$

where f is expressed in GHZ. Chaff normally falls within a motion, which randomizes the orientation of individual dipoles resulting in insensitivity to the radar polarization.

4.4.3 POINT (DISCRETE) CLUTTERS

Radar echoes obtained from regions of atmosphere where no apparent reflecting sources seem to exist are commonly called *ghosts* or *angels*. Angel echoes are classified as dot angels, usually known as point targets due to birds and insects, or distributed angels, due to the inhomogeneous atmosphere. Birds and insects in substantial number can also appear as distributed angels, and can have a degrading effect on radar. For an MTI radar it is difficult to remove clutter due to flying birds by Doppler filtering. Operation at UHF has proven to be a satisfactory method for reducing the effects of such clutter as birds and insects. An inhomogeneous atmosphere does not produce enough reflectivity to be a source of clutter to most radar.

Birds

Although the radar cross section of a single bird is small, the birds that travel in flocks may produce a significantly large total cross section. Birds that fly at speeds around 90 km/hour are usually difficult to reject by MTI radars. Some examples of radar cross section taken at three frequencies with vertical polarization[14] are shown in Table 4.3. As observed, the largest values occur at S-band. Other examples of radar cross section are given in Figure 4.12, which plots the average radar cross section as a function of the bird's weight. The solid circles are the average over a ±20° sector around the broadside aspects.[15] The x's are the average of the ±20° head-on and ±20° tail-on aspects.

The backscatter from birds fluctuates over quite a range of values. Thus it should be properly described statistically. Fluctuations in the RCS of a single bird in flight have been measured to have a log-normal distribution for mid-microwave bands, and an empirical formula that relates the wing beat frequency f (Hz) of birds to their length ℓ (mm) of the wing is found[16] to be expressed by $f\ell^{0.827} = 572$.

TABLE 4.3
Radar Cross Section of Birds

Bird Type	Frequency	Mean RCS (cm²)	Median RCS (cm²)
Grackle	X	16	6.9
	S	25	12
	UHF	0.57	0.45
Sparrow	X	1.6	0.8
	S	14	11
	UHF	0.02	0.02
Pigeon	X	15	6.4
	S	80	32
	UHF	11	8

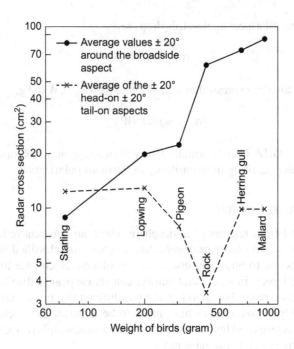

FIGURE 4.12 Radar cross sections of birds at S-band with vertical polarization (From Skolnik, 1980).

Insects

Insects in sufficient number are readily detected by radar, and can clutter the display, thus reducing the capability of a radar to detect desired targets. At X-band the RCS of a variety of insects showed a variation from 0.02 to 9.6 cm² with longitudinal polarization and between 0.01 and 0.95 cm² for transverse polarization. A desert locust or a honeybee might have a cross section of about 1 cm² at X-band. The cross section of insects below X-band is approximately proportional to the fourth of the frequency.[17] Appropriate echoes are obtained only when insect body lengths exceed $\lambda/3$.

Insects viewed broadside have RCS values between 10 and 1,000 times larger than when viewed end-on. Insect echoes are more likely to be found at lower altitudes, near dawn or twilight. Large concentration of insect angel echoes would not be expected at temperatures below 40°F or above 90°F. Sensitivity time control (STC) can mitigate the adverse effects of clutter due to insects. A bee would have a broadside RCS of about 1 cm² at X-band. This would not increase significantly up to W-band as the bee moved from the Mie to the optical region. RCS of different insects, including aphids, mosquitoes, ladybugs, honeybees, moths, and locusts were discussed by Riley.[18] He demonstrated the relationship between the RCS and the mass of an insect, with the variation of water droplet shown for comparison. Similar comparisons have been made for both birds and insects.[19]

4.5 CLUTTER STATISTICAL DISTRIBUTIONS

Radar clutter within a resolution cell is composed of a large number of scatterers with random phases and amplitude, and is characterized by statistical fluctuations that must be described by probability density function. These distributions describing the clutter depend on the character of the clutter itself, the frequency, and the grazing angle. Several distributions have been suggested, including the Raleigh distribution, the log-normal distribution, the K distribution, the Weibull distribution, and others.

Targets and Interference

Rayleigh distribution: If the sea and the land clutter are composed of many small scatterers within the resolution cell, then the statistical fluctuation of the clutter voltage envelope r_c may be modeled using a Rayleigh probability density function,

$$p(r_c) = \frac{2r_c}{x_0} \exp\left(\frac{-r_c^2}{x_0}\right), \quad \text{for } r_c \geq 0$$
$$= 0, \quad \text{for } r_c < 0 \tag{4.55}$$

where x_0 is the mean squared value of the envelope r_c.

Log-normal distribution: The log-normal distribution empirically fits some low-angle land clutter, also high-resolution samples of real clutter in the plateau region. The probability density function of the voltage envelope r_c is given by

$$p(r_c) = \frac{1}{r_c\sqrt{4\pi \ln(\bar{r}_c / r_m)}} \exp\left[-\frac{[\ln(r_c / r_m)]^2}{4\ln(\bar{r}_c / r_m)}\right], \quad \text{for } r_c \geq 0$$
$$= 0, \quad \text{for } r_c < 0 \tag{4.56}$$

where $\bar{r}_c$ and r_m are the mean and median values of the voltage envelope r_c, respectively.

K-distribution: This distribution has been suggested by Ward and Watts[20] as fitting sea clutter given by

$$p(r_c) = \frac{4}{\Gamma(M)}[M / 2\bar{\sigma}]^{(M+1)/2} r_c^M K_{M-1}\left(r_c\sqrt{\frac{2M}{\bar{\sigma}}}\right) \quad \text{for } r_c \geq 0$$
$$= 0 \quad \text{for } r_c < 0 \tag{4.57}$$

where

$\quad\quad r_c$ = the clutter amplitude envelope
$\quad\quad M$ = shape parameter
$\quad K_{M-1}(\cdot)$ = modified Bessel function of order $(M - 1)$, $M > 0$
$\quad\quad \bar{\sigma}$ = average power in the bandpass clutter signal of voltage envelope r_c
$\quad\quad \Gamma(M)$ = gamma function of the shape parameter M.

Weibull distribution: The Weibull distribution has also been examined for modeling sea clutter. The Weibull[21] distribution empirically fits low-angle clutter (less than 5°) for frequencies between 1 and 10 GHz. The Weibull clutter has an envelope r_c, described by the probability density function,

$$p(r_c) = \frac{b}{\alpha}\left(\frac{r_c}{\alpha}\right)^{b-1} \exp\left[-\left(\frac{r_c}{\alpha}\right)^b\right] \quad \text{for } r_c \geq 0$$
$$= 0 \quad \text{for } r < 0 \tag{4.58}$$

where b is called a shape parameter and

$$\alpha = \frac{\bar{r}_c}{\Gamma(1 + (1 / b))}. \tag{4.59}$$

is a scale factor with $\bar{r}_c$, which is the mean value of r_c.

4.6 CLUTTER SPECTRUM

The clutter is not always stationary. Consequently, the clutter spectrum exhibits some Doppler frequency spread because of wind, motion of the radar scanning antenna, and transmitter frequency drift rate. Since most of the clutter power is concentrated around zero Doppler with some spreading at higher frequencies, it is customary to model clutter by Gaussian-shaped power spectral density as

$$S_c(\omega) = \frac{P_c}{\sqrt{2\pi\sigma_\omega^2}} \exp\left(\frac{-(\omega - \omega_0)^2}{2\sigma_\omega^2}\right) \tag{4.60}$$

where σ_ω represents the Doppler frequency spread in radians per second, P_c is the clutter power, and ω_0 is the center Doppler frequency. The main reason for clutter spreading is actual clutter motion because of wind. The frequency standard deviation σ_{wind} (in radians per second) due to wind is given by

$$\sigma_{\text{wind}} = \frac{2\pi}{\lambda} V_{\text{rms}} \tag{4.61}$$

where λ is the radar signal wavelength and V_{rms} is the RMS value of the clutter velocity. A second source spectral spreading is the motion of the radar scanning antenna.[22] Accordingly, the standard deviation of the effective Doppler radial frequency due to the scanning antenna with a uniform aperture with width D is given by

$$\sigma_\theta = \frac{1.4D}{\lambda} \dot{\theta} \tag{4.62}$$

where $\dot{\theta}$ is the antenna scan rate in radians per second. The third source of spectral spreading is the transmitter frequency drift rate $\dot{f}$ (Hertz per second). According to Schleher (1978), the resulting standard deviation is

$$\sigma_f = \frac{2.67}{B} \dot{f} \tag{4.63}$$

where B is the transmitted pulse bandwidth. Assuming that the different effects are independent and disturb the return echo in a multiplicative way, the resulted spectrum is obtained by convolving the different Gaussian spectra, each with its respective σ. More precisely,

$$\sigma_\omega^2 = \sigma_{\text{wind}}^2 + \sigma_\theta^2 + \sigma_f^2. \tag{4.64}$$

4.7 RADAR RECEIVER NOISE

In radar system noise sets, thresholds below which desired target echoes are obscured is the chief factor limiting radar sensitivity. Much of the radar receiver is designed to mitigate or otherwise overcome the debilitating effects of noise. Therefore, it is essential to have a thorough understanding of noise, which is crucial to maximizing radar performance.

Noise is a random, usually unwanted, electromagnetic energy that interferes with the ability of the radar receiver to detect the wanted signal. Sources of noise for a radar system usually include thermal emissions from the target, thermal motion of the conduction electrons in resistors and semiconductors in the receiver, atmospheric phenomena, any interference both intentional and unintentional, shot noise due to the discrete nature electric charge, and flicker noise in active devices. Noise is mostly apparent in regions with a low signal level, such as the weak received echo signal corrupting the desired signal in a radar receiver. The interference noise also results from extraterrestrial

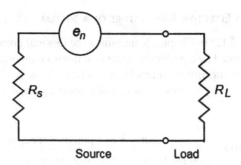

FIGURE 4.13 Thevenin equivalent circuit of thermal noise source.

radiation sources (galactic or cosmic nose), mainly in the Milky Way, absorption of electromagnetic radiation in the atmosphere, and the noise temperature of the earth. Noise is characterized by its statistical properties. Noise that is directly proportional to the temperature of the ohmic portions of the circuit and the receiver bandwidth is called thermal noise or Johnson noise or white noise. It occurs at all frequencies with equal amplitudes and contains equal power within a fixed bandwidth at any center frequency. Because noise cannot be separated from the returned echo signals, the receiver noise is also amplified along with the desired signal.

Thermal noise or white noise was first measured by John B. Johnson, and the theoretical explanation was given by Harry Nyquist[23] both at Bell Labs. Any electrical system containing thermal noise can be modeled as the Thevenin equivalent circuit with an equivalent voltage source in series with a noiseless input resistor, as shown in Figure 4.13. The voltage source is a random noise signal with voltage variance within some frequency band, and the mean-square voltage $\langle e_n^2 \rangle$ is given by

$$\langle e_n^2(f) \rangle = \int_{-B}^{B} (4R_s hf) \left[\exp\left(\frac{hf}{KT}\right) - 1 \right]^{-1} df \tag{4.65}$$

where f is assumed to be positive frequency in Hertz, $h = 6.625 \times 10^{-34}$ is the Plank's constant, T is the temperature of the receiver, K is the Boltzmann's constant, B is the noise bandwidth of the receiver, and R_s is the resistance of the voltage source. The noise bandwidth is expressed as

$$B = \frac{\int_{-\infty}^{\infty} |H(f)|^2 df}{|H(f_0)|^2} \tag{4.66}$$

where $H(f)$ is the frequency response of the filter, and $H(f_0)$ is the peak response of the filter. Maximum power transfer occurs when the source resistance is equal to the load resistance, that is, when $R_L = R_S$. Thus the power delivered to the load resistance is given by

$$P_n = \frac{\langle e_n^2(f) \rangle}{4R_s} = \int_{-B}^{B} (hf) \left[\exp\left(\frac{hf}{KT}\right) - 1 \right]^{-1} df. \tag{4.67}$$

For typical radar frequencies and reasonable anticipated temperatures, it is observed that $f \ll KT/h \approx 6{,}000$ GHz, assuming $T = T_0$. This implies that P_n can be reasonably approximated to

$$P_n \approx KT_0 B. \tag{4.68}$$

4.7.1 NOISE FACTOR AND EFFECTIVE TEMPERATURE OF A SYSTEM

Noise factor, as defined by (3.12) in Chapter 3, measures the thermal noise generated in the receiver compared to the noise generated by a perfect receiver at standard temperature $T = 290°K$. This ideal receiver has a unity noise factor. But for a practical receiver, the noise factor must be greater than unity. Noise factor, denoted by F, is a power ratio, while noise figure, denoted by NF, is the decibel equivalent implying:

$$NF = 10\log\left(\frac{\text{output noise of practical receiver}}{\text{output noise of ideal receiver at std. temp } T_0}\right). \qquad (4.69)$$

More precisely,

$$NF = 10\log\frac{N_0}{KT_0 BG_a} \qquad (4.70)$$

where the symbols have their usual meanings, assuming the network at T_0.

Noise factor takes into account all noise sources and specifies the performance of the total system, and can be expressed in terms of the effective noise temperature T_e and the source temperature T_0 of the network. Consider the single component amplifier model as shown in Figure 4.14, where

$$N_G = KT_e B, \quad \text{and} \quad N_i = KT_0 B. \qquad (4.71)$$

Therefore, the output noise power is written as

$$N_0 = (N_i + N_G)G_a = KTG_a(T_0 + T_e). \qquad (4.72)$$

Equation (3.13) can be written as

$$F = \frac{S_i/N_i}{S_0/N_0} = \frac{N_0}{N_i(S_0/S_i)}. \qquad (4.73)$$

Using (4.71) and (4.72) yields

$$F = \frac{KBG_a(T_0 + T_e)}{KT_0 BG_a}. \qquad (4.74)$$

More precisely,

$$F = 1 + \frac{T_e}{T_0}. \qquad (4.75)$$

Equivalently, we can write

$$T_e = (F-1)T_0. \qquad (4.76)$$

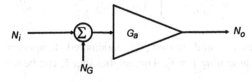

FIGURE 4.14 Single component amplifier model.

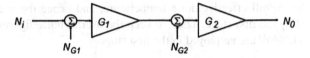

FIGURE 4.15 Cascaded two-stage amplifiers.

Now consider a two-stage cascaded network, as shown in Figure 4.15. In this model, network 1 is defined by power gain G_1, effective temperature T_{e1}, noise figure F_1, and bandwidth B. Similarly, network 2 is defined by power gain G_2, effective temperature T_{e2}, noise figure F_2, and bandwidth B. The output noise power is then given by

$$N_o = KT_0 B G_1 G_2 + KT_{e1} B G_1 G_2 + KT_{e2} B G_2 \tag{4.77}$$

where

$$N_o = K(T_0 + T_e) B G_1 G_2.$$

In other words,

$$N_0 = G_1 G_2 \left(N_i + N_{G1} + \frac{N_{G2}}{G_1} \right). \tag{4.79}$$

This leads to substitute the model of Figure 4.15 by an equivalent model with equivalent temperature T_e of Figure 4.16. Combining (4.77) and (4.78), simplifies the result in an expression of the overall equivalent noise temperature as

$$T_e = T_{e1} + \frac{T_{e2}}{G_1}. \tag{4.80}$$

Application of (4.76) in (4.80) yields

$$F_0 = F_1 + \frac{F_2 - 1}{G_1}. \tag{4.81}$$

Equation (4.80) can be extended to determine the overall effective temperature of an n-stage cascaded system as

$$T_e = T_{e1} + \frac{T_{e2}}{G_1} + \frac{T_{e3}}{G_1 G_2} + \cdots + \frac{T_{en}}{G_1 G_2 G_3 \cdots G_{n-1}}. \tag{4.82}$$

Similarly, (4.81) can be extended to determine the overall noise figure of an n-stage cascaded system as

$$F_0 = F_1 + \frac{F_2 - 1}{G_1} + \frac{F_3 - 1}{G_1 G_2} + \cdots + \frac{F_n - 1}{G_1 G_2 G_3 \cdots G_{n-1}}. \tag{4.83}$$

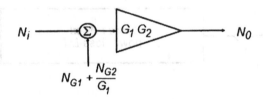

FIGURE 4.16 Equivalent model of two-stage cascaded amplifiers.

It is observed that the overall effective noise temperature, and hence the overall noise figure, is essentially dominated by the first stage. Thus, to keep the overall noise figures as low as possible, low-noise amplifiers (LNA)[24] are employed in the first stage.

Example 4.3

For a four-stage cascade amplifier, $G_1 = 6.02$ dB, $G_2 = 7.78$ dB, $G_3 = 9.03$ dB, $G_4 = 10.00$ dB, $T_{e1} = 145$ K, $T_{e2} = 200$ K, $T_{e3} = 300$ K, and $T_{e4} = 400$ K, find the overall effective noise temperature and the operating noise figure of the amplifier assuming the source at standard temperature T_0.

Solution: We have

$$G_1 = 6.02 \text{ dB} = 4, \quad G_2 = 7.78 \text{ dB} = 6, \quad G_3 = 6.02 \text{ dB} = 8, \quad G_4 = 6.02 \text{ dB} = 10,$$

$$T_{es} = T_{e1} + \frac{T_{e2}}{G_1} + \frac{T_{e3}}{G_1 G_2} + \frac{T_{e4}}{G_1 G_2 G_3} = 145 + \frac{200}{4} + \frac{300}{(4)(6)} + \frac{400}{(4)(6)(8)} = 209.58$$

$$F_{0p} = 1 + \frac{T_{es}}{T_0} = 1 + \frac{209.58}{290} = 1.72 = 2.36 \text{ dB}$$

4.7.2 Noise Temperature of Absorptive Network

An *absorptive network* contains passive and lossy elements. These introduce losses by absorbing energy from the signal and then converting it to heat. Resistive attenuators, transmission lines, and waveguides, and even rainfall, are all examples of absorptive networks.

Consider an absorptive network[25] having a power loss, denoted by L, which is defined as the ratio of input power to output power, and is always greater than unity. As such, the gain is less than unity. Assume that the lossy network is matched at both ends, terminating at a resistor at one end and an antenna at the other. Also assume that the system is at some ambient temperature T_x, as shown in Figure 4.17. The noise power radiated by the antenna is

$$P_{rad} = \frac{KT_x B}{L} + KT_{out} B. \tag{4.84}$$

Since the antenna is matched to the resistive source at T_0, the available noise power is $P_{rad} = KT_0 B$. Using this relation (4.84) yields

$$T_{out} = T_x \left(1 - \frac{1}{L}\right). \tag{4.85}$$

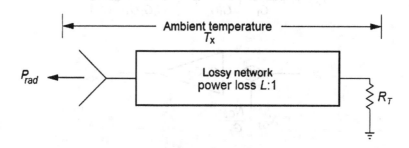

FIGURE 4.17 Lossy network matched at both ends.

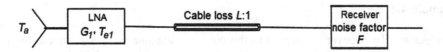

FIGURE 4.18 A system illustrating the overall noise temperature with the LNA followed by the lossy cable.

This is the equivalent noise temperature of the lossy network referred to the output terminals. Thus the equivalent noise temperature of the lossy network referred to the input is

$$T_{in} = T_{out}L = T_x(L-1). \tag{4.86}$$

For the assumption that the network operates at standard temperature $T_x = T_0$ and $T_{in} = T_e$, it is observed, by comparing (4.76) and (4.86), that

$$F = L. \tag{4.87}$$

This indicates that at standard room temperature the noise factor of a lossy network is equal to the loss factor.

4.7.3 Overall Effective Temperature of a Composite System

A typical radar receiving system consisting of amplifiers and a lossy network is considered, as shown in Figure 4.18, in order to illustrate the effects in the overall temperature. The average system noise temperature referred to the input at the antenna is

$$T_s = T_a + T_{e1} + \frac{(L-1)T_0}{G_1} + \frac{L(F-1)T_0}{G_1}. \tag{4.88}$$

Consider another system similar to Figure 4.18 but the components are rearranged, as shown in Figure 4.19. The system temperature referred to input at the antenna is

$$T_s = T_a + T_{e1} + T_0(L-1) + T_{e1}L + \frac{L(F-1)T_0}{G_1}. \tag{4.89}$$

For the systems shown in Figures 4.18 and 4.19, assuming $F = 12$ dB, $L = 5$ dB, LNA gain $G_1 = 50$ dB, and its temperature $T_{e1} = 150°$K, it can be easily computed that T_s is 185°K in the case of Figure 4.18, and 1136° K in the case of Figure 4.19. This indicates that the LNA must be placed in front of the lossy cable. To work with the noise figure rather than the system noise temperature, the conversion can be easily made. If T_{es} represents the average effective input noise temperature of the entire receiving system at point A, and the antenna is considered the source for the system, then

$$T_{es} = T_s - T_a, \tag{4.90}$$

and the average system operating noise figure, denoted by F_{op}, is expressed as

$$F_{op} = 1 + \frac{T_{es}}{T_a}.$$

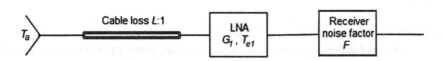

FIGURE 4.19 A system illustrating the overall noise temperature with the lossy cable followed by the LNA.

70

Fundamental Principles of Radar

Example 4.4

Consider a radar receiver connected to the output of antenna via a lossy transmission line. Assume that the antenna temperature $T_a = 175$ K, transmission loss $L = 1.61$ dB at physical temperature $T_x = 255$ K, receiver effective input noise temperature $T_e = 200$ K, noise bandwidth of the receiver B = 10 MHz.

a. Find the average system noise temperature, average effective input noise temperature of the entire receiving system, and the average system operating noise figure.
b. Find the necessary available power gain G to obtain an average output power gain of 0.5 W.

Solution:

a. We have L = 1.61 dB = 1.45.

Then the average system noise temperature T_s and the average effective input noise temperature of the entire receiving system T_{es} are calculated as

$$T_s = T_a + T_x(L-1) + T_e L = 175 + 255(1.45-1) + (200)(1.45) = 579.75 \text{ K}$$
$$T_{es} = T_x(L-1) + T_e L = 255(1.45-1) + (200)(1.45) = 404.75 \text{ K}$$

The average system operating noise figure is then

$$F_{op} = 1 + \frac{T_{es}}{T_a} = 1 + \frac{404.75}{175} = 3.31 = 5.20 \text{ dB}$$

b.

$$N_0 = KT_s B \frac{G}{L} \Rightarrow G = \frac{N_0 L}{KT_s B} = \frac{(0.5)(1.45)}{(1.38 \times 10^{-23})(579.75)(10 \times 10^6)} = 9.06 \times 10^{12} = 129.57 \text{ dB}$$

4.8 SYSTEM LOSSES

The losses that occur throughout the radar system are generally termed as *system losses*. The losses reduce the SNR at the receiver output. A system loss is associated in each stage of signal processing in both transmitting and receiving portions of the radar system. In addition to losses due to signal processing, electromagnetic energy in traveling through the atmosphere suffers another loss known as *atmospheric attenuation*. Other losses caused by field use of the radar system and display losses are qualitative in nature. These loss factors are usually estimated on the basis of past experience and experimental observations of the radar designer.

The loss factors are not statistical in the true sense of the word. Some of these may be very hard to determine precisely. Although the loss associated with any one factor may be small, there are many possible loss mechanisms, and their total contributions in the system losses can be significant. All these loss factors can be broadly classified into three categories such as design losses, operational losses, and propagation losses.

4.8.1 Design Losses

Design losses occur at the design stage, which may be overlooked. Among the most important are the following.

Plumbing loss: There are always some small power losses that occur in the transmission line, waveguides, and duplexers connecting output of the transmitter to the antenna. In addition to the transmission losses, an extra loss occurs at each connection or bends in the line. Connector losses are usually small in the order of a few decibels, but if the connection is

Targets and Interference

poorly designed and implemented, it may contribute significant attenuation. The radar signal suffers attenuation as it passes through the duplexer. The greater the isolation required from the duplexer on the transmission line, the larger the insertion loss will be due to introducing the component into the transmission line.

MTI loss: An MTI loss is introduced when a radar ID is designed to discriminate moving and stationary targets. The MTI discrimination technique results in a complete loss of sensitivity for blind speeds corresponding to certain relative velocities of the target. The blind speed problem and the loss resulting therefrom are discussed in more details in Chapter 5.

Antenna beam-shape loss: In the radar equation the antenna gain was assumed to be a constant equal to the maximum value. But, in practice, the train of received pulses is amplitude-modulated by the shape of the antenna beam. The antenna gain is less if the pulses are received at the beam edges. In case of a fan beam or pencil beam, if the target passes through the center of the beam, the signal received gives the maximum value. But if the target passes through any other point of the beam, the signal received will not obviously correspond to the maximum value resulting in beam-shape loss. When the antenna scans rapidly enough so that the gain on transmit is not the same as the gain on receive, an additional loss, called the *scanning loss*, occurs.

Collapsing loss: The collapsing loss occurs when the unwanted noise from other sources converges in the vicinity of the actual target return. The convergence of these returns may occur in the detection circuitry or on the C-scope display. A collapsing loss can occur when the output of a high-resolution radar is displayed on a device whose resolution is coarser. It also results if the outputs of two or more radar receivers are combined while only one contains a useful signal while the other contains noise. Similar effects may occur in inadequate receiver bandwidth.

Limiting loss: Although a well-designed and engineered radar receiver will not limit the received signal under normal circumstances, the intensity-modulated displays such as the PPI or the B-scope have limited dynamic range. Some receivers sometimes employ limiting for some special purpose. Limiting results in a loss of only a fraction of decibel for a large number of integrated pulses.

Radome loss: Radar dome or radome is a protective dielectric enclosure that protects the radar installation from inclement weather. Because the transmitted and received energy are intercepted by the radome, there is two-way radome loss whenever radomes are used. This loss is a function of the frequency of operation and radome thickness, and sometimes of the radome material.

Integration loss: For radar detection, a train of pulses is received from each target and processed to obtain the high-detection probability. The process by which the pulses are combined is known as *integration*. Integration loss occurs when there is a failure in integrating all signal pulses coherently in the receiver filter. In general, noncoherent integration is not as efficient as coherent integration, but noncoherent integration and detection are most easily implemented at the cost of integration loss. This integration loss is caused by the passage of the radar signal through the envelope detector at a lower level. The loss is the difference between the total energy contained in the returned pulse and that obtained from the integration of these pulses.

4.8.2 OPERATIONAL LOSSES

The second category comprises operational losses, some of which are discussed next.

Operator loss: Crossing of a threshold in the receiver indicates the presence of a target. In many cases this threshold crossing is displayed to an operator for visual identification. The performance will decrease if the operator is distracted, overloaded, or not well-trained. When the operator introduces loss into the system, it is easy to select a proper value to compensate for the loss.

Field degradation: When a radar is operated under field conditions, the performance usually deteriorates even more especially if the equipment is operated and maintained by inexperienced or unmotivated personnel. This loss may even apply to equipment operated by experienced and motivated professionals under adverse field conditions. Factors contributing to field degradation include equipment age, errors in equipment setting and maintenance, water in transmission lines, incorrect mixer-crystal current, loose cable connections, etc. Careful observation of performance monitoring instruments and timely preventive maintenance can keep radar performance up to design level. A good estimate of the field degradation is difficult to obtain because it cannot be predicted and is dependent on the conditions under which it is operating. Because of environmental conditions, a loss of 1 to 10 decibels may occur between electronic and human detection.

4.8.3 PROPAGATION LOSSES

The third category system losses are due to propagation of radar signals in the atmosphere. In an analysis of radar performance, it is convenient to assume that radar and target are both located in free space. However, the presence of other objects in the universe modifies the free-space behavior, which can have a significant effect on radar performance. The modifying objects that are considered are the earth and atmosphere. Some of the mechanisms that affect radar performance are as follows.

Attenuation: The attenuation of radar waves in the lower atmosphere occurs as it propagates through the earth's atmosphere, and is due primarily to the presence of both free molecules and suspended particles such as dust grains and water drops, which condense in fog and rain. In the absence of condensation, the attenuation is due to oxygen and water vapor. The effect of atmospheric attenuation is to absorb and scatter power in the radar wave, thus attenuating both the signal that strikes the target and the signal that returns to the radar after being reflected from the target.

A radar wave passing through an ionized layer of the atmosphere also results in attenuation, which is primarily due to the absorption of signal energy when the vibrating electrons collide with each other. There are two types of absorption in the atmosphere: nondeviating and deviating. Nondeviating absorption results in a medium where the refractive index is approximately unity, while when the refractive index is less than 1, deviation absorption results.

Attenuation of radar waves in the normal, clear atmosphere is negligible at lowest radar frequencies, becoming significant in the microwave bands, and imposing severe limits on radar performance in the millimeter-wave bands.

Reflection: Radar signals propagate directly from the radar to the target. But there can also be energy that travels to the target via a path that includes a reflection from the earth's surface. The direct and ground-reflected waves interfere at the target either constructively or destructively to produce reinforcements or nulls, respectively. The lobe structure that results causes nonuniform illumination of the coverage and influences the normal performance of the radar.

Refraction: Refraction is the change in the direction that radar waves travel due to spatial change in the index of refraction in the atmosphere. The refractive index is defined as the ratio of the speed of light in free space to the phase velocity of the wave in the medium. Refraction impacts radar waves by introducing an error in measuring elevation angle due to the change in the refractive index as a function of altitude, and by extending the horizon thereby increasing the measured value of the range. The most common way to deal with refraction is to replace the actual earth with an imaginary larger earth whose effective radius is four thirds the radius of the actual earth. At times, atmospheric conditions can cause more than usual bending, resulting in an increased range.

Targets and Interference

PROBLEMS

4.1 A C-band monostatic radar operating at 5 GHz is to track a conducting sphere calibration target. Find the sphere's cross section if the diameter of the sphere is 20 cm.

4.2 A hollow aluminum sphere with a diameter of 20 cm is the target at 10.5 GHz. (a) Find the radar cross section of the target. (b) What is the minimum frequency that makes the sphere optical?

4.3 Find the target RCS of a sphere with a diameter of 1 m. Find the target RCS of a flat plate with the same projected area as the sphere for C-band radar operating at 6 GHz.

4.4 For an X-band radar operating at 10 GHz, find the target RCS of 1.2 m × 1.2 m flat plate and of a corner reflector with 1.2 m sides.

4.5 A low-noise RF amplifier is connected to an antenna with $T_s = T_a = 250$ K. The amplifier has a noise figure of $F = 10$ dB. Find the overall operating noise figure F_s of the system.

4.6 An LNA is connected to a radar receiver that has a noise figure of 10 dB. The available power gain of the LNA is 40 dB, and its effective noise temperature is 150 K. Find the overall noise temperature and the noise figure referred to the input.

4.7 For the system shown in Figure 4.18, the radar receiver noise figure $F = 10$ dB, the cable loss $L = 3$ dB, the LNA gain $G_1 = 40$ dB, and its equivalent noise temperature $T_{el} = 170$ K. The antenna noise temperature is 40 K. Find the overall noise temperature of the radar receiver system referred to the input.

4.8 Repeat the calculation when the system of Figure 4.1 is modified to the system as shown in Figure 4.19.

4.9 An LNA is connected to an antenna having a noise temperature of 40 K through a lossy feeder cable having a loss of 3 dB. The LNA has a noise temperature of 100. Find the system noise temperature referred to (a) the cable feeder input and (b) the LNA input.

4.10 A radar receiver system consists of an antenna having a noise temperature of 70 K, feeding directly into an LNA. The LNA has a noise temperature of 110 K and a gain of 50 dB. The coaxial feeder between the LNA and the receiver has a loss of 3 dB, and the main receiver has a noise figure of 10 dB. Find the system noise temperature referred to input.

4.11 A radar receiver consisting of RF amplifier with a noise figure of 10 dB and a gain of 30 dB, and a combination of an IF mixer and amplifier with noise figure of 12 dB and a gain of 50 dB.

(a) Find the overall noise figure referred to the input.

(b) If the antenna temperature is 800 K, find the operating noise figure for the system.

REFERENCES

1. G. T. Ruck et al., *Radar Cross-Section Handbook* (New York: Plenum Press, 1970).
2. J. W. Crispin and K. M. Siegel, *Method of Radar Cross-Section Analysis* (New York: Academic Press, 1968).
3. R. F. Harrington and J. R. Mautz, "Straight Wires with Arbitrary Excitation and Loading," *IEEE Trans, Antennas and Propagation*, AP-15 (1967).
4. E. F. Knot, J. F. Shaeffer, and M. T. Tuley, *Radar Cross Section* (Norwood, MA: Artech House, 1985).
5. L. N. Ridenour, *Radar System Engineering, MIT Radiation Laboratory Series* (New York: McGraw-Hill Book Company,1947).
6. J. I. Marcum, "A Statistical Theory of Target Detection by Pulsed Radar: Mathematical Appendix," *IRE Transaction on Information Theory,* IT-6, no. 2 (1960).
7. Peter Swerling, "Detection of Fluctuating Pulsed Signal in the Presence of Noise," *IRE Transaction on Information Theory,* IT-3, no. 3 (1957).
8. H. Goldstein, "A Primer of Sea Echo," *Report No. 157,* U.S. Navy Electronics Laboratory, San Diego, 1950.
9. R. M. Long, *Radar Reflectivity of Land and Sea* (Dedham, MA: Artech House, 1983).

10. R. D. Hayes and F. B. Dyer, "Computer Modeling for the Fire Control Radar Systems," *Technical Report* no. 1, Contract DAA25-73-C-0256, Georgia Institute of Technology, Atlanta, December 1974.
11. M. I. Skolnik, "Sea Echo," in *Radar Handbook,* ed. M. I. Skolnik (New York: McGraw-Hill Book Company, 1970).
12. P. M. Austin, "Radar Measurement of the Distribution of Precipitation in New England Storm," *Proceedings of the 10th Weather Conference* (1965): 247–254.
13. K. L. S. Gunn and T. W. R. East, "The Microwave Properties of Precipitation Particles," *Quarterly Journal of the Royal Meteorological Society,* 80 (1954): 522–545.
14. T. G. Konrad, J. J. Hicks, and E. B. Dobson, "Radar Characteristics of Birds in Flight," *Science,* 159 (January 19, 1968), 274–280.
15. E. W. Houghton, F. Blackwell, and T. A. Willmot, "Bird Strike and Radar Properties of Birds," *International Conference on Radar—Present and Future,* IEE Conference Pub 05 (1973): 257–262.
16. W. L. Flock and J. L. Green, "The Detection and Identification of Birds in Flight, Using Coherent and Noncoherent Radars," *Proceedings of the IEEE,* 53 (1974): 745–753.
17. K. M. Glover, K. R. Hardy, T. G. Konrad, W. N. Sullivan, and A. S. Michaels, "Radar Observations of Insects in Free Flight," *Science,* 254 (1966), 967–972.
18. J. R. Riley, "Radar Cross Section of insects," *Proceedings of the IEEE,* 73 (February 1985): 228–232.
19. C. R. Vaughn, "Birds and Insects as Radar Targets: A Review," *Proceedings of the IEEE,* 73 (February 1985): 205–227.
20. K. D. Ward and S. Watts, "Radar Sea Clutter," *Microwave Journal,* 29, no. 6 (1985): 109–121.
21. R. R. Boothe, "The Weibull Distribution Applied to the Ground Clutter Backscatter Coefficient," *USAMC Report* RE-TR-69-15, June 1969.
22. D. C. Schleher, ed., *MTI Performance Analysis in MTL Radar* (Dedham, MA: Artech House, 1978), 183–251.
23. H. Nyquist, "Thermal Agitation of Electric Charge in Conductors," *Physical Review,* 32, no. 1 (1928): 110–113.
24. O. Holt, "Technology Survey—A Sampling of Low Noise Amplifiers," *The Journal of Electronic Defense* (2015).
25. D. Roddy, *Satellite Communications* (New York: McGraw-Hill Book Company, 2001).

5 Propagation of Radar Waves

"If you travel a lot, if you like roaming about in order to lose yourself, you can end up in the strangest places. I think it must be a kind of built-in radar, which often takes me to places that are either peculiarly quiet or peculiar in a quiet sort of way."

—Wim Wenders

5.1 INTRODUCTION

The concept of EM energy traveling in straight lines applies only when the radar system exists in an otherwise empty universe (free space) without any interference from the earth and its atmosphere, which are mainly responsible for modifying the free space behavior. Although the free space analysis is only an approximation for radar detection calculations, an accurate prediction of the radar performance can be achieved by including the effects of *reflection* from the surface of the earth, *refraction* caused by an inhomogeneous atmosphere, *diffraction* caused by interaction of the radar waves with the earth's curvature beneath the radar LOS, and *attenuation* of radar waves by gases constituting the atmosphere. In addition to these effects, clutters and meteorological conditions including rain, dust, and fog can also affect the propagation of radar waves. Despite accurate quantitative predictions are not always easy to obtain because of the difficulty in acquiring the necessary knowledge of the environment in which the radar operates; some concepts and ideas may result in possible areas of investigation in order to mitigate or eliminate those effects.

The atmosphere can be divided roughly into three layers: the troposphere, the stratosphere, and the ionosphere, as illustrated in Figure 5.1. The first layer, which extends from the earth's surface to an altitude of 10 km is known as the troposphere. The radar wave refracts as it travels through this layer.

Refraction occurs due to spatial change in the index of refraction, which is a function of frequency and dielectric constant of the layer. The layer above the troposphere extending from 10 km to 60 km behaves like free space and is called the stratosphere. The stratosphere has no effect on radar waves. The ionosphere extends from about 60 km to about 600 km, and is composed of a significant amount of ionized free electrons. The presence of free electrons in the ionosphere affects the propagation of the EM wave in different ways including reflection at low frequency, refraction, absorption, noise emission, and polarization rotation. When considering the effects of the earth and its atmosphere, two different regions can be distinguished: the interference region, which lies above the line of sight, and the diffraction region, which lies below the line of sight of the radar, as depicted in Figure 5.2.

5.2 REFLECTION EFFECTS DUE TO MULTIPATH PHENOMENA

Multipath[1] is the propagation of an EM wave from one point to another by more than one path. It usually consists of a direct path and one or more indirect paths by reflection from the surface of the earth or sea or from large man-made structures, as depicted in Figure 5.3. At frequencies below approximately 40 MHz, it may also include more than one path through the ionosphere.

5.2.1 REFLECTION FROM A PLANE FLAT EARTH

It is worthwhile first to consider a greatly simplified geometry of multipath propagation over a plane, flat, and reflecting surface of the earth, as shown in Figure 5.4, and then extend the results to account for the actual round earth. The radar antenna is located at a height h_r. The target is

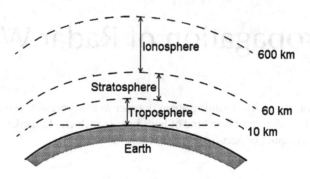

FIGURE 5.1 Layers of the atmosphere.

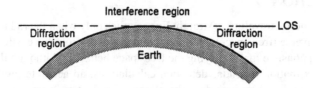

FIGURE 5.2 Interference and diffraction regions.

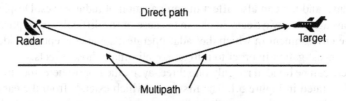

FIGURE 5.3 Multipath phenomena.

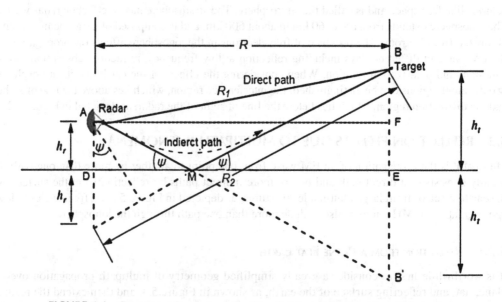

FIGURE 5.4 Geometry of specular reflection of a plane reflecting the earth's surface.

Propagation of Radar Waves

at height h_t and at a distance R from the radar. The incident and reflected angles at point M are both ψ, also known as the grazing angle. The EM energy radiated from the antenna arrives at the target via two separate paths: the direct path AB $= R_1$ and the indirect path AMB = AMB' $= R_2$ where B' is the image of B. Using the geometry of Figure 5.3, the direct and indirect paths are expressed as

$$R_1 = \sqrt{R^2 + (h_t - h_r)^2}, \tag{5.1}$$

and

$$R_2 = \sqrt{R^2 + (h_t + h_r)^2} \tag{5.2}$$

where the approximations are valid when $R \gg h_t$, h_r. Equations (5.1) and (5.2) can be approximated using the binomial series expansion as

$$R_1 \approx R + \frac{(h_t - h_r)^2}{2R}, \tag{5.3}$$

and

$$R_2 \approx R + \frac{(h_t + h_r)^2}{2R}. \tag{5.4}$$

The difference between the reflected path and the direct path is

$$\Delta R = R_2 - R_1 \approx \frac{2h_t h_r}{2R}, \quad h_t \gg h_r. \tag{5.5}$$

Any difference in relative phase between the direct and the reflected waves is the difference in path length and the change in phase that occurs in reflection at point M. Thus the phase difference caused by the two paths is given by

$$\Delta\phi = \frac{2\pi \Delta R}{\lambda} = \frac{4\pi h_t h_r}{\lambda}. \tag{5.6}$$

The reflection coefficient of the flat, smooth surface may be considered as a complex quantity $\Gamma = \rho e^{j\varphi}$, where ρ describes the change in amplitude and φ is the phase shift on reflection. It is assumed—for a smooth surface with good reflecting properties—that the complex reflection coefficient $\Gamma = -1$, implying that the reflected signal suffers no change in amplitude, but its phase is shifted by 180°. The total phase difference between the direct and the surface reflected signals as measured at the target is

$$\Delta\Phi = \Delta\phi + \pi = \frac{4\pi h_t h_r}{\lambda} + \pi. \tag{5.7}$$

The resultant amplitude of the two signals, each with unity amplitude, is expressed as

$$E_T = \left|1 + e^{-j\Delta\Phi}\right| = \left[2(1 + \cos\Delta\Phi)\right]^{1/2} = \left[2(1 - \cos\Delta\phi)\right]^{1/2}. \tag{5.8}$$

The overall signal strength is then modified at the target located at B by the ratio of the signal strength in the presence of the earth to the signal strength at the target in free space. The modulus of this ratio, known as the propagation factor F, is also expressed by (5.8). Therefore, the ratio of the power incident on the target to that, which would be incident if the target were located in free space, is

$$F^2 = E_T^2 = 2(1 - \cos \Delta\phi) = 4\sin\left(\sin\frac{\Delta\phi}{2}\right)^2. \tag{5.9}$$

Because of the reciprocity, the path from radar to target is the same as from target to radar. The expression of the power ratio at the radar is

$$F^4 = 16\left(\sin\frac{\Delta\phi}{2}\right)^4. \tag{5.10}$$

Substituting (5.6) in (5.10) yields

$$F^4 = 16\sin^4\frac{2\pi h_t h_r}{\lambda R}. \tag{5.11}$$

Finally, the radar equation describing the received echo power must be modified by the propagation factor F^4 given by (5.11). It is observed that the fourth power relation between the range and echo signal results in a variation of radar range from 0 to 2 times range of the same radar in free space.

The maxima of F^4 in (5.11) occur when the argument of the sine term is equal to odd multiples of $\pi/2$, and are, therefore, defined by

$$\frac{4h_r h_t}{\lambda R} = 2n + 1 \tag{5.12}$$

where $n = 0, 1, 2, \ldots$ The minima or nulls occur when the sine term is zero. More precisely,

$$\frac{2h_r h_t}{\lambda R} = n. \tag{5.13}$$

Thus the maxima are produced in the propagation factor at target heights $h_t = (2n+1)$ $(\lambda R/4h_r)$, $n = 0,1,2,3,\cdots$, and the maxima are produced at target heights $h_t = n(\lambda R/2h_t)$, $n = 0,1,2,3,\cdots$. Therefore, the antenna elevation coverage due to the presence of plane reflecting surface is represented by a lobed structure, as indicated in Figure 5.5. A target located at the maximum of a particular lobe can be detected at a range twice that of the same radar located in free space. However, targets at other heights can be less than the free space range. For targets located in the null, no echo signal is received.

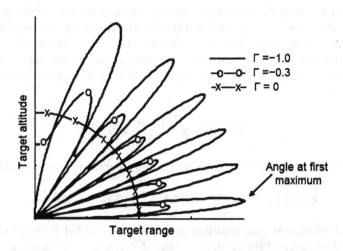

FIGURE 5.5 Vertical lobe structures due to the plane, reflecting surface.

Propagation of Radar Waves

Finally, the radar equation, expressed in (3.10) in Chapter 3, can be modified by incorporating the effects of multipath reflections due to the presence of a plane, reflecting surface as

$$P_r = \left[\frac{P_t G^2 \lambda^2 \sigma}{(4\pi)^3 R^4} \right] \left[16 \sin^4 \frac{2\pi h_t h_r}{\lambda R} \right]. \tag{5.14}$$

For targets at low angles, (5.14) can be approximated by

$$P_r \approx \frac{4\pi P_t G^2 \sigma (h_t h_r)^4}{\lambda^2 R^8}. \tag{5.15}$$

The power of the received echo from targets at low angles varies inversely as the eighth power of the range instead of the usual fourth power. The accelerated drop of the received echo power as the range is increased is the result of the direction to the target getting closer to a null at a zero grazing angle.

Example 5.1

Consider an X-band airborne radar operating at 10 GHz at a height of 4 km and a target at a height of 7 km with a radar-target horizontal range $R = 40$ km. Calculate (a) the length of the direct and indirect paths, (b) the angle of incidence (grazing angle), (c) the time for the radar signal to intercept the target via the direct and indirect paths, and (d) the Doppler shift along the direct path assuming a radar velocity of 150 m/s and the approaching target with a velocity of 300 m/s.

Solution:

a. Using (5.1):

$$R_1 = \sqrt{R^2 + (h_t - h_r)^2} = \sqrt{(40 \times 10^3)^2 + [(7 \times 10^3) - (4 \times 10^3)]^2} = 40.11 \text{ km}$$

Using (5.2) gives:

$$R_2 = \sqrt{R^2 + (h_t + h_r)^2} = \sqrt{(40 \times 10^3)^2 + [(7 \times 10^3) + (4 \times 10^3)]^2} = 41.48 \text{ km}$$

b. It can be easily shown from Figure (5.4) that

$$\sin \psi = \frac{h_t + h_r}{R_2} \Rightarrow \psi = \sin^{-1} \left[\frac{(7 \times 10^3) + (4 \times 10^3)}{(41.48 \times 10^3)} \right] = 15.38°$$

c.

$$t_1 = \frac{40.11 \times 10^3}{3 \times 10^8} = 0.134 \text{ ms}, \quad \text{and} \quad t_2 = \frac{41.48 \times 10^3}{3 \times 10^8} = 0.138 \text{ ms}$$

d. Both the target and the airborne radar make an angle θ with the direct path. The angle θ is given by

$$\tan \theta = \frac{(7-4)(10^3)}{40 \times 10^3} \Rightarrow \theta = 4.289°$$

The Doppler shift is

$$f_d = \frac{2v_r}{\lambda} = \frac{(20)(300 - 150) \cos(4.289°)}{(3 \times 10^8 / 10^{10})} = 99.72 \text{ kHz}$$

The reflecting coefficient of smooth surface depends largely on the type of polarization of the wave. The reflection coefficients for horizontal and vertical polarizations are expressed by

$$\Gamma_h = \frac{\sin\psi - \sqrt{\varepsilon - (\cos\psi)^2}}{\sin\psi + \sqrt{\varepsilon - (\cos\psi)^2}} \quad (5.16)$$

and

$$\Gamma_v = \frac{\varepsilon\sin\psi - \sqrt{\varepsilon - (\cos\psi)^2}}{\varepsilon\sin\psi + \sqrt{\varepsilon - (\cos\psi)^2}} \quad (5.17)$$

where ψ is the grazing angle and ε is the complex permittivity (dielectric constant) given by

$$\varepsilon = \varepsilon' - j\varepsilon'' = \varepsilon' - j60\sigma\lambda. \quad (5.18)$$

where σ is the conductivity of the lossy dielectric and λ is the wavelength. The typical values of $\varepsilon = \varepsilon' - j\varepsilon'' = \varepsilon' - j60\sigma\lambda$, where ε' and ε'' can be found in the literature.

Figure 5.6 displays the magnitude plots of reflection coefficients corresponding to horizontal and vertical polarizations for seawater at 28°C where ε'= 65, and ε''= 30.7 at X-band. Likewise, Figure 5.7 displays the phase plots of reflection coefficients corresponding to horizontal and vertical polarizations for seawater at 28°C where ε'= 65, and ε''= 30.7 at X-band. The magnitude of the reflection coefficients of horizontally polarized energy behave differently from waves with vertical polarization in both cases. It is observed that the reflection coefficient, in both cases, for the vertically polarized wave is less than that for the horizontally polarized wave. The different reflection coefficients with the two polarizations result in different coverage patterns. Vertical polarization is preferred when complete vertical coverage is desired.

5.2.2 Reflection from a Smooth Spherical Earth

To account for the complete treatment of the multipath propagation on radar performance, the curvature of the earth cannot be ignored. This is especially true for coverage at low elevation angles near the horizon. The direct and the reflected waves interfere with each other to produce a lobed

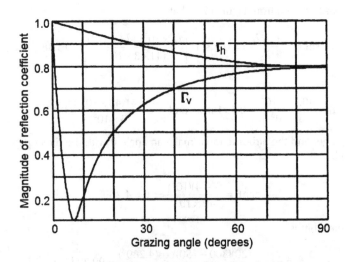

FIGURE 5.6 Magnitude of reflection coefficients for sea water (From Mahafza, 1998).

Propagation of Radar Waves

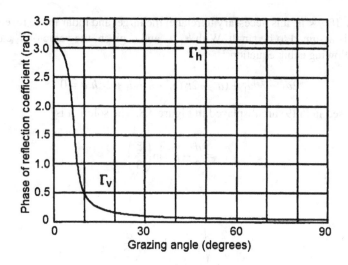

FIGURE 5.7 Phase of reflection coefficients for sea water (From Mahafza, 1998).

radiation pattern similar to that described for the flat earth. Lobing is not as pronounced as in the case of spherical earth.[2]

The analysis of surface reflection from the spherical earth requires a determination of the geometrical specular reflection point at a distance d_1, shown in Figure 5.8, from either the radar antenna or the target. Approximate solutions[3] are available for smaller angular distance between the radar and the target near the surface of the earth, and also for very large distances such as the case of the earth station and the geostationary satellite.

For smooth spherical earth, the reflected wave diverges because of the earth's curvature. As a result, the reflected energy is defocused and decreased in field strength, which can be accounted for by a factor called the *divergence factor*, denoted by D. The divergence factor can be derived using the geometrical configuration in Figure 5.8. In this analysis it is assumed that the grazing angle ψ is

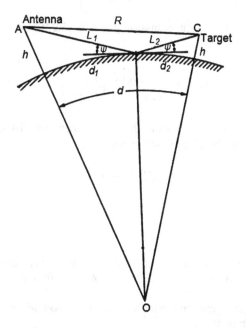

FIGURE 5.8 Multipath propagations over the spherical earth.

small, and $h_r \geq h_t$. If $h_r < h_t$, it is necessary to relabel the target and radar in the reverse manner. The grazing angle ψ is assumed to be small. With h_r, h_t, and d given, the value of d_1 can be calculated by solving the following cubic equation:[4]

$$2d_1^3 - 3dd_1^2 + (d^2 - 2r_e(h_r + h_t))d_1 + 2r_e h_r d = 0. \tag{5.19}$$

The parameters used in (5.19) are displayed in Figure 5.8. The solution is

$$d_1 = \frac{d}{2} - p \, \sin\left(\frac{\xi}{3}\right) \tag{5.20}$$

where

$$p = \frac{2}{\sqrt{3}} \sqrt{r_e(h_t + h_r) + \left(\frac{d}{2}\right)^2}, \tag{5.21}$$

$$\xi = \sin^{-1}\left[\frac{2r_e(h_t - h_r)d}{p^3}\right]. \tag{5.22}$$

For a small grazing angle, it can be shown that

$$\psi = \sin^{-1}\left[\frac{2r_e h_r + h_r^2 - d_1^2}{2r_e d_1}\right] \approx \frac{1}{d_1}\left[h_r - \frac{d_1^2}{2r_e}\right]. \tag{5.23}$$

A widely accepted approximation for the divergence factor is given by Kerr[5] as

$$D = \left[1 + \frac{2d_1(d - d_1)}{r_e d \sin\psi}\right]^{-1/2}. \tag{5.24}$$

By using the small grazing angle approximation given by (5.23) in the expression for D in (5.24) we get

$$D \approx \left[1 + \frac{2d_1^2(d - d_1)}{r_e d(h_r - (d_1^2 / 2r_e))}\right]^{-1/2}. \tag{5.25}$$

For all these results to be valid, it is necessary that d not exceed the distance of the radio horizon of the radar. This is true if it satisfies the following relation:

$$d \leq \sqrt{2r_e h_r} + \sqrt{2r_e h_t}. \tag{5.26}$$

Example 5.2

Consider a shipboard radar at an altitude of 25 m above the spherical smooth sea to observe a target at an altitude of 1.2 km and a surface distance of 50 km from the radar.

   a. Show that the target appears to the radar above the radio horizon.
   b. Find the grazing angle ψ.
   c. Find the divergence factor D.

Propagation of Radar Waves

83

Solution:

a. The radio horizon between the radar and target with the given altitudes is

$$d = \sqrt{(2)(4/3)(6371)(0.025)} + \sqrt{(2)(4/3)(6371)(1.2)} = 20.61 + 142.78 = 163.39 \text{ km}$$

where

$$h_r = 25 \text{ m} = 0.025 \text{ km}$$

where all distance lengths are expressed in km. The surface distance of 50 km is less than 448.96, implying that the target appears to the radar above the horizon.

b. Using (5.23) by reversing labeling (since $h_r < h_t$) gives

$$\psi \approx \frac{1}{d_1}\left[h_t - \frac{d_1^2}{2r_e}\right]$$

where

$$d_1 = \frac{d}{2} - p\sin\left(\frac{\xi}{3}\right)$$

where

$$p = \frac{2}{\sqrt{3}}\sqrt{r_e(h_t + h_r) + \left(\frac{d}{2}\right)^2} = \frac{2}{\sqrt{3}}\sqrt{\frac{4}{3}(6371)(1.2 + 0.025) + \left(\frac{50}{2}\right)^2} = 121.27 \text{ km}$$

$$\xi = \sin^{-1}\left[\frac{2r_e(h_r - h_t)d}{p^3}\right] = \sin^{-1}\left[\frac{(2)(4/3)(6371)(0.025 - 1.2)(50)}{(121.27)^3}\right] = -.594 \text{ rad}$$

leading to

$$d_1 = \frac{d}{2} - p\,\sin\left(\frac{\xi}{3}\right) = \frac{50}{2} - (121.27)\sin\left(\frac{-.594}{3}\right) = 48.855 \text{ km}$$

$$\psi \approx \frac{1}{d_1}\left[h_t - \frac{d_1^2}{2r_e}\right] = \frac{1}{48.855}\left[1.2 - \frac{(48.855)^2}{(2)(4/3)(6371)}\right] = 0.0217 \text{ rad} = 1.24°$$

c. Using (5.25) gives:

$$D \approx \left[1 + \frac{2d_1^2(d - d_1)}{r_e d(h_t - (d_1^2/2r_e))}\right]^{-1/2} = \left[1 + \frac{(2)(48.855)^2(50 - 48.855)}{(4/3)(6371)(50)\left[1.2 - \frac{48.855^2}{(2)(4/3)(6371)}\right]}\right]^{-1/2} = .9924$$

Roughness Factor

Roughness of the surface also lowers the effective magnitude of the reflection coefficient by scattering radiation in nonspecular directions. The effect of roughness on the magnitude of the reflection coefficient has been accounted for by use of *scalar roughness factor S*, having values between 0 and 1. The Rayleigh roughness criterion[6] S_r is commonly used to estimate the maximum surface irregularity that will not significantly lower the reflection coefficient, and expressed as

$$S_r = 4\pi(h_s/\lambda)\sin\psi \tag{5.27}$$

where h_s is the maximum peak-to-trough variations within the Fresnel zone and λ is the free space wavelength. In general, a surface is considered smooth for $S_r < 0.3$. For rough surfaces the reflected signal has two components: one is a *specular* component, which is coherent with the incident signal, and the other is a *diffuse* component, which fluctuates in magnitude and phase with a Rayleigh distribution. The specular component is described by a reflection coefficient $\Gamma_s = \rho_s \Gamma_0$ where ρ_s is the reduction factor due to the specular component. For a slightly rough surface with a random height distribution we have

$$\rho_s = \exp\left(-\frac{S_r^2}{2}\right).$$

(5.28)

Equation (5.28) tends to underestimate ρ_s for very rough surfaces. A recent derivation[7] of ρ_s for the sea surface suggests a better estimate given by the expression

$$\rho_s = \exp\left(-\frac{S_r^2}{2}\right)I_0\left(\frac{S_r^2}{2}\right)$$

(5.29)

where I_0 is the modified Bessel function of zero order. This expression is in good agreement with experimental results. The diffuse component of the reflected field results from scattering over a large area with the major contribution from the region well outside the first Fresnel zone. The region contributing to diffuse scatter is known as the *glistening surface*, from which signals are scattered randomly. The diffuse amplitude reflection coefficient Γ_d is expressed as

$$\Gamma_d = \rho_d \Gamma_0$$

(5.30)

where ρ_d is the coefficient depending solely on surface irregularities. There is no simple expression for ρ_d in the literature, and its value varies between 0.2 and 0.4. The total field above a reflecting surface is the resultant of the direct component, the coherent specular component, and the random diffuse component, showing what is popularly known as the *Nakagami-Rice distribution*. The total scattered power is constant given by

$$\langle \rho_s \rangle^2 + \langle \rho_d \rangle^2 = \text{constant}$$

(5.31)

where $\langle . \rangle$ indicates the rms values. This representation is sometimes useful in estimating the reflection from rough surfaces.

5.3 REFRACTION OF EM WAVES

Refraction is the deviation of an electromagnetic wave from a straight line as it passes through the atmosphere due to a spatial variation in the index of refraction as a function of height. The index of refraction, denoted by the symbol n, is defined as

$$n = \frac{c}{v_p}$$

(5.32)

where c is the velocity of light in free space and v_p is the velocity of propagation of the wave in the medium characterized by

$$v_p = \frac{1}{\sqrt{\mu\varepsilon}} = \frac{1}{\sqrt{\mu_0\mu_r\varepsilon_0\varepsilon_r}} = \frac{c}{\sqrt{\varepsilon_r}}$$

(5.33)

where μ_o, ε_o are the absolute values of permeability and permittivity of the medium, respectively, and μ_r, ε_r are the relative permeability and permittivity of the medium, respectively.

Propagation of Radar Waves 85

For a nonmagnetic medium, $\mu_r = 1$ and $c = 1/\sqrt{\mu_0 \varepsilon_0}$. Using (5.33) in (5.32) yields

$$n = \frac{1}{\sqrt{\varepsilon_r}}. \tag{5.34}$$

Averaged over many locations and over long periods of time, it has been found that the refractive index n of the troposphere decreases with increasing altitude. The refractivity N is related to n by the relation given by

$$N = (n-1) \times 10^6. \tag{5.35}$$

The refractivity depends on meteorological conditions such as temperature, atmospheric pressure, and water vapor pressure, and an approximate formula is expressed as

$$N = \frac{77.6}{T} \left(p + \frac{4810e}{T} \right) \tag{5.36}$$

where T is the air temperature in Kelvin, p is the total pressure in millibars, and e is the partial pressure of water vapor in millibars. Equation (5.33) shows that v_p increases as n decreases. As a result, the EM wave travels faster as a direct function of the altitude. Thus the wave tends to bend downward under normal atmospheric conditions.

The previous discussion assumes that the index of refraction decreases in a smooth monotonic fashion as a function of height, but this leads to errors in the measurements of range and time delay. To account for such errors, the atmosphere, for a first approximation, is modeled as a series of planar slabs, each of a constant refractive index $\{n_m : m = 0,1,2,3,\cdots\}$. Then the angle $\{\alpha_m : m = 0,1,2,3,\cdots\}$ at which the wave emerges from each layer is related to its corresponding refractive index by Snell's law:

$$n_0 \cos\alpha_0 = n_1 \cos\alpha_1 = n_2 \cos\alpha_2 \ldots\ldots = n_m \cos\alpha_m. \tag{5.37}$$

However, a better approximation is obtained by considering a radially stratified atmosphere with M spherical layers, each of a constant refractive index $\{n_m : m = 0,1,2,3,\cdots\}$ between concentric spheres with a corresponding height of the layer from the center of the earth $\{r_m : m = 0,1,2,3,\cdots\}$ and an angle emergence $\{\alpha_m : m = 0,1,2,3,\cdots\}$. Again, Snell's law applies as

$$n_0 r_0 \cos\alpha_0 = n_1 r_1 \cos\alpha_1 = n_2 r_2 \cos\alpha_2 \ldots\ldots = n_m r_m \cos\alpha_m. \tag{5.38}$$

If α_0 and the index of refraction n_0 are known as a function of height, the remaining angles can be easily found from (5.38). The approximation improves as the thickness of each layer is made narrower. In reality the atmosphere is not radially stratified. The information about the index of refraction as a function of height for various locations on the earth's surface can be combined with the generalized ray tracing techniques to evaluate refraction effects more accurately. Normally this is required only for very high-precision radar systems. For most applications the model of radially stratified atmosphere is an adequate approximation.

5.3.1 Four-Thirds Earth Model

To account for atmospheric refractions, a classical way of dealing with it is to replace the actual earth of radius r_0 by an imaginary equivalent earth of radius $r_e = \kappa r_0$, with the actual atmosphere replaced by a homogeneous atmosphere in which the EM waves propagate in starlight lines rather than curved lines, as depicted in Figure 5.9. The constant κ is the ratio of the effective radius to the actual earth radius, and is written as

$$\kappa = \frac{r_e}{r_o}. \tag{5.39}$$

FIGURE 5.9 Illustration of the effects of equivalent earth: (a) refraction due to actual earth and (b) refraction due to the equivalent earth.

For a radially stratified atmosphere, the curvature of the ray[8] relative to the surface of the earth is given by

$$\frac{1}{r_e} = \frac{1}{r_o} + \frac{dn}{dh} \qquad (5.40)$$

where the symbols have their usual meanings, h represents the distance above the earth's surface, and (dn/dh) represents the refractivity gradient, and is normally negative. For this new earth, the ray path is a straight line with the introduction of the ratio κ. Substitution of (5.39) into (5.40) results in an explicit expression of κ given by

$$\kappa = \frac{1}{1 + r_o(dn/dh)}. \qquad (5.41)$$

Assuming r_o = 6,370 km, and a typical value of the refractivity gradient $(dn/dh) = 3.9 \times 10^{-8}$ m at microwave frequencies, a typical value κ becomes 4/3. By this the inhomogeneous atmosphere is replaced by a homogeneous atmosphere over a slightly larger earth through which the EM waves are assumed to travel in straight lines. It is only an approximation, and may not yield correct results if precise radar measurements are desired.

5.3.2 Anomalous Propagation

Nonstandard propagation occurs when κ is not equal to 4/3. Even though the four-thirds earth model provides only an approximation of the effects of atmospheric refraction, propagation, which deviates from this approximation, is called *anomalous propagation*. Three forms of deviation that commonly occur are *sub-refraction, super-refraction* and *ducting*. These are illustrated in Figure 5.10.

When $(dn/dh) \geq 0$, a condition called sub-refraction occurs, which causes the wave to bend away from, instead of toward, the earth. This results in a decrease of the ground coverage. For (dn/dh) more negative waves bend more strongly toward the earth. This condition is called super-refraction resulting in the increase of ground coverage. An effect called ducting occurs when $(dn/dh) < -16 \times 10^{-8}/m$. In this case the radius of curvature of a transmitted wave will be less than or equal to the radius of curvature of the earth. Ducting can dramatically increase or decrease the radar coverage.

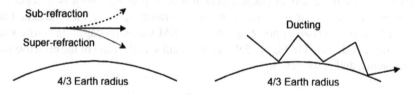

FIGURE 5.10 Illustrations of anomalous propagation.

Propagation of Radar Waves

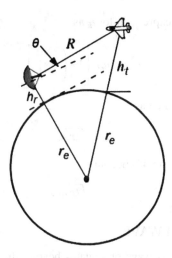

FIGURE 5.11 Target height in the four-thirds earth model.

Target Height in the Four-Thirds Earth Model

For a four-third earth model, Blake[9] derives the target height finding equation in terms of height of the radar h_r, elevation angle θ and the range R as

$$h_t = h_r + 6{,}076\, R \sin\theta + 0.6625\, R^2 \cos^2\theta \tag{5.42}$$

where h_t and h_r are expressed in feet, and R in nautical miles. The parameters used in (5.42) are shown in Figure 5.11.

Range to the Radar Horizon

Range to the radar horizon r_h can be obtained as a function of radar height h_r using the geometrical consideration of Figure 5.12. In this figure the radius of the earth r_0 is assumed to be constant. The right-angle triangle OBA will result in

$$r_h = \sqrt{(r_o + h_r)^2 - r_o^2}. \tag{5.43}$$

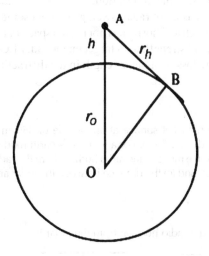

FIGURE 5.12 Range to the radar horizon.

Solving (5.43), we can write the expression for r_h as

$$r_h^2 = 2h_r r_o \left[1 + \left(\frac{h_r}{r_o} \right)^2 \right]. \tag{5.44}$$

Since $h_r \ll r_o$, (5.44) is approximated as

$$r_h \approx \sqrt{2r_o h_r}. \tag{5.45}$$

When refraction is accounted for, (5.45) becomes

$$r_h \approx \sqrt{2r_e h_r}. \tag{5.46}$$

5.4 DIFFRACTION OF EM WAVE

The mechanism by which the EM wave propagates beyond the geometrical horizon is generally termed as *diffraction*. In the diffraction mechanism the EM wave curves around the edges and penetrates into the region beyond the geometrical horizon. Diffraction depends on the frequency and the size of the obstacle compared to the wavelength.

The effect of the earth and its atmosphere can be roughly separated into two regions: the interference region and the diffraction region, as illustrated before in Figure 5.2. The straight line that separates these two regions is just a tangent to the surface of earth, and is usually referred to as *LOS*.

The effect of diffraction is explained by Huygens's principle, which states that every elementary area of a wavefront serves as a center that radiates in all directions on the front side of the wavefront.[10] The diffraction not only permits EM energy to penetrate into the shadow region but also influences the field outside that region. The lower the frequency, the more the wave is diffracted. The mechanism of diffraction is especially useful at very low frequencies. At radar frequencies little energy is diffracted. Thus the radar coverage cannot be extended much beyond the LOS by diffraction. Therefore, it can be concluded that if a low-altitude radar coverage is desired beyond the geometrical horizon in the diffraction region, the frequency should be as low as possible.

The distance between the radar and the target along the LOS is

$$R = \sqrt{2\kappa r_o h_r} + \sqrt{2\kappa r_o h_t} \tag{5.47}$$

where the symbols have their usual meanings. Equation (5.47) for calculating the distance along the LOS should not be used as a measure of radar coverage without some reservation. It has been shown that a target located at the geometrical horizon is not in free space but is definitely within the diffraction region of the radar. The field strength for a target on the radar LOS might vary from 10 to 30 dB below that in free space.[11] The loss of signal strength in the diffraction region can be quite high.

Example 5.3

Consider a radar over a smooth flat surface at an altitude of 350 m and an aircraft approaching the radar at a constant altitude of 1.5 km. For a four-thirds earth model with the radius of the earth $r_0 = 6371$ km, find (a) the surface range to the radio horizon from the radar, (b) the surface range to the radio horizon from the target, and (c) the distance between the radar and the target along the LOS.

Solution:

a. The surface range to the radio horizon from the radar is

$$d_r = \sqrt{2r_e h_r} = \sqrt{(2)(4/3)(6371 \times 10^3)(350)} = 77.112 \text{ km}$$

Propagation of Radar Waves

b. The surface range to the radio horizon from the target is

$$d_t = \sqrt{2r_e h_t} = \sqrt{(2)(4/3)(6371\times 10^3)(1500)} = 159.637 \text{ km}$$

c. The distance between the radar and the target along the LOS is

$$d = \sqrt{2r_e h_r} + \sqrt{2r_e h_t} = (77.112 + 159.637) \text{ km} = 236.749 \text{ km}$$

5.5 ATTENUATION BY ATMOSPHERIC GASES

The attenuation of EM waves in the lower atmosphere is due primarily to the presence of oxygen and water vapor when no condensation or dust particles are present. Attenuation results when a portion of the energy incident on molecules of these atmospheric gases is absorbed as heat or used in molecular transformation of the atmospheric particles. Atmospheric attenuation increases significantly in the presence of fog, dust, rain, and clouds. The attenuation of the atmosphere may be expressed in terms of two parameters such as attenuation coefficient, α, in nepers per kilometer, and the effective path length of propagation L in kilometers as

$$A = A_0 10^{-0.05\alpha L} \qquad (5.48)$$

where A_0 is the value of A at $L = 0$. Attenuation results when a portion of the incident energy on the atmospheric gases is absorbed or lost.

Figure 5.13[12,13] shows the typical values of atmospheric attenuation due to oxygen and water vapor as a function of wavelength. Atmospheric attenuation is relatively insignificant above 10 cm and becomes much more pronounced at MMW frequencies. As illustrated, relative resonance peaks occur at 22.24 GHz and at 184 GHz due to water vapor absorption, while oxygen

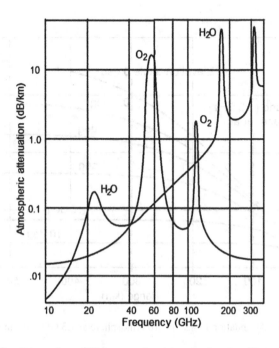

FIGURE 5.13 Attenuation of EM energy by atmospheric gases (From Blake, 1972).

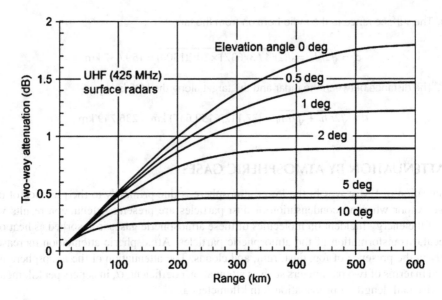

FIGURE 5.14 Attenuation in a standard atmosphere at 425 MHz (From Blake, 1972).

resonance peaks occur at 60 GHz and 118 GHz.[14] A relative minimum in the absorption occurs at 35 GHz and between 70 GHz and 110 GHz, making these desirable frequency bands to operate. However, the windows of minimum attenuation are not entirely clear as shown in Figure 5.13. The attenuation of the atmospheric gases decreases with increasing altitude. Consequently, the attenuation experienced by radar will depend on the altitude of the target as well as the range. Atmospheric attenuation is a function of range, frequency, and elevation. Blake[15] prepared plots for attenuation as a function of range, for various antenna elevation angles and frequencies.

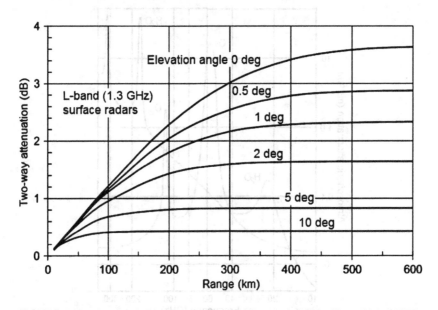

FIGURE 5.15 Attenuation in a standard atmosphere at 1.3 GHz (From Blake, 1972).

Propagation of Radar Waves

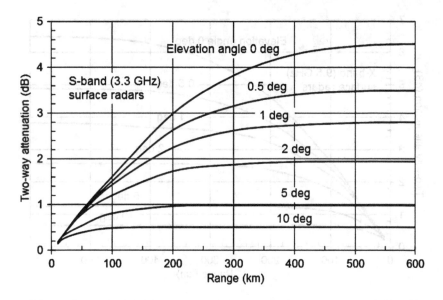

FIGURE 5.16 Attenuation in a standard atmosphere at 3.3 GHz (From Blake, 1972).

Two-way attenuations in clear atmosphere for surface radars are given as functions of range and elevation angle for UHF, L-, S-, C-, and X-bands in Figures 5.14 through 5.18 (Blake 1972), which were also replotted on logarithmic scales using custom radar functions by Curry.[16] For targets in space, the atmospheric loss in penetration of the entire atmosphere is depicted in Figure 5.19 (Blake 1972). Further details on atmospheric attenuation affecting the radar range performance are available in the literature (Blake 1972).

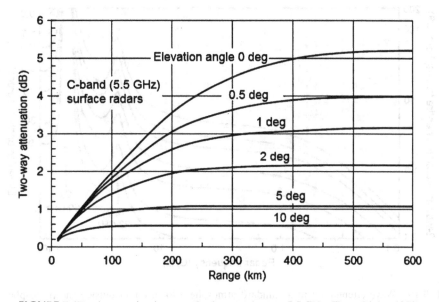

FIGURE 5.17 Attenuation in a standard atmosphere at 5.5 GHz (From Blake, 1972).

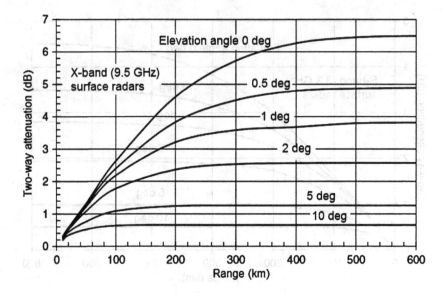

FIGURE 5.18 Attenuation in a standard atmosphere at 9.5 GHz (From Blake, 1972).

5.6 IONOSPHERIC ATTENUATION

When an EM wave passes through the ionosphere, there results a periodic force on the ionized electrons causing some signal energy to be absorbed. The total attenuation A, in decibels, due to the collision of the vibrating electrons with other particles is given by the integral[17]

$$A = 20\log\left[\exp\left(-\int_{s_1}^{s_2} v\,ds\right)\right] \qquad (5.49)$$

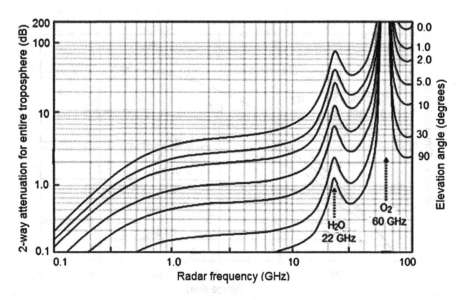

FIGURE 5.19 Wave attenuation in a standard atmosphere for a target outside the troposphere (From Blake, 1972).

Propagation of Radar Waves

or

$$A = -8.68 \int_{s_1}^{s_2} v \, ds \qquad (5.50)$$

where v is the absorption coefficient of the medium, and s_1 and s_2 are the limits of the path. Two types of ionospheric absorption are observed: *nondeviating* and *deviating*. Nondeviating absorption occurs when the index of refraction is unity. But when the index of refraction is much less than 1, the resulting absorption is called deviating absorption, which causes considerable deviation of the wave from the original direction of propagation.

Attenuation in the ionosphere is a problem for radars operating in HF and VHF bands, but seldom exceeds 1 dB at frequencies above 100 MHz. The attenuation from the ionosphere (in dB) varies inversely with the square of frequency. It is usually significant only at frequencies below about 300 MHz. The normal daytime two-way attenuation for signal paths passing through the ionosphere is shown by Millman[18] as a function of elevation angle for VHF and UHF frequencies. At nighttime, the ionospheric attenuation is less than 0.1 dB for all elevation angles.

A more serious problem may be due to the *Faraday rotation* as the signal penetrates the ionosphere. The free electrons in the ionosphere are not uniformly distributed but form layers. Furthermore, clouds of electrons may travel through the ionosphere and give rise to fluctuations in the radar signal. One of the effects of the ionosphere is to produce a rotation of the polarization of the signal known as the Faraday rotation. When a linearly polarized wave travels through the ionosphere, it sets motion in the free electrons in the ionized layers. These electrons move in the earth's gravitational field thereby experiencing a force. The direction of the electron motion is no longer parallel to the electric field of the radar wave; and as the electrons react back on the wave, the net effect is to shift the polarization. The angular shift in polarization, and hence the Faraday rotation, is dependent on the length of the path in the ionosphere, the strength of the earth's magnetic field in the ionized region, and the electron density in the region. Faraday rotation is inversely proportional to the square of the frequency. Consequently, the effect due to the Faraday rotation is not considered to be a serious problem for frequencies above about 10 GHz.

PROBLEMS

5.1 A radar mounted at a height of 30 m on a smooth planar surface transmits a peak power of 1 MW, and operates with a target at a height of 50 m and of RCS 4.0 m² at a range of 40 km. The antenna's directive gain toward the target is 34.77 dB. Assume a C-band radar operating at a frequency of 5.4 GHz, and the radar and the target are above the smooth flat surface. For multipath propagation, find (a) the propagation factor and (b) the signal power at the radar receiver.

5.2 Repeat Problem 5.1 for targets at small angles.

5.3 Find the angles corresponding to the first lobe and null for the radar-target system described in Problem 5.1.

5.4 Referring to Figure 5.4, show that the following relationship is true:

$$\psi = \tan^{-1} \left[\frac{(h_t + h_r)^2}{R_1^2 - (h_t - h_r)^2} \right]^{1/2}$$

94   Fundamental Principles of Radar

5.5 Using the law of cosines to appropriate triangles in Figure 5.7 relevant to a smooth spherical surface, derive the following relationships:

$$L_1 = \left[h_r^2 + 4r_e(r_e + h_r)\sin^2\left(\frac{d_2}{2r_e}\right) \right]^{1/2}$$

$$L_2 = \left[h_t^2 + 4r_e(r_e + h_t)\sin^2\left(\frac{d_1}{2r_e}\right) \right]^{1/2}$$

$$R_d = \left[(h_t - h_r)^2 + 4(r_e + h_r)(r_e + h_t)\sin^2\left(\frac{d}{2r_e}\right) \right]^{1/2}$$

5.6 Using (5.23) for the expression of ψ, show that the expressions for d_1 and d_2 are given by

$$d_1 \approx -r_e\psi + \sqrt{(r_e\psi)^2 + 2r_e h_r}$$

$$d_2 \approx -r_e\psi + \sqrt{(r_e\psi)^2 + 2r_e h_t}$$

5.7 Consider a radar is at a height of 12 m above a smooth spherical earth, and the target at a height of 450 m. Find the values of d_1, d_2, d, and radial range R_d when the grazing angle is 3° for acceptable performance.

5.8 Consider a radar located 50 m above a smooth spherical ocean surface and a target at a height of 1.2 km tracked by the radar. If the grazing angle due to multipath is 2°, find (a) distance to the reflection point from the radar, (b) distance to the target over the ocean surface, and (c) the divergence factor.

REFERENCES

1. Institute of Electrical and Electronics Engineers, "IEEE Standard Radar Definition," *IEEE Standard 686-1982m* (New York: IEEE, 1982).
2. C. Demb and M. H. L. Pryer, "The Calculation of Field Strengths over a Spherical Earth," *Journal of the Institution of Electrical Engineers—Part III: Radio and Communication Engineering,* 94, no. 31(1947): 325–339.
3. L. Boithias, *Radio Wave Propagation* (New York: McGraw-Hill Book Company, 1987).
4. M. I. Skolnik, ed., "Prediction of Radar Range," *Radar Handbook,* Second Edition (New York: McGraw-Hill Book Company, 1990).
5. D. E. Kerr, ed., *Propagation of Short Wave* (Lexington, MA: Boston Technical Publishers, Inc., 1964).
6. P. Beckmann and A. Spizzichino, *The Scattering of Electromagnetic Waves from Rough Surfaces* (Oxford, United Kingdom: Pergamon Press, 1963).
7. A. R. Miller et al., "New Derivation for the Rough Surface Reflection Coefficient and for the Distribution of Sea Wave Elevation," *IEEE Pro*ceedings, *Part H,* 131, no. 2 (1984): 114–116.
8. H. R. Reed, and C. M. Russell, *Ultra High Frequency Propagation* (London: Chapman and Hall Ltd., 1966).
9. L. V. Blake, *Radar Range performance Analysis* (Norwood, MA: Artech House, 1986).
10. C. R. Burrows and S. S. Attwood, *Radio Wave Propagation* (New York: Academic Press, Inc., 1949).
11. M. I. Skolnik, "Radar Horizon and Propagation Loss," *Proceedings of the IRE, 45* (1957): 697–698.
12. J. H. Van Vleek, "The Absorption of Microwaves by Oxygen," *Physical Reviews,* 71 (1947): 413–424.

13. J. H. Van Vleek, "The Absorption of Microwaves by Uncondensed Water Vapor," *Physical Reviews,* 71 (1947): 425–433.
14. A. W. Straiton and C. W. Tolbert, "Anomalies in the Absorption of Radio Waves by Atmospheric Gases," *Proceedings of the IRE,* 48 (1960): 898–903.
15. L. V. Blake, "Radar/Radio Tropospheric Absorption and Nosie Temperature," *U.S. Naval Research Laboratory Report, 7461* (1972), Washington, D.C., also in NTIS documents, AD753197.
16. G. R. Curry, *Radar Essentials—A Concise Handbook for Radar Design and Performance Analysis* (Rayleigh, NC: SciTech Publishing, Inc., 2012).
17. J. M. Kelso, *Radio Ray Propagation in the Atmosphere* (New York: McGraw-Hill Book Company, 1964).
18. G. H. Millman, "Atmospheric Effects on Radio Wave Propagation," in *Modern Radar,* ed. R. S. Berkowitz (New York: John Wiley and Sons, 1965).

6 Continuous Wave (CW) Radars

"In our culture of constant access and nonstop media, nothing feels more like a curse from God than time in the wilderness. To be obscure, to be off the beaten path, to be in the wilderness feels like abandonment, it seems more like exile than a vacation. To be so far off of everyone's radar that the world might forget about us for a while? That's almost akin to death...[But] far from being punishment, judgment, or a curse, the wilderness is a gift. It's where we can experience the primal delight of being fully known and delighted in by God."

—Jonathan Martin

6.1 INTRODUCTION

The first radar and many earlier and modern versions are based on the transmission of a continuous wave (CW) of EM energy and then the reception of this CW energy when reflected from a moving target. If the target is in motion relative to the radar, the received signal will be shifted in frequency from the transmitted frequency by an amount known as Doppler shift, which is essentially the basis of CW radar. Detailed descriptions of Doppler phenomenon are given in most physics texts, and a particularly thorough discussion emphasizing the moving target discrimination and resolution in simple CW radar systems has already been discussed, to some extent, in Chapter 2. The returned echo not only indicates that a target is present, but also the time that elapses between the transmission of the signal and receipt of the echo is a measure of the distance of the target. The amount of target information that can be extracted from the received waves is a function of the characteristics of the transmitted wave. Targets may be ascertained by modulating either the amplitude, frequency, or phase of the transmitted signal, and observing the round-trip time required for the signal to return to the radar after being reflected from targets.

During the early development of radar, it was recognized that a CW would have advantages in the measurement of the Doppler effect. The CW radar's advantages are its simplicity and ability to handle, without velocity ambiguity, targets at any range and with nearly any velocity. It can be safely mentioned that the CW radar has all the advantages without corresponding disadvantages. However, the spillover, the direct leakage of the transmitter and its accompanying noise into the receiver, is a major problem. Particular applications dictate whether CW radars or pulsed radars are to be utilized. There are several advantages of CW radars over pulsed radars. Hardware required for the CW radars is simpler, lighter, and smaller. The transmitted power level of a CW radar is lower than the peak power level of a pulsed radar. CW radars detect targets at shorter range than many pulsed radars.

The type of radar that employs continuous transmission, either unmodulated or modulated, finds wide applications. Some of the common applications are in weapon fuzzes, weapon seekers, aircraft altimeters, and police radars. Historically, the early radar experiments were carried out with continuous rather than pulsed waves. The CW radar is of interest not only because of its many applications, but it helps us understand the nature and the use of Doppler information contained in the echo signal. Also, the CW radar provides a measurement of relative velocity that may be used to distinguish moving targets from stationary clutters.

There are two types of CW radars: *unmodulated CW* and *modulated CW*.

6.2 UNMODULATED CONTINUOUS WAVE (CW) RADAR

Unmodulated CW radars may utilize a *homodyne* receiver for a very simple case or a *superheterodyne* receiver if flicker noise is of major concern.

6.2.1 HOMODYNE RECEIVER

A block diagram of the simplest CW radar possible is illustrated in Figure 6.1. The signal generated by the transmitter is an unmodulated CW at frequency f_0, and is routed through a duplexer to the antenna. A portion of the radiated energy is intercepted by the moving target and scattered, and some of it is received back at the antenna. If the target is in motion with a radial velocity v_r relative to the radar, the received signal will be shifted in frequency from the transmitted frequency by an amount $\pm f_d$ as given by $f_d = 2v_r / \lambda$ [see (2.18) in Chapter 2]. The plus sign applies if the distance between the target and radar is decreasing (closing target), and the minus sign applies if the distance is increasing (opening target).

The received echo signal is heterodyned in the detector (mixer) with a portion of the transmitted signal f_0 to produce a beat signal with frequency f_d. The sign of f_d is not lost. The signal is then amplified by a low-frequency amplifier to a level suitable for operating the indicator. This type of receiver, which beats the received signal with the transmitted signal, is called homodyne. The low-frequency amplifier has a frequency response characteristic similar to that of Figure 6.2. It eliminates echoes from stationary targets and amplifies only the Doppler echo signal. The upper cut-off frequency is selected to pass the highest Doppler frequency expected.

6.2.2 DOPPLER SHIFT AND RANGE RATE

The Doppler frequency, as discussed in Chapter 2, is defined as the difference between the received frequency and the transmitted frequency. The Doppler frequency $f_d = 2v_r / \lambda$, derived in Chapter 2, can also be alternatively derived by considering the unmodulated CW radar with a homodyne receiver of Figure 6.1. Consider the radar transmitted signal consisting of a sine wave of frequency f_0 in the form

$$s_t(t) = \sin(2\pi f_0 t + \varphi_0) \tag{6.1}$$

where φ_0 is the phase at $t = 0$. The signal is reflected from a moving target (assuming approaching or closing target) whose range at any time t from the radar is given by

$$R(t) = R_0 - \dot{R}t \tag{6.2}$$

where R_0 is the range at $t = 0$. Equation (6.2) ignores the additive effect of acceleration and higher range derivatives. The signal received from the target at time t would have been transmitted Δt seconds prior to time t where $\Delta t = 2R(t) / c$, and is given by

$$s_r(t) = \sin[2\pi f_0(t - \Delta t) + \varphi_0]. \tag{6.3}$$

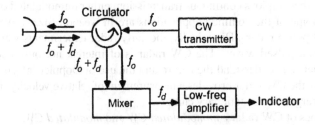

FIGURE 6.1 A simple CW radar system with a homodyne receiver.

Continuous Wave (CW) Radars

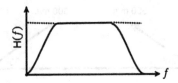

FIGURE 6.2 Frequency-response characteristic of a low-frequency amplifier.

The value of the delay Δt can be written as

$$\Delta t = \frac{2R(t)}{c} = \frac{2(R_0 - \dot{R}t)}{c}. \tag{6.4}$$

Substituting (6.4) in (6.3) yields

$$s_r(t) = \sin\left[2\pi f_0 t + 2\pi f_0 \frac{2\dot{R}}{c} t - 4\pi f_0 \frac{R_0}{c} + \varphi_0\right]. \tag{6.5}$$

Denoting $f_d = 2\dot{R}f_0/c$, the received signal can be written as

$$s_r(t) = \sin\left[2\pi(f_0 + f_d)t - 4\pi f_0 \frac{R_0}{c} + \varphi_0\right]. \tag{6.6}$$

Thus the frequency of the received signal is shifted by f_d called the Doppler shift, given by

$$f_d = \frac{2\dot{R}f_0}{c}. \tag{6.7}$$

For a receding or opening target, the range is given by

$$R(t) = R_0 + \dot{R}t. \tag{6.8}$$

and consequently the Doppler shift will be

$$f_d = -\frac{2\dot{R}f_0}{c}. \tag{6.9}$$

Thus we can also write the general expression of the Doppler shift as

$$f_d = \pm\frac{2\dot{R}f_0}{c}. \tag{6.10}$$

The plus sign applies if the target is approaching, and the minus sign applies if the target is receding.

Example 6.1

Consider two aircrafts, A and B as shown in the figure that follows. Aircraft A carrying a CW radar transmits at a frequency of 350 MHz, and tracks an aircraft B. Aircraft A travels at the speed of 500 m/s and makes an angle of 40° with an AB line of sight. Aircraft B approaching aircraft A at a speed of 800 m/s makes an angle of 30° with a BA line of sight.

    a. Find the Doppler shift as recorded by the CW radar in aircraft A.
    b. Find the direction of aircraft B for the Doppler frequency shift to be zero.

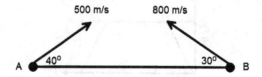

Solution:

a. The relative velocity between the two aircrafts is

$$v_r = v_B \cos\theta_B - (-v_A \cos\theta_A) = 800\cos 30° + 500\cos 40° = 1075.84 \text{ m/s}$$

The Doppler shift is positive for the approaching target and given by

$$f_d = \frac{2v_r f_A}{c} = \frac{(2)(1075.84)(350 \times 10^6)}{3 \times 10^8} = 2.51 \text{ kHz}$$

b. The direction of aircraft B for zero Doppler shift is calculated by setting the relative velocity equal to zero. This yields

$$v_r = 800\cos\theta_B - 500\cos 40° = 0 \Rightarrow \theta_B = \cos^{-1}\left[\frac{500\cos 40°}{800}\right] = \pm 61.39°$$

6.2.3 SUPERHETERODYNE RECEIVER

The receiver in a simple homodyne CW radar is not as sensitive because of increased *flicker noise*, which occurs in electronic devices within the radar. The noise power produced by the flicker effect varies with frequency as $1/f$. Thus the detector of the CW receiver can introduce a considerable amount of flicker noise resulting in reduced receiver sensitivity. One way to overcome the effects of flicker noise is to amplify the received signal at an intermediate frequency (IF) high enough to render the flicker noise small, and then heterodyne the signal down to lower frequencies. This is achieved by using a simple dual-antenna configuration of the CW radar system shown in Figure 6.3. The receiver of this system is called a *superheterodyne receiver*. Instead of the usual local oscillator, a portion of the transmitted signal is shifted in frequency by an amount equal to the IF before it is mixed with the received signal. Since the output of the mixer consists of two sidebands on either side of the carrier plus the carrier, a narrow bandpass filter is used to remove all the components except the lower sideband at $f - f_{IF}$.

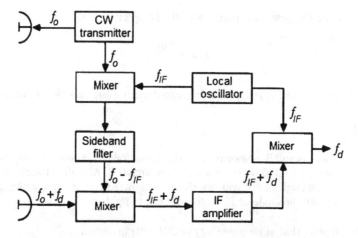

FIGURE 6.3 A simple CW radar with a superheterodyne receiver.

Continuous Wave (CW) Radars

The improvement in receiver sensitivity with the IF superheterodyne might be in the order of 30 dB over the homodyne receiver. The output of the sideband filter is then mixed with the received signal, and then amplified by an IF amplifier. The output of the IF amplifier is then mixed in the second mixer with the local oscillator signal to obtain a Doppler frequency signal that is directed toward the Doppler amplifier followed by an indicator.

6.2.4 Sign of the Radial Velocity

The sign of the radial velocity can be determined with separate filters, as shown in Figure 6.4. Let the transmitter signal $s_t(t)$ be represented by

$$s_t(t) = A_0 \cos(2\pi f_t), \tag{6.11}$$

then the echo signal $s_r(t)$ received by the receiving antenna is given by

$$s_r(t) = C_1 A_0 \cos(2\pi f_t \pm 2\pi f_d + \varphi) \tag{6.12}$$

where f_t and f_d are the transmitter and Doppler frequency, respectively, A_0 is the amplitude of the transmitted signal, φ is a constant phase shift depending on the range of initial detection, and C_1 is a constant determined from the radar equation. As shown in Figure 6.4, channel A generates a difference signal $s_A(t)$ represented by

$$s_A(t) = C_2 A_0 \cos(\pm 2\pi f_d + \varphi). \tag{6.13}$$

Likewise, the output $s_B(t)$ of the channel B mixer is

$$s_B(t) = C_2 A_0 \cos(\pm 2\pi f_d + \varphi + 90°). \tag{6.14}$$

When the target is approaching (closing target), the Doppler becomes positive resulting in a pair of two channel outputs as

$$\begin{aligned} s_A^+(t) &= C_2 A_0 \cos(2\pi f_d + \varphi) \\ s_B^+(t) &= C_2 A_0 \cos(2\pi f_d + \varphi + 90°). \end{aligned} \tag{6.15}$$

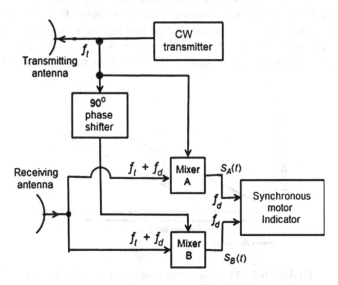

FIGURE 6.4 Measurement of the sign of Doppler frequency.

Similarly, if the target is receding (opening target), outputs of the two channels are

$$s_A^-(t) = C_2 A_0 \cos(2\pi f_d - \varphi)$$
$$s_B^-(t) = C_2 A_0 \cos(2\pi f_d - \varphi - 90°).$$
(6.16)

Outputs of the two channels are applied to a two-phase synchronous motor.[1] The direction of motor rotation is an indication of whether the target is approaching or receding. Electronic methods may also be used instead of a synchronous motor to determine the sign of the radial velocity.

6.3 FREQUENCY MODULATED CW RADARS

Simple unmodulated CW radars cannot measure the target range. This limitation is overcome by modulating the transmitted CW signal where the carrier signal may be modulated in amplitude, frequency, or phase. *Frequency modulation* may be linear or sinusoidal. To this end, multiple-frequency and phase-coded CW radar techniques are useful in many applications.

6.3.1 LINEAR FREQUENCY-MODULATED CW RADARS

CW wave radars may use the *linear frequency modulation* (LFM) technique to measure both the range and the Doppler information. In the frequency-modulated CW radar, the transmitter frequency is made to increase linearly with time as shown by the solid line in Figure 6.5. If the target is at range R, an echo signal will return after a delay time $\Delta t = 2R/c$. The dotted line represents the echo signal. The correlation of frequencies of transmitted and received echo signals at any given instant of time produces a *beat frequency* f_b. If there is no moving target, the beat frequency is a measure of the target's range, and $f_b = f_r$, where f_r is the beat frequency due to range only. The beat frequency as a function of the rate change of carrier frequency f_c is given by

$$f_r = \dot{f}_c t_d = \frac{2R}{c} \dot{f}_c.$$
(6.17)

In practical CW radars the LFM waveform cannot be continually changed in one direction only, and thus a periodicity in the modulation is necessary. This modulation does not need to be triangular, but may be sawtooth, sinusoidal, or some other form. The triangular frequency modulation waveform is shown in Figure 6.6 for a stationary target at range R. In this figure the solid line represents the transmitted signal, while the dashed line represents the received signal. The beat frequency f_b, defined as the difference between the transmitted and received signal, is depicted also in Figure 6.6. The time delay Δt is a measure of the range of the target given by $R = c\Delta t/2$.

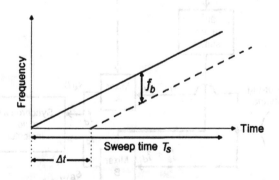

FIGURE 6.5 Frequency-time relationships in FMCW radar.

Continuous Wave (CW) Radars

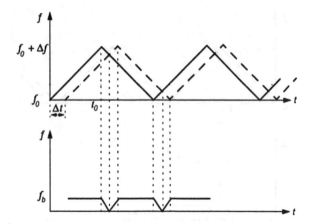

FIGURE 6.6 Triangular LFM and beat frequency for a stationary target.

Simple geometrical considerations reveal that the modulating frequency is $f_m = 1/2t_0$, which results in the rate of frequency change $\dot{f}_c$ given by

$$\dot{f}_c = \frac{\Delta f}{t_0} = \frac{\Delta f}{1/2f_m} = 2f_m\Delta f \tag{6.18}$$

where Δf is the maximum frequency deviation. Then the beat frequency is readily written as

$$f_b = f_r = \dot{f}_c \Delta t = (2f_m \Delta f)(2R/c) = \frac{4Rf_m\Delta f}{c}. \tag{6.19}$$

Equation (6.19) can be written in terms of the measured beat frequency to determine the range of the target as

$$R = \frac{cf_b}{4f_m\Delta f}. \tag{6.20}$$

Now, if the target is not stationary and approaching the radar at a radial velocity $v_r = \dot{R}$, a Doppler frequency shift will be superimposed on the received echo signal resulting in the frequency-time plot of the echo signal to be shifted up or down as shown in Figure 6.7. In this case, the shift due to Doppler frequency will subtract from the frequency shift due to the range during the positive slope and will add during the negative slope.

Designating the beat frequencies during the positive and negative slope portions by f_b^+ and f_b^-, we get

$$f_b^+ = \frac{2R}{c}\dot{f}_c - \frac{2\dot{R}}{\lambda} \tag{6.21}$$

where $\dot{R}$ is the range rate, termed as target radial velocity as seen by the radar, and

$$f_b^- = \frac{2R}{c}\dot{f}_c + \frac{2\dot{R}}{\lambda}. \tag{6.22}$$

Adding the previous two equations, we get the expression for range R,

$$R = \frac{c}{4\dot{f}_c}(f_b^+ + f_b^-), \tag{6.23}$$

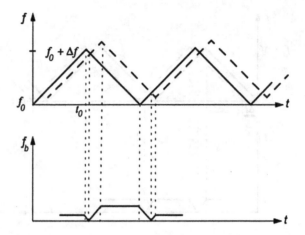

FIGURE 6.7 Triangular LFM and beat frequency for a moving target.

and subtracting (6.21) from (6.22), we get the expression for the range rate $\dot{R}$,

$$\dot{R} = \frac{\lambda}{4}(f_b^- - f_b^+). \tag{6.24}$$

Clearly, the CW radar using triangular LFM can extract both range and radial velocity of the target.

Example 6.2

Consider an X-band linear frequency-modulated continuous wave (LFM-CW) radar to measure both range and velocity, which operates at a frequency of 10 GHz. Assume a target at a range of 300 km that is approaching the radar with a velocity of 230 m/s. A modulating frequency of 20 Hz is selected for this radar with a peak frequency deviation of 250 KHz. Determine the beat frequencies f_b^+ and f_b^- during the positive and negative slope of the frequency excursion, respectively.

Solution:

f_b^+ can be determined by using (6.21):

$$f_b^+ = \frac{2R}{c}\dot{f} - \frac{2\dot{R}}{\lambda}$$

where

$$\lambda = \frac{3 \times 10^8}{10 \times 10^9} = 0.03 \text{ m}, \quad \dot{f} = \frac{\Delta f}{1/2f_m} = 2f_m \Delta f = (2)(20)(250 \times 10^3) = 10 \text{ MHz/s}$$

$$f_b^+ = \frac{2R}{c}\dot{f} - \frac{2\dot{R}}{\lambda} = \frac{(2)(300 \times 10^3)(10 \times 10^6)}{3 \times 10^8} - \frac{(2)(240)}{0.03} = 20000 - 16000 = 4 \text{ kHz}$$

Similarly, f_b^- can be determined by using (6.22):

$$f_b^- = \frac{2R}{c}\dot{f} + \frac{2\dot{R}}{\lambda}$$

Thus,

$$f_b^- = 20000 + 16000 = 36 \text{ kHz}$$

Continuous Wave (CW) Radars

6.3.2 Sinusoidal Frequency-Modulated CW Radars

From a practical point of view a sinusoidal FM can be used to measure the target range instead of a linear FM. The transmitted signal of the FMCW can be written as

$$s(t) = A_t \sin \psi(t) \tag{6.25}$$

where

$$\psi(t) = 2\pi \int f(t)\,dt \tag{6.26}$$

The instantaneous frequency of the transmitted FM signal is expressed as

$$f(t) = f_0 + \Delta f \cos(2\pi f_m t) \tag{6.27}$$

where f_0 is the frequency of the unmodulated carrier, Δf is called the maximum frequency deviation, and f_m is the frequency of the sinusoidal modulating signal. A fundamental characteristic of an FM is that the frequency deviation Δf is proportional to the amplitude of the modulating signal, and is independent of the modulating frequency.

Using (6.27), the angle $\psi(t)$ of the FM signal is obtained as

$$\psi(t) = 2\pi \int_0^t f(t)\,dt = 2\pi f_0 t + \frac{\Delta f}{f_m} \sin(2\pi f_m t). \tag{6.28}$$

Thus the FM signal itself is given by

$$s(t) = A_t \sin\left[2\pi f_0 t + \frac{\Delta f}{f_m} \sin(2\pi f_m t)\right]. \tag{6.29}$$

The received signal from the target (assumed to be stationary) will be the result of the signal transmitted T seconds earlier, where $T = 2R/c$, with R and c representing target range and velocity of propagation, respectively, as before. Thus the received signal $r(t)$, as shown in Figure 6.8, will be a replica of the transmitted signal but displaced in time by the transit time T, and is given by

$$\begin{aligned} r(t) &= A_r \sin\left[2\pi f_0 (t-T) + \frac{\Delta f}{f_m} \sin[2\pi f_m (t-T)]\right] \\ &= A_r \sin[\psi(t-T)] \end{aligned} \tag{6.30}$$

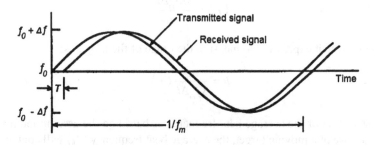

FIGURE 6.8 Sinusoidal frequency modulation in CW radars.

where A_r represents the amplitude of the received signal. The received signal of (6.30) and the transmitted signal of (6.25) are heterodyned in a mixer followed by a lower sideband filter resulting in

$$s_0(t) = A_0 \sin[\psi(t) - \psi(t-T)]. \tag{6.31}$$

Substituting the value of $\psi(t)$ from (6.28) and simplifying yield

$$s_0(t) = A_0 \sin\left(\frac{2\Delta f}{f_m}\sin(\pi f_m T)\cos\left[2\pi f_m\left(t-\frac{T}{2}\right)\right] + 2\pi f_0 T\right). \tag{6.32}$$

If we assume that $T \ll 1/f_m$, we get

$$\sin(\pi f_0 T) \cong \pi f_0 T. \tag{6.33}$$

leading to the following result:

$$\frac{2\Delta f}{f_m}\sin(\pi f_m T) \cong 2\pi \Delta f T. \tag{6.34}$$

Using this approximation (6.34) in (6.32), we get

$$s_0(t) = A_0 \sin\left(2\pi f_0 T + (2\pi \Delta f T)\cos\left[2\pi f_m\left(t-\frac{T}{2}\right)\right]\right). \tag{6.35}$$

The beat frequency f_b is obtained by differentiating the argument of the sine wave of (6.35) with respect to t, and is given by

$$f_b = (\pi f_m \Delta f T)\sin(2\pi f_m t - \pi f_m T + \pi) \tag{6.36}$$

where π in the argument of the sign function accounts for the positive sign. The average beat frequency $\langle f_b \rangle$ over one half of a modulating cycle is obtained by integrating the previous expression as

$$\langle f_b \rangle = \frac{1}{1/2 f_m}\int_0^{1/2 f_m} f_b(t)\,dt = (4\pi f_m \Delta f T)\cos(\pi f_m T). \tag{6.37}$$

Since $(\pi f_m T) \ll 1$, then $\cos(\pi f_m T) \cong 1$, and $T = 2R/c$, (6.37) is simplified to

$$\langle f_b \rangle = \frac{8Rf_m \Delta f}{c}. \tag{6.38}$$

This results in an explicit expression to measure the range of the target as

$$R = \frac{c}{8 f_m \Delta f}\langle f_b \rangle. \tag{6.39}$$

The previous derivation of the average beat frequency is based on the assumption that the target is stationary. In the case of a moving target, the average beat frequency $\langle f_b \rangle$ will contain a component of Doppler frequency shift f_d in (6.38).

Continuous Wave (CW) Radars

6.4 MULTIPLE-FREQUENCY CW RADAR

The radar range measurement can also be obtained under certain circumstances by measuring the phase of the echo signal relative to the phase of the transmitted signal. Consider a CW radar transmitting a single-frequency sine wave with a frequency f_0 of the form

$$s(t) = \sin(2\pi f_0 t). \tag{6.40}$$

The amplitude of the signal is considered to be unity, since it does not influence the phase. The echo signal received by the radar from a target at a distance R, which returns after a time delay $T = 2R/c$, is expressed as

$$r(t) = \sin[(2\pi f_0 (t - T)] = \sin[(2\pi f_0 t - 2\pi f_0 T)]. \tag{6.41}$$

If the transmitted and the received are mixed in a phase detector, the output is proportional to the phase difference $\Delta\phi$ of the two signals, and is given by

$$\Delta\phi = 2\pi f_0 T = \frac{4\pi f_0 R}{c}. \tag{6.42}$$

This phase difference is directly proportional to the range of the target, which is given by

$$R = \frac{c\Delta\phi}{4\pi f_0} = \frac{\lambda}{4\pi} \Delta\phi. \tag{6.43}$$

This equation indicates that an unambiguous range is measured as long as $\Delta\phi$ remains less than 2π. This unambiguous range is much too small for most practical applications.

The maximum unambiguous range can be extended by transmitting two separate CW signals differing slightly in frequency. Consider the case where the transmitted signal consisting of two continuous sine waves of frequencies f_1 and f_2 separated by an amount Δf. Ignoring the amplitude factors again, the transmitted signals with constant arbitrary phase angles ϕ_1 and ϕ_2 may be written as

$$s_1(t) = \sin(2\pi f_1 t + \phi_1), \text{ and}$$
$$s_2(t) = \sin(2\pi f_2 t + \phi_2). \tag{6.44}$$

The return signal from the target is shifted in frequency by the Doppler shift, and shifted in phase proportional to range R. The received signals are then expressed as

$$r_1(t) = \sin\left[2\pi(f_1 \pm f_{1d})t - \frac{4\pi f_1 R}{c} + \phi_1 \right], \text{ and}$$
$$r_2(t) = \sin\left[2\pi(f_2 \pm f_{2d})t - \frac{4\pi f_2 R}{c} + \phi_2 \right] \tag{6.45}$$

where f_{1d} and f_{2d} are Doppler frequency shifts associated with transmitted frequencies f_1 and f_2, respectively. Since the frequencies f_1 and f_2 are approximately the same, the Doppler frequency shifts f_{1d} and f_{2d} are consequently equal to each other. Thus $f_{1d} = f_{2d} = f_d$. Separating the two components of the received signal and heterodyning each received signal component with corresponding component of the transmitted signal, we obtain the resulting two Doppler-frequency components as

$$s_{1d}(t) = \sin\left(\pm 2\pi f_{1d}t - \frac{4\pi f_1 R}{c}\right), \text{ and}$$

$$s_{2d}(t) = \sin\left(\pm 2\pi f_{2d}t - \frac{4\pi f_2 R}{c}\right). \tag{6.46}$$

The phase difference between these components of (6.46) is

$$\Delta\phi = \frac{4\pi(f_2 - f_1)R}{c}. \tag{6.47}$$

Hence, denoting $\Delta f = f_2 - f_1$, we get

$$R = \frac{c\Delta\phi}{4\pi\Delta f}. \tag{6.48}$$

Again, the maximum unambiguous range corresponds to $\Delta\phi = 2\pi$, and is given by

$$R_{\max} = \frac{c}{2\Delta f}. \tag{6.49}$$

6.5 PHASE-MODULATED CW RADAR

Limitations of unmodulated CW radars for not being able to measure target range can also be overcome by modulating the carrier signal in phase. In *phase-modulated CW* (PMCW) radars, the target range is measured by applying a discrete phase shift every τ second to the transmitted CW signal, thus producing a phase-coded waveform. The signal after being reflected from the target is received by the radar receiver where it is correlated with the stored version of the transmitted signal. The time delay between the transmitted signal and the occurrence of maximum correlation provides the target range information.

There are two types of phase-coding techniques: *binary phase codes* and *polyphase codes*. The principle of range determination is the same for both coding techniques, but the binary phase codes are simpler than the other. A binary phase code and the resulting PM-CW waveform are shown in Figure 6.9. A binary phase code is generated by dividing the CW carrier into N segments of equal duration τ, and coding each segment randomly with either 1 or 0. A 1 designation represents no phase shift (in-phase) with respect to the unmodulated carrier, while 0 designation represents a phase shift of π radians (out-of-phase). The codes in CW radars are periodic in nature.

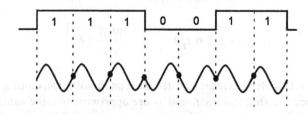

FIGURE 6.9 Binary phase code and the resulting PMCW.

Continuous Wave (CW) Radars

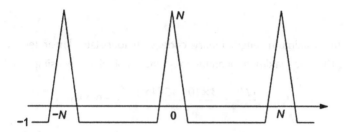

FIGURE 6.10 Autocorrelation function of an MLS code of length N.

Pseudorandom binary phase codes are also known as *maximal length sequence* (MLS) codes in which the statistics associated with their occurrence are similar to that associated with the coin-toss sequence. An MLS code has a length of $N = 2^n - 1$, where n is the number of stages in shift register generating the code. Since the code is periodic, its autocorrelation function has a peak value at $t = 0$ and at an integer multiple of $N\tau$, and a value of -1 everywhere else as shown in Figure 6.10. The most common method of generating MLS codes is to use linear shift registers. Figure 6.11 shows the four-stage linear feedback register generating the MLS code. The output from 3 and 4 are summed, which is fed into stage 1. The output results in a sequence of length $N = 2^4 - 1 = 15$. Several different feedback connections to different stages of the shift register are possible to generate different maximum length codes. For example, a maximum of 18 different maximal length codes, each of length $N = 127$, can be generated for a scheme of seven-stage shift register.

A PMCW radar can be used to measure target range by correlating the received waveform with a stored version of the transmitted code. An alternative method of implementing a PMCW radar is by correlating the delayed version of the transmitted code. If the delay time is adjusted to maximize the autocorrelation function, it provides the measurement of target range. The transmitted waveform must return to radar in $N\tau$ seconds or less in order to result in the maximum unambiguous range of PMCW radar given by

$$R_{max} = \frac{cT_{max}}{2} = \frac{cN\tau}{2} \tag{6.50}$$

where T_{max} is the maximum permissible round-trip time of the signal between the radar and the target.

Example 6.3

A PMCW radar uses maximal length (pseudo-random) binary codes to measure the target range by applying a discrete phase shift every 2 μs to the transmitted CW signal. The maximal length code is implemented by using a four-stage linear feedback register. Find the maximum unambiguous range of the target.

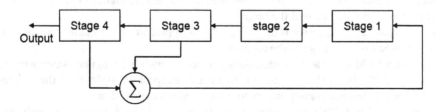

FIGURE 6.11 Four-stage linear feedback register.

Solution:

The length of the maximum length binary code with four-stage linear feedback register is $N = 2^4 - 1 = 15$. Thus the maximum unambiguous range is obtained by using (6.50):

$$R_{max} = \frac{cN\tau}{2} = \frac{(3 \times 10^8)(15)(4 \times 10^{-6})}{2} = 9.0 \text{ km}$$

Codes that use harmonically related phases based on certain fundamental phase increments are called *polyphase codes*. Polyphase codes are complex, and allow for any M possible phase shifts, where M is called the order of the code. The possible phase states are

$$\phi_k = \frac{2\pi}{M}k, \qquad k = 0,1,2,...,M-1. \tag{6.51}$$

This enables you to code a long, constant amplitude pulse so that the output resulting from the matched system has particular desired properties. Many different polyphase codes—described very briefly next—are employed depending on system designs and applications.

The *Frank polyphase codes* are nothing but discrete approximations to a linear FM waveform.[2] In Frank polyphase codes, for each integer M, there is a Frank code of length $N = M^2$ that employs the phase shifts given also by (6.51). A Frank code of M^2 is referred to as an M-phase Frank code. Note that the size of the fundamental phase increment decreases as the number of groups is increased, and because of phase stability, this may degrade the performance of very long codes. The Frank codes provide an interesting alternative to pseudo-random codes when long codes are needed for a radar application in which extended clutter or a high-density target environment is expected; they also should be considered in applications in which the Doppler sensitivity of binary codes poses a concern. Codes similar to Frank codes with improved Doppler tolerance have been investigated by Krestschmer and Lewis.[3]

The Welti[4] and Golay[5] codes are sidelobe-cancelling codes that may seem to represent an ideal solution to the sidelobe suppression problem, but the difficulties involved in their implementation as well as their sensitivities to echo fluctuations and Doppler shift make them less than ideal. The Welti codes form a large, general set of polyphase codes exhibiting the property that the sum of the autocorrelations of the two codes in a pair adds to twice the single autocorrelation at the peak and to zero elsewhere. The Welti codes form a large, general set of polyphaser codes, and the Golay codes form the subset of these sidelobe-cancelling codes that are binary.

PROBLEMS

6.1 Assuming an LFM-CW radar using a triangular modulation as shown in Figure 6.6 with a maximum frequency deviation of 250 kHz and a modulating frequency of 20 Hz. For a stationary target located at a range of 150 km, find the beat frequency of the radar system.

6.2 Consider an LFM-CW radar that uses the triangular modulation as shown in Figure 6.6 with $\Delta F = 200$ kHz and $f_m = 30$ Hz, and measures a beat frequency of 12 kHz for a stationary target. Find the range of the target.

6.3 In Problem 6.2, if the target is an aircraft that is approaching the radar at a radial velocity of 100 m/s, find the range of the aircraft.

6.4 A certain LFM-CW radar is modulating a carrier with a triangular waveform at a frequency of 200 Hz with a maximum frequency sweep of 45 MHz. Find the average beat frequency difference corresponding to a range increment of 12 m.

6.5 Consider an LFM-CW radar operating at a frequency of 12 GHz uses a triangular modulation scheme with a maximum frequency sweep of 220 kHz at a frequency rate of 10 MHz

Continuous Wave (CW) Radars

per second. Find the beat frequencies during the positive and negative slopes of the triangular waveform when the target initially located at a range of 360 m is approaching the radar at a radial velocity of 230 m/s.

6.6 An X-band LFM-CW radar is used to measure both the range and the velocity. The beat frequencies f_b^+ and f_b^- during the positive and negative slope of the frequency excursion are measured to provide 6.667 kHz and 40 kHz, respectively. Assume that a triangular baseband signal with a frequency of 20 Hz is frequency modulating the radar signal of 10 GHz with a maximum frequency sweep of 250 kHz.

    a. Determine the range and velocity of the target.
    b. Determine the maximum range if the maximum time delay is selected as 10% of t_0.
    c. Determine the maximum unambiguous range.

6.7 An X-band CW radar operates at a frequency of 10 GHz and triangularly sweeps a bandwidth of frequency excursion of 2 MHz at a 200 Hz rate. The frequency difference between the transmit and receive signals on the up-sweep is 65.51 kHz and on the down-sweep it is 82.65 kHz. Find the target range and radial velocity.

6.8 Consider a sinusoidal FMCW radar that operates at a frequency of 3 GHz, and measures the average beat frequency of 160 Hz for a stationary target. Assume that the maximum frequency deviation is 15 kHz and the sinusoidal modulating signal has a frequency of 50 Hz. Find the target range.

6.9 A certain radar transmitter is modulating the carrier with a sinusoidal triangular waveform with a frequency of 250 Hz with a frequency sweep of 40 kHz. Find the average beat frequency difference if the target is at a range of 100 km.

6.10 Consider an X-band radar operating at 10.4 GHz uses sinusoidal FM for ranging. The modulating is a sinusoidal signal having the frequency of 100 Hz. The phase difference between the received modulated signal and that transmitted is 3.5°. Find the range of the target.

6.11 The unambiguous range can be increased by using multiple frequency transmission of sinusoidal signals. Consider transmitted sinusoidal signals of frequency $f_1 = 88.5$ kHz and $f_2 = 90$ kHz. Find the maximum unambiguous range.

6.12 If the transmitted signal of a multiple frequency of a CW radar consists of two continuous sinewaves of frequencies $f_1 = 100$ kHz and $f_2 = 105$ kHz, find the corresponding maximum unambiguous range.

REFERENCES

1. H. P. Kalmus, "Direction Sensitive Doppler Device," *Proceedings of the IRE,* 43 (1955): 690–700.
2. R. L. Frank, "Polyphase Codes with Good Nonperiodic Correlation Properties," *IEEE Transaction on Information Theory,* 9 (1963): 43–45.
3. F. F. Krestschmer, Jr. and B. L. Lewis, "Doppler Properties of Polyphase Coded Pulse Compression Waveforms," *NRL Report 8635, Naval Research Laboratory*, Washington D.C., September 1982.
4. G. R. Welti "Quaternary Codes for Pulsed Radar," *IRE Transaction on Information Theory,* 7 (1960): 82–87.
5. M. J. E. Golay, "Complementary Series," *IRE Transaction on Information Theory,* 7, no. 2 (1961): 82–87.

7 MTI and Pulse Doppler Radars

"But by providing the background picture—the universal situational awareness that we desire—by showing the anomalies, the Space-Based Radar will change the nature of how we do our analysis and our intelligence."

—**Stephen Cambone**

7.1 INTRODUCTION

Relative velocity between the target and the radar creates a Doppler shift of the transmitted frequency. It was shown previously that this Doppler shift is proportional to the radial speed of the target. Thus a measurement of Doppler frequency affords a means of measuring radial speed, which is more accurate than other methods. In addition, the Doppler shift can be used in radar system applications for several advantages that include separating desired target returns from those of fixed targets, and extracting information concerning radial velocity of the target. A pulse radar that utilizes such advantages is called a *moving target indication* (MTI) or *pulse Doppler radar.* The physical principles of both these radars are the same but they differ in their mode of operation. For instance, the MTI radar operates on low pulse repetition frequencies, and uses a delay line canceller filter to isolate moving targets from stationary targets, thus causing ambiguous Doppler velocity but unambiguous range measurements. On the other hand, the pulse Doppler radar operates on high pulse repetition frequency, and is the one in which the Doppler measurement is unambiguous but the range measurement can be either ambiguous or unambiguous, and the Doppler data are extracted by the range gates and Doppler filters.

7.2 DESCRIPTION OF OPERATION

A simple CW radar, consisting of a transmitter, receiver, indicator, and the transmitting and receiving antennas, is shown in Figure 7.1. In principle the CW radar can be converted to a pulse radar, as shown in Figure 7.2, by providing a power amplifier and a pulse modulator to turn the amplifier on and off to generate pulses. There is no local oscillator here since the reference signal is diverted to the receiver directly from the CW oscillator to serve also as the coherent reference needed to detect the Doppler frequency shift. By coherent it is meant that the phase of the transmitted signal is preserved in the reference signal. This kind of reference signal is the distinguishing feature of coherent MTI radar.

Let the oscillator voltage (transmitted signal) at any time t be represented by

$$v_{osc} = v_t(t) = A_t \sin(2\pi f_t t) \tag{7.1}$$

which results in a reference signal $v_{ref}(t)$ and an echo signal $v_{rec}(t)$, assuming a Doppler shit of $f_d = (2v_r f_t / c)$ as

$$v_{ref} = A_1 \sin(2\pi f_t t) \tag{7.2}$$

and

$$v_{rec}(t) = A_r \sin\left[2\pi f_t \left(1 \pm \frac{2v_r}{c}\right)t - \frac{4\pi f_t R_0}{c}\right] \tag{7.3}$$

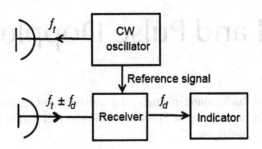

FIGURE 7.1 Simple CW radar.

where

A_t = amplitude of the oscillator signal
A_1 = amplitude of the reference signal
A_r = amplitude of the echo signal
f_t = transmitted frequency
f_d = Doppler frequency shift
v_r = radial velocity of the target
c = velocity of propagation
R_0 = initial range of the target.

The received echo signal is then heterodyned with the transmitted signal in the mixer stage of the receiver resulting in a difference signal $v_d(t)$ given by

$$v_d(t) = A_d \sin\left[2\pi f_d t - \frac{4\pi f_t R_0}{c}\right] \qquad (7.4)$$

where A_d is the amplitude of the difference signal at the output of the mixer; and it is assumed that the target is approaching toward the radar at a velocity of v_r.

For stationary targets the Doppler frequency $f_d = 0$, v_d is no longer a function of time, but may take on any constant value from $+A_d$ to $-A_d$ including zero. However, when the target is in motion relative to the radar, f_d assumes a value other than zero and the voltage corresponding to the difference frequency from the mixer will vary with time. Note that all these frequencies are with reference to the carrier waveform and, have nothing to do with the pulse repetition frequency. Figure 7.3(a) shows the reflected signal from the target. The frequency of this signal may have been changed due

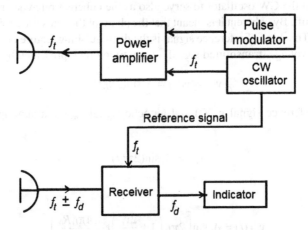

FIGURE 7.2 Pulse radar using Doppler information.

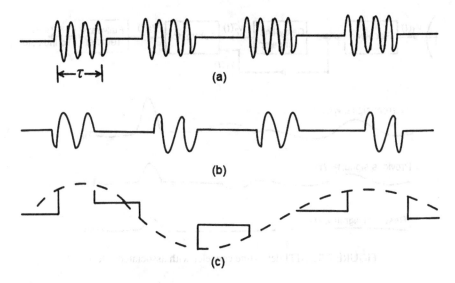

FIGURE 7.3 (a) RF pulse train, (b) video pulse train for Doppler frequency $f_d > 1/\tau$, (c) video pulse train for Doppler frequency $f_d < 1/\tau$, as depicted (From Skolnik, 1980).

to the motion of the target. In Figure 7.3(b), the difference signal is shown in presence of a moving target for the case when the resultant Doppler frequency is such that $f_d \tau > 1$, and in Figure 7.3(c) for the case $f_d \tau < 1$, where τ is the width of the pulse. When $f_d \tau > 1$, the Doppler shift can be found easily from the information contained in one pulse only, whereas, when $f_d \tau < 1$ the Doppler shift can be extracted from the information contained in many more pulses. The difference signal is the output of the mixer and is also called the bipolar video output, which is displayed on the A-scope in successive sweeps, as depicted in Figure 7.4. Arrows indicate the positions of the moving targets.

Note that the amplitudes of the signals from stationary targets do not change with the number of sweeps. But the echo signals from moving targets will change in amplitude over successive sweeps according to (7.4). If one sweep, say Figure 7.4(b), is subtracted from the previous sweep in Figure 7.4(c), echoes from stationary objects will be canceled leaving only moving targets, as shown in Figure 7.4(a).

This bipolar video is not good enough for a PPI since the screen display will show bright patches for all stationary targets, and spots of fluctuating brightness for moving targets. But what we actually require is the Doppler information regarding moving targets only. One method to extract this information is to employ delay-line cancelers. The delay-line canceler acts as a filter to eliminate d-c components due to stationary targets and to pass a-c components due to moving targets. Subtraction of two echoes from two successive sweeps is accomplished in a delay-line canceler, as indicated

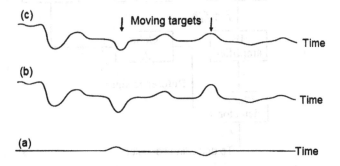

FIGURE 7.4 Successive sweeps of an MTI radar A-scope display.

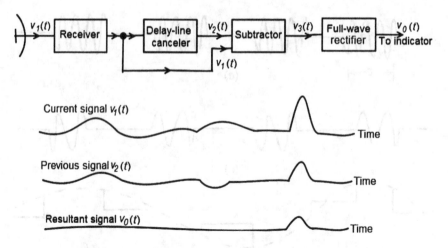

FIGURE 7.5 MTI delay-line canceler with associated effects.

in Figure 7.5. In this situation, the current signal is delayed by one pulse time period (reciprocal of the pulse repetition frequency) and subtracted from the signal coming next. The outputs of the two channels are subtracted from one another, the resultant of which is also bipolar. This is converted to unipolar video in a full-wave rectifier.

The block diagram of the simple MTI radar illustrating the reference signal shown in Figure 7.2 is not necessarily the most typical. MTI radar with a power amplifier controlled by a stable local oscillator is shown in Figure 7.6. The characteristic feature of coherent MTI radar is that the transmitted signal must be coherent in phase with the reference signal in the receiver. The coherent reference signal is supplied by an oscillator called *coho* whose frequency is the same as the IF used in the receiver. The output of the coho f_c is also mixed with the local oscillator frequency f_l, which must also be a stable local oscillator known as *stalo* for stable local oscillator. The IF signal is produced

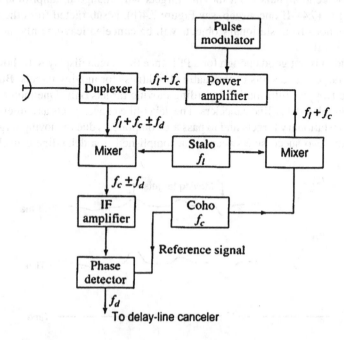

FIGURE 7.6 Coherent MTI radar with power amplifier transmitter.

MTI and Pulse Doppler Radars

by heterodyning the RF echo signal with the stalo signal. The function of the stalo is to provide the necessary frequency translation from the IF to the transmitted frequency. Any stalo phase deviation that may occur is canceled on reception because the stalo that generates the transmitted signal also acts as a local oscillator in the receiver. The reference and the IF echo signals are both fed into a mixer called the *phase detector*.

7.3 SINGLE DELAY-LINE CANCELER

As mentioned earlier, the delay-line canceler acts as a filter, which rejects the d-c components of fixed unwanted targets such as buildings, hills, and trees, and retains for detection the a-c components of moving targets such as aircraft. The delay-line canceler with associated effects on waveforms is depicted in Figure 7.5. For the purpose of mathematical analysis, the single delay-line canceler represented in Figure 7.5 is modified to Figure 7.7.

The signal $v_1(t)$ received from a particular target at range R_0 is reproduced here from (7.4) with some modifications as

$$v_1(t) = V_0 \sin(2\pi f_d t - \phi_0) \tag{7.5}$$

where V_0 represents the amplitude of the video signal, and $\phi_0 = 4\pi f_t R_0 / c$. The signal received from the previous transmission is similar in its format, except it is delayed by a time T = pulse repetition interval, and is

$$v_2(t) = V_0 \sin[2\pi f_d(t-T) - \phi_0]. \tag{7.6}$$

The amplitude V_0 is assumed to be constant for both the pulses over the interval T. The two signals are subtracted from one another resulting in a signal $v_3(t)$ given by

$$\begin{aligned} v_3(t) &= v_1(t) - v_2(t) \\ &= V_0 \sin(2\pi f_d t - \phi_0) - V_0 \sin[2\pi f_d(t-T) - \phi_0]. \end{aligned} \tag{7.7}$$

Using the trigonometric identity:

$$\sin\alpha - \sin\beta = 2\sin\left(\frac{\alpha-\beta}{2}\right)\cos\left(\frac{\alpha+\beta}{2}\right)$$

we get

$$v_3(t) = 2V_0 \sin(\pi f_d T)\cos[2\pi f_d(t-T/2) - \phi_0] \tag{7.8}$$

Equation (7.8) indicates that the output from the canceler consists of a cosine wave at the Doppler frequency f_d with an amplitude $2V_0 \sin(\pi f_d T)$. It means that the amplitude of the canceled video

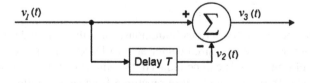

FIGURE 7.7 A single delay-line canceler.

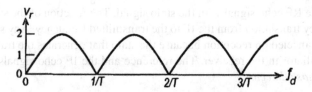

FIGURE 7.8 Frequency response of the single delay-line canceler.

output is a function of the Doppler shift and the pulse repetition interval T or pulse repetition frequency f_r. The output of the subtractor is then fed into the full-wave rectifier, which provides the magnitude of $v_3(t)$, given by

$$|v_3(t)| = v_0(f_d) = 2V_0|\sin(\pi f_d T)|. \tag{7.9}$$

Thus the magnitude of the relative frequency response $v_r = |v_3(t)/v_1(t)|$ of the delay-line canceler can be demonstrated in Figure 7.8.

Blind Speeds

Figure 7.8 indicates that the frequency response of the single delay-line canceler becomes zero if

$$f_d = \frac{n}{T} = nf_r \tag{7.10}$$

where $n = 0, 1, 2, \ldots$, and the pulse repetition frequency (PRF) $f_r = 1/T$. Therefore, the delay-line canceler, according to (7.10), not only eliminates stationary targets ($n = 0$), but also rejects any moving targets whose Doppler frequency happens to be the same as the PRF or a multiple thereof. Radial velocities of the target corresponding to which the MTI response is zero are known as *blind speeds* and are given by

$$v_{bn} = \frac{n\lambda}{2T} = \frac{n\lambda f_r}{2} \qquad n = 1, 2, 3, \cdots \tag{7.11}$$

where v_{bn} is the *nth* blind speed of the target. If λ is expressed in meters, f_r in Hz, and the relative radial velocity in mph, the blind speeds are

$$v_{bn} \approx 1.12n\lambda f_r. \tag{7.12}$$

Only the first blind speed is significant, since others are the integer multiple of the first, and is obtained from (7.12), given by

$$v_{b1} = \frac{\lambda}{2T} = \frac{\lambda f_r}{2}. \tag{7.13}$$

The blind speeds can be one of the serious limitations, since they cause desired moving targets to be canceled out along with the undesired clutter at zero Doppler frequency. Based on (7.13), the detrimental effects of blind speeds can be reduced, to some extent, by some methods that include operating at high radar frequency, high pulse repetition frequency, multiple radar frequencies, or multiple pulse repetition frequencies.

MTI and Pulse Doppler Radars

Example 7.1

The first blind speed can be used to determine the maximum unambiguous range at the radar frequency of operation. For a radar operating at a frequency of 300 MHz (UHF), find the maximum unambiguous range if the first blind speed is 700 mph.

Solution:

The first blind speed v_1 when converted into MKS unit is

$$v_1 = \frac{(700)(1609)}{(60)(60)} = 312.86 \text{ m/s}$$

The maximum unambiguous range is obtained from $R_{un} = \frac{cT}{2}$, where the corresponding first blind speed is given by $v_1 = \frac{\lambda}{2T}$, which leads to the explicit expression for R_{un} as

$$R_{un} = \frac{c\lambda}{4v_1} = \frac{c^2}{4v_1 f}$$

Thus

$$R_{un} = \frac{c^2}{4v_1 f} = \frac{(3 \times 10^8)^2}{(4)(312.86)(300 \times 10^6)} \approx 240 \text{ km}$$

7.4 DOUBLE DELAY-LINE CANCELER

In most radar applications the response of a single delay-line canceler is not acceptable since it does not have a broad notch in the top band. A double delay-line canceler has a better response through widening the clutter-rejection notches. This is accomplished by the output of the delay-line canceler through a second delay-line canceler. The output of the two single delay-line cancelers connected in cascade is the square of a single delay-line canceler given by $4\sin^2(\pi f_d T)$. Two basic configurations of such a double delay-line canceler are shown in Figure 7.9(a) and 7.9(b).

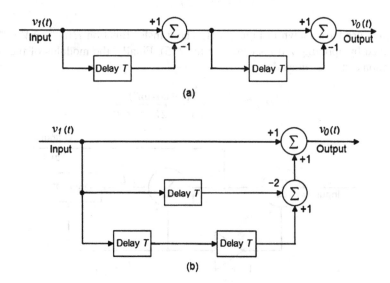

FIGURE 7.9 Two configurations of a double delay-line canceler (From Mahafza, 2005).

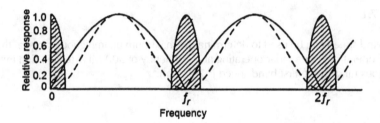

FIGURE 7.10 Realtive frequency response of the single delay-line ccanceler (solid curve) and the double delay-line canceler (dashed line). Shaded area represents clutter (From Skolnik, 1980).

The relative response of the double delay-line canceler compared with that of a single delay-line canceler is shown in Figure 7.10. To demonstrate the additional cancellation, the finite width of the clutter spectrum is also shown in this figure. The double delay-line canceler configuration of Figure 7.9(b), commonly called three pulse canceler, has the same frequency response characteristic as that of configuration Figure 7.9(a). The outputs of both the configurations are the same, and can be written as

$$v_0(t) = v_i(t) - 2v_i(t-T) + v_i(t-2T) \tag{7.14}$$

where $v_i(t)$ and $v_0(t)$ are the input and overall output in both the configurations.

7.5 MTI RECURSIVE FILTER

The typical configuration of a time-domain filter with feedback loops is illustrated in Figure 7.11. This is commonly known as a *recursive filter*. Using z-transforms as the basis for design, it is possible to synthesize any frequency function.[1] The synthesis technique described by White and Ruvin[2] may be applied with any known low-pass filter characteristic, whether it is a Butterworth or Chevyshev filter. In the time domain, we can write, using Figure 7.11, as

$$\begin{aligned} v_0(t) &= v_i(t) - (1-K)w(t) \\ v(t) &= v_0(t) + w(t) \\ w(t) &= v(t-T) \end{aligned} \tag{7.15}$$

The parameters used are shown in Figure 7.11. The transfer function $H(e^{j\omega T}) \triangleq V_0(e^{j\omega T})/V_i(e^{j\omega T})$ can be obtained by applying z-transformation to (7.15). Finally, the modulus of the square of the transfer function can be evaluated as

$$\left| H(e^{j\omega T}) \right|^2 = \frac{2(1-\cos\omega T)}{1+K^2 - 2K\cos\omega T}. \tag{7.16}$$

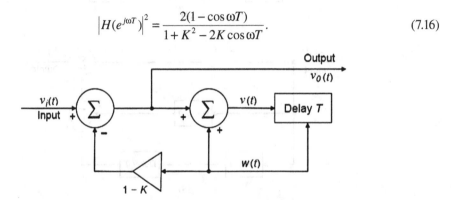

FIGURE 7.11 MTI recursive filter (From Mahafza, 2005).

MTI and Pulse Doppler Radars

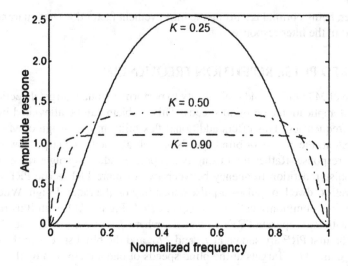

FIGURE 7.12 Frequency response of a recursive filter (From Mahafza, 1998).

The normalized frequency response $|H(e^{j\omega T})|$ of such a filter is plotted from (7.16) as a function of a normalized frequency (f / f_r), as shown in Figure 7.12. By changing the gain factor K, the filter response can be controlled. The value of K should be selected less than unity in order to avoid oscillation due to positive feedback.

7.6 MTI NONRECURSIVE FILTER

The *Nonrecursive filter*, sometimes known as a *tapped delay-line* filter, in its general form consists of N pulses and $(N-1)$ delay lines, and is shown in Figure 7.13. The weights for a nonrecursive filter with N-stage tapped delay lines, that give a response proportional to $\sin^N(\pi f_d T)$, are chosen so that they are the binomial coefficients of the expansion $(1-x)^N$ with alternating signs. In general, the binomial coefficients are given by

$$w_i = (-1)^{i-1} \frac{N!}{(N-i+1)!(i-1)!}, \qquad i = 1, 2, \ldots, N+1. \tag{7.17}$$

The nonrecursive filter with alternating binomial weights maximizes the improvement factor as well as the probability of maximum detection. The difference between an optimal filter with optimal weights and one with binomial weights is so insignificant that the latter one is considered to be

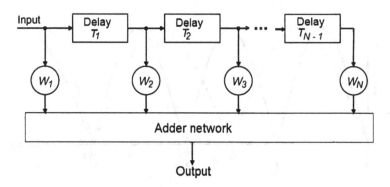

FIGURE 7.13 General form of MTI nonrecursive filter.

optimal. However, being optimal regarding the improvement factor does not guarantee a deep notch or a flat passband in the filter response.

7.7 STAGGERED PULSE REPETITION FREQUENCIES

The performance of MTI radars and their ability to perform adequate target detection is dependent, in many practical applications, on blind speeds. This problem can be alleviated by staggering the pulse repetition frequencies. This offers additional flexibility in the design of MTI filters. It mitigates, to some extent, the effects of blind speeds as well as allows a sharper low-frequency cutoff in the frequency response. Rather than using two separate radars, a single radar is employed that time shares its pulse repetition frequency between two or more PRFs. The PRF can be switched scan-to-scan, dwell-to-dwell, or pulse-to-pulse depending on the radar design. When the switching is pulse-to-pulse, it is commonly called a *staggered PRF*. Figure 7.14 is an illustration of an MTI radar operating with two separate PRFs f_{r1} and f_{r2}, where $f_{r1} = 1.5 f_{r2}$, on a time-shared basis. The blind speeds of the first PRF are shown at f_{r1} and $2 f_{r1}$, and the blind speeds of the second PRF are shown at f_{r2}, $2 f_{r2}$, and $3 f_{r2}$. Targets at the blind speeds of one are covered by the other radar with a different PRF, except when $2 f_{r1} = 3 f_{r2}$, where both PRFs have the same blind speed. In fact, zero response occurs only when the blind speeds of each PRF coincide.

Consider a radar system with two PRIs T_1 and T_2, such that

$$\frac{T_1}{T_2} = \frac{n_1}{n_2} \qquad (7.18)$$

where n_1 and n_2 are integers. The first blind speed occurs when

$$\frac{n_1}{T_1} = \frac{n_2}{T_2}. \qquad (7.19)$$

This is illustrated in Figure 7.14, where $n_1 = 2$ and $n_2 = 3$. The ratio $k_s = T_1 / T_2$ is called the *staggering ratio*. The closer this ratio approaches unity, the greater the value of the first blind speed will be. Thus the choice of the staggering ratio is a compromise between the value of the first blind speed and the depth of the nulls within the filter passband. An example of the frequency response with a four-period (five-pulse) stagger is shown in Figure 7.15, which might be used with a long-range air traffic control radar.[3] In this example the periods are in the ratio 25:30:27:31 and the first blind speed is 28.25 times that of a constant PRF waveform with the same average period. In general, if there are N PRFs related by

$$\frac{n_1}{T_1} = \frac{n_2}{T_2} = \cdots\cdots = \frac{n_N}{T_N} \qquad (7.20)$$

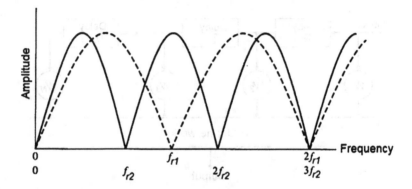

FIGURE 7.14 Frequency response of two PRFs f_{r1} (dash line) and f_{r2} (solid line) $f_{r1} = 1.5 f_{r2}$.

MTI and Pulse Doppler Radars

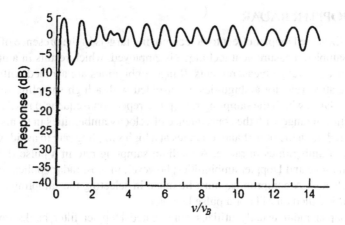

FIGURE 7.15 Frequency response of five-pulse (four-period) stagger (From Skolnik, 1980).

and if v_{b1} is equal to the first blind speed corresponding to f_{r1} of a nonstaggered waveform, then the first true blind speed v_{B1} for the staggered waveform is

$$\frac{v_{B1}}{v_{b1}} = \frac{n_1 + n_2 + \cdots + n_N}{N}. \tag{7.21}$$

Example 7.2

Consider a staggered X-band MTI radar operating at 12 GHz that uses two different PRFs with a stagger ratio of 33:34 to alleviate the problem of blind speeds. If the first PRF is 408 Hz, find the blind speeds for both PRFs and the first true blind speed of the resultant staggered waveform.

Solution:

The wavelength of the MTI radar is $\lambda = (3 \times 10^8 / 12 \times 10^9) = 0.25$ m. The stagger ratio for this radar is 33:34 implying that

$$\frac{n_1}{n_2} = \frac{T_1}{T_2} = \frac{33}{34} \Rightarrow \frac{33}{T_1} = \frac{34}{T_2}$$

This can be written as

$$33 f_{r1} = 34 f_{r2} \Rightarrow f_{r2} = \frac{33 f_{r1}}{34} = \frac{(33)(408)}{34} = 396 \text{ Hz}$$

Then the blind speeds corresponding to the two PRFs are

$$v_{b1} = \frac{\lambda f_{r1}}{2} = \frac{(0.025)(408)}{2} = 5.1 \text{ m/s} = 18.36 \text{ km/h}$$

$$v_{b2} = \frac{\lambda f_{r2}}{2} = \frac{(0.025)(396)}{2} = 4.95 \text{ m/s} = 17.82 \text{ km/h}$$

The first true blind speed of the resultant staggered waveform is then

$$v_{B1} = \left(\frac{n_1 + n_2}{2}\right) v_{b1} = \left(\frac{33 + 34}{2}\right)(18.36) = 615.06 \text{ m/s} \approx 2214.22 \text{ km/h}$$

7.8 PULSE DOPPLER RADAR

A pulse radar utilizes the Doppler effect to detect moving targets in the presence of fixed targets. In a pulse radar, a sampled measurement technique is employed, which results in ambiguities in both the relative velocity and range measurements. Range ambiguities are avoided with a low sampling rate, while the relative velocity ambiguities are avoided with a high sampling rate. MTI usually refers to a pulse radar in which the sampling rate (pulse repetition frequency) is selected low enough to avoid ambiguities in range with the consequence of velocity ambiguities in terms of blind speeds. A pulse Doppler radar, on the other hand, operates at a high sampling rate to avoid blind speeds with the consequence of ambiguities in range. A medium sampling rate in a pulse radar can theoretically resolve both range and Doppler ambiguities; however, in most radar applications the sampling rate cannot be selected. A compromise must be made in selecting the sampling rate to determine whether the radar is called an MTI or a pulse Doppler.

The pulse Doppler radar usually utilizes range-gated Doppler filter banks rather than delay-line cancelers. A simplified pulse Doppler radar block diagram is shown in Figure 7.16. The range gates are implemented as filters that operate corresponding to the detection range. The width of the detection range determines the desired range resolution. The width of each gate in a series of the range gates configuration is matched to the radar pulse width. The narrow-band filter bank is usually implemented using fast Fourier transform (FFT), where the bandwidth of the individual filters corresponds to the FFT frequency resolution. Both MTI and pulse Doppler radars use a high-power amplifier of some sort, and both use digital-signal processing. Thus the equipment differences are no longer significant enough to distinguish one from the other. The basic difference between an MTI and the pulse Doppler radar is the PRF and duty cycle that each employ.

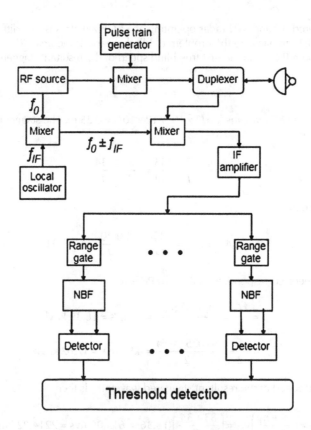

FIGURE 7.16 Simplified pulse Doppler radar.

MTI and Pulse Doppler Radars

7.9 RANGE AND DOPPLER AMBIGUITIES

Some types of measurements introduce unavoidable ambiguities in the measurement process. This happens when the PRF is too low to sample Doppler frequency directly and too high to sample the range directly. The line spectrum of a train of pulses has a $(\sin x)/x$ (sinc x) envelope where the line spectra are separated by the PRF f_r, as illustrated in Figure 7.17. The Doppler filter bank is capable of resolving target Doppler as long as the pulse radar is designed so that $f_r = (2v_{r\max}/\lambda)$, where $v_{r\max}$ is the maximum anticipated target radial velocity. If the Doppler frequency is high enough to make an adjacent spectral line move inside the Doppler band, the radar can be Doppler ambiguous. Thus the radar systems require high PRF when detecting a high-speed target to avoid Doppler ambiguities. But when a long-range radar is required to detect a high-speed target, it may not be possible to be both range and Doppler unambiguous, which can be resolved by employing multiple PRFs.

7.9.1 Resolution of Range Ambiguity

As mentioned earlier, the range and Doppler ambiguities can be resolved by using multiple PRFs. Theoretically, a large number of PRFs can be used to approach the unambiguous range of the basic PRF of a multiple-PRF ranging system. However, in practice, two or three PRFs are sufficient to resolve the most ambiguous range problem. As an illustration, consider a radar that uses two PRFs f_{r1} and f_{r2} on transmit to resolve range ambiguities, as shown in Figure 7.18. Let R_{u1} and R_{u2} be the unambiguous ranges corresponding to f_{r1} and f_{r2}, respectively. And let $R_u(R_u \gg R_{u1}, R_{u2})$ be the desired unambiguous range corresponding to a desired basic PRF f_B.

To construct an effective two-PRF ranging system, we may start by selecting a basic PRF f_B, and an integer N. The two ranging PRFs can then be calculated by

$$f_{r1} = Nf_B$$
$$f_{r2} = (N+1)f_B. \tag{7.22}$$

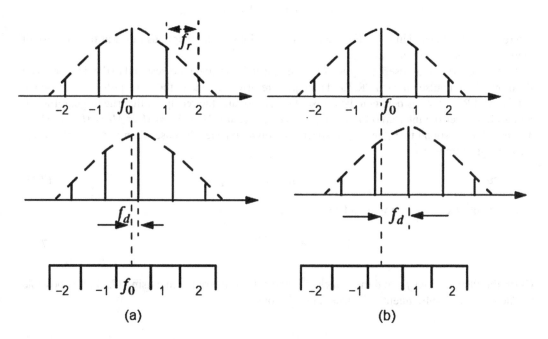

FIGURE 7.17 Spectra of transmitted, received, and Doppler bank (From Mahafza, 2005).

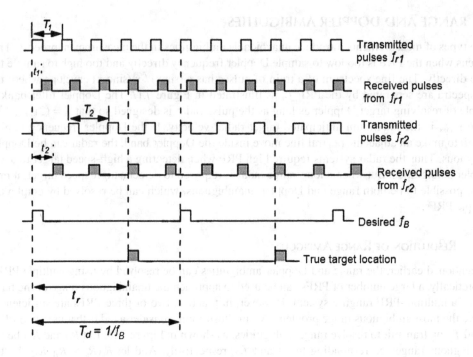

FIGURE 7.18 Resolving range ambiguity using two PRFs (From Mahafza, 2005).

The multipliers N and $(N+1)$ are relatively *prime* numbers. Similarly, a three-PRF system can be constructed using the relations

$$f_{r1} = N(N+1)f_B$$
$$f_{r2} = N(N+2)f_B$$
$$f_{r3} = (N+1)(N+2)f_B$$

where N is an integer. Three-PRF systems provide a longer unambiguous range at the expense of more complicated configuration.

In relation to symbols used in Figure 7.18, the time delays t_1 and t_2 correspond to the time between the transmit of a pulse on each PRF and receipt of return echo due to the same pulse. Let M be the number of PRF intervals between the transmit of a pulse and the receipt of the true target return. It follows that, over the interval 0 to T_d, the possible results are: $M_1 = M_2 = M$ or $M_2 = M_1 + 1$, where M_1 and M_2 correspond to f_{r1} and f_{r2}, respectively. First consider the case when $t_1 < t_2$. In this case, the round-trip time t_r is

$$t_r = t_1 + MT_1 = t_2 + MT_2. \qquad (7.23)$$

The value of M can now be written from (7.23) as

$$M = \frac{t_2 - t_1}{T_1 - T_2}. \qquad (7.24)$$

Given the PRFs f_{r1} and f_{r2} and the values of t_1 and t_2 from the radar, and using (7.23), it is possible to calculate t_r and subsequently the true range R from

$$R = \frac{ct_r}{2}. \qquad (7.25)$$

MTI and Pulse Doppler Radars

127

Now, in the case of $t_1 > t_2$, the round-trip time t_r is

$$t_r = t_1 + MT_1 = t_2 + (M+1)T_2 \tag{7.26}$$

which results in an expression of M given by

$$M = \frac{(t_2 - t_1) + T_2}{T_1 - T_2}. \tag{7.27}$$

Using (7.25)–(7.27), we can readily write the expression for the range. The multiple PRF method of range allows us to utilize a high PRF radar and calculate the range in addition to the relative velocity.

7.9.2 Resolution of Doppler Ambiguity

The analysis of the resolution of Doppler ambiguity is analogous to that of range ambiguity. We can write the expression for M in terms of f_{d1} and f_{d2} corresponding to f_{r1} and f_{r2}, respectively, as

$$M = \frac{(f_{d2} - f_{d1}) + f_{r2}}{f_{r1} - f_{r2}}, \qquad \text{if } f_{d1} > f_{d2} \tag{7.28}$$

and

$$M = \frac{(f_{d2} - f_{d1})}{f_{r1} - f_{r2}}, \qquad \text{if } f_{d1} < f_{d2}. \tag{7.29}$$

Thus the true Doppler is calculated from

$$\begin{aligned} f_d &= Mf_{r1} + f_{d1}, \text{ and} \\ f_d &= Mf_{r2} + f_{d2}. \end{aligned} \tag{7.30}$$

This result of (7.30) can be extended to the three-PRFs scheme as follows:

$$f_d = n_i f_{ri} + f_{di}, \quad i = 1, 2, 3. \tag{7.31}$$

Example 7.3

An X-band pulsed Doppler radar operating at a frequency of 12 GHz uses the three-PRFs scheme to resolve Doppler ambiguities for a target that is approaching the radar at a radial velocity of 500 m/s. The PRFs employed are: $f_{r1} = 11$ kHz, $f_{r2} = 14$ kHz, and $f_{r3} = 17$ kHz.

    a. Find the Doppler frequency position of the target corresponding to each PRF.
    b. Find the true Doppler frequency for another closing target appearing at Doppler frequencies of 5 kHz, 22 kHz, and 19 kHz for each PRF.
    c. Find the radial velocity corresponding to the Doppler frequency obtained in part (b).

Solution:

    a. The Doppler frequency for the target that is approaching the radar at a radial velocity of 500 m/s is

$$f_d = \frac{2v_r}{\lambda} = \frac{(2)(500)}{(3 \times 10^8 / 12 \times 10^9)} = \frac{1000}{0.025} = 40 \text{ kHz}$$

From (7.31):

$$f_d = n_i f_{ri} + f_{di}, \quad i = 1, 2, 3$$

we can write the desired Doppler frequency as a function of each set of PRF and corresponding Doppler frequency as

$$40 = n_1 f_{r1} + f_{d1} = 11 n_1 + f_{d1} \tag{7.32}$$

$$40 = n_2 f_{r2} + f_{d2} = 14 n_2 + f_{d2} \tag{7.33}$$

$$40 = n_3 f_{r3} + f_{d3} = 17 n_3 + f_{d3} \tag{7.34}$$

where all frequencies are in kHz. To satisfy that the Doppler frequency cannot be greater than the corresponding PRF, we can write the previous relations as

$$f_{d1} = 40 - 11 n_1 < 11 \tag{7.35}$$

$$f_{d2} = 40 - 14 n_2 < 14 \tag{7.36}$$

$$f_{d3} = 40 - 17 n_3 < 17 \tag{7.37}$$

From (7.35) we can easily show that $f_{d1} = 7$ kHz for $n_1 = 3$. Similarly from (7.36) and (7.37), we can find that $f_{d2} = 12$ kHz for $n_2 = 2$, and $f_{d3} = 6$ kHz for $n_3 = 2$.

b. Again, the Doppler frequency for another closing target appearing at 5 kHz, 22 kHz, and 19 kHz for each PRF can be obtained by using (7.31):

$$f_d = n_i f_{ri} + f_{di}, \quad i = 1, 2, 3$$

This generates the following expressions for the given Doppler frequencies:

$$f_d = 11 n_1 + 5 \tag{7.38}$$

$$f_d = 14 n_2 + 22 \tag{7.39}$$

$$f_d = 17 n_3 + 19 \tag{7.40}$$

where all frequencies are in kHz. From (7.38) we can obtain a series of f_d for $n_1 = 0, 1, 2, 3, \cdots$ as

$$f_d = 5, 16, 27, 36, \cdots \text{kHz, for } n_1 = 1, 23 \cdots$$

Similarly, from (7.39) and (7.40), we obtain the following:

$$f_d = 22, 36, 50, 64, \cdots \text{kHz, for } n_2 = 0, 1, 2, 3, \cdots$$

$$f_d = 17, 36, 53, 70, \cdots \text{kHz, for } n_3 = 0, 1, 2, 3, \cdots$$

By inspection of the previous three expressions for f_d, it reveals that $f_d = 36$ kHz is the true target Doppler frequency for $n_1 = 3$, and $n_2 = n_3 = 1$.

c. It follows that the radial velocity of the target corresponding to $f_d = 36$ kHz is

$$v_r = \frac{\lambda f_d}{2} = \frac{(0.025)(36 \times 10^3)}{2} = 450 \text{ m/s}$$

MTI and Pulse Doppler Radars

PROBLEMS

7.1 Find the maximum unambiguous range for an X-band radar operating at a frequency of 10 GHz corresponding to a first blind speed of 500 mph.

7.2 An S-band radar operating at a frequency of 3 GHz is used to detect a target in an unambiguous maximum range of 30 km. What is the corresponding first blind speed in mph that limits this desired range?

7.3 A pulse Doppler radar is used to resolve the Doppler ambiguity, and employs two different PRFs to obtain a desired unambiguous range of 120 km. Determine f_{r1}, f_{r2}, R_{u1}, and R_{u2}. Assume $N = 63$.

7.4 Consider a pulse Doppler radar that uses two PRFs to resolve range ambiguity. If the desired unambiguous range is 200 km, find the unambiguous ranges for the two PRFs. Select the integer $N = 7$.

7.5 A longer unambiguous range can be obtained by constructing a three-PRFs MTI system along with the provision of resolving range ambiguity. Design an MTI radar for this purpose that uses three PRFs to attain a desired unambiguous range of 100 km. Determine the unambiguous ranges corresponding to the three PRFs selected.

7.6 A pulsed Doppler radar operating at a frequency of 10 GHz uses a three-PRF scheme to resolve Doppler ambiguities for a target that is approaching the radar at a radial velocity of 540 m/s. The PRFs employed are: $f_{r1} = 14$ kHz, $f_{r2} = 17$ kHz, and $f_{r3} = 20$ kHz.

   a. Find the Doppler frequency position of the target corresponding to each PRF.

   b. Find the true Doppler frequency for another closing target appearing at Doppler frequencies of 8 kHz, 16 kHz, and 10 kHz for each PRF.

   c. What is the radial velocity corresponding to the true Doppler frequency as obtained in part (b)?

REFERENCES

1. H. Urkowitz, "Analysis and Synthesis of Delay Line Periodic Filters," *IRE Transactions on Circuit Theory,* 4, no. 2 (1957): 41–53.

2. W. D. White and A. E. Ruvin, "Recent Advances in Synthesis of Comb Filters," *IRE National Convention Record,* 5, no. 2 (1957): 186–199.

3. W. W. Shrader, "MTI Radar," in *Radar Handbook,* ed. M. I. Skolnik (New York: McGraw-Hill Book Company, 1970).

8 Pulse Compression Radar

"I never got tired of watching the radar echo from an aircraft as it first appeared as a tiny blip in the noise on the cathode-ray tube, and then grew slowly into a big deflection as the aircraft came nearer. This strange new power to 'see' things at great distances, through clouds or darkness, was a magical extension of our senses. It gave me the same thrill that I felt in the early days of radio when I first heard a voice coming out of a horn..."

—**Robert Hanbury Brown**

8.1 INTRODUCTION

Pulse compression involves the transmission of a long coded pulse and the processing of the received echo to obtain a relatively narrow pulse. The average transmitted power of a radar is augmented by increasing the pulse width, which results in an undesirable effect of decreasing the range resolution capability of the radar. It is often desirable to increase the pulse width while simultaneously maintaining adequate range resolution. This is accomplished by using pulse compression techniques where a long pulse containing some sort of frequency or phase modulation is transmitted, which, upon reception, must be compressed to permit the separation of adjacent range resolution cells. Simply speaking, the pulse compression allows a radar to utilize a relatively long pulse to achieve a large amount of transmitted energy, while simultaneously attaining the range resolution of a short pulse. Transmission of long pulses permits a more efficient use of the average-power capability of the radar. Generation of high peak power signals is avoided. The radar is less vulnerable to interfering signals that differ from the coded transmitted signal.

If the relative phases of the frequency components of a narrow pulse are changed by a phase-distorting filter, the frequency components combine to generate an expanded pulse, which is transmitted. The received echo is processed in the receiver in a compression filter where the relative phases of the frequency components are readjusted to produce a compressed pulse again. A pulse compression radar is a practical implementation of a matched filter system. It is, therefore, necessary to discuss the matched filter followed by the ambiguity function as background information to study the pulse compression radar.

8.2 THE MATCHED FILTER

The matched filter, sometimes known as a North filter or the conjugate filter, was developed, by Dwight North.[1] The matched filter is a filter that produces the maximum attainable SNR at the output when both useful signal and additive white noise are present at the input. In radar applications matched filters are used extensively to improve the SNR, and are of paramount importance. The maximum SNR at the output is achieved by matching the radar receiver transfer function to the received signal. Figure 8.1 shows the basic block diagram of a matched filter radar system. The transmitted waveform is generated by a signal generator designated as $s(t)$. The signal output $s(t)$ is amplified, fed to the antenna, radiated, reflected from a target, and then received by the receiver. The output of the receiver is fed into the matched filter after suitable amplification. The matched filter impulse response $h(t)$ is simply a scaled, time-reversed, and delayed form of the input signal. The shape of the impulse response is related to the signal and, therefore, matched to the input. The matched filter has the property of being able to detect the

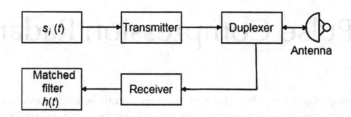

FIGURE 8.1 Basic block diagram of a matched filter radar system.

signal even in the presence of noise. It yields a higher output peak signal to mean noise power ratio for the input than for any other signal shape with the same energy content.

Consider that the transmit signal $s(t)$ with white additive white Gaussian noise with a two-sided uniform spectral density $N_0 / 2$ that is passed through a matched filter with a frequency transfer function $H(\omega)$. We want to find the $H(\omega)$ that will maximize the SNR at a given observation time t_0. More precisely,

$$SNR = \frac{|s_0(t_0)|^2}{\overline{n_0^2(t)}} \tag{8.1}$$

where $s_0(t_0)$ is the filter output corresponding to the input signal $s(t)$ at the observation time t_0 and $\overline{n_0^2(t)}$ is the average power of the white noise at the output of the filter. Let the Fourier transform of $s(t)$ be $S(\omega)$. Then the output signal of the filter output prior to envelope detection is

$$s_0(t_0) = \frac{1}{2\pi} \int_{-\infty}^{\infty} S(\omega) H(\omega) \exp(j\omega t_0) \, d\omega. \tag{8.2}$$

The power spectral density (in W/Hz) of the noise $n(t)$ at the filter output, denoted by $N(\omega)$, is

$$N(\omega) = \frac{N_0}{2} |H(\omega)|^2. \tag{8.3}$$

The average noise output power is then

$$\overline{n_0^2(t)} = \frac{1}{2\pi} \int_{-\infty}^{\infty} N(\omega) \, d\omega = \frac{1}{2\pi} \int_{-\infty}^{\infty} \frac{N_0}{2} |H(\omega)|^2 \, d\omega. \tag{8.4}$$

When simplified, this yields

$$\overline{n_0^2(t)} = \frac{N_0}{4\pi} \int_{-\infty}^{\infty} |H(\omega)|^2 \, d\omega. \tag{8.5}$$

Substituting (8.2) and (8.5) into (8.1) gives

$$SNR = \frac{|s_0(t_0)|^2}{\overline{n_0^2(t)}} = \frac{\left|\frac{1}{2\pi} \int_{-\infty}^{\infty} S(\omega) H(\omega) \exp(j\omega t_0) \, d\omega\right|^2}{\frac{N_0}{4\pi} \int_{-\infty}^{\infty} |H(\omega)|^2 \, d\omega} = \frac{\left|\int_{-\infty}^{\infty} S(\omega) H(\omega) \exp(j\omega t_0) \, d\omega\right|^2}{\pi N_0 \int_{-\infty}^{\infty} |H(\omega)|^2 \, d\omega}. \tag{8.6}$$

Pulse Compression Radar 133

We will now make use of the Schwartz inequality, which states that for any two complex signals $A(\omega)$ and $B(\omega)$, the following inequality is true

$$\left| \int_{-\infty}^{\infty} A(\omega)B(\omega)\,d\omega \right|^2 \le \int_{-\infty}^{\infty} |A(\omega)|^2\,d\omega \int_{-\infty}^{\infty} |B(\omega)|^2\,d\omega. \tag{8.7}$$

This equality holds if and only if

$$A(\omega) = A_0 B^*(\omega) \tag{8.8}$$

where∗ denotes the complex conjugate, and A_0 is an arbitrary constant and can be assumed to be unity. Applying the Schwartz inequality to (8.6) yields

$$SNR \le \frac{1}{\pi N_0} \int_{-\infty}^{\infty} |S(\omega)|^2\,d\omega. \tag{8.9}$$

If we assume that equality occurs at $t = t_0$ and $A_0 = 1$, then

$$SNR = \frac{2E}{N_0}, \tag{8.10}$$

where we have used the expression of the energy of the signal given by

$$E = \frac{1}{2\pi} \int_{-\infty}^{\infty} |S(\omega)|^2\,d\omega. \tag{8.11}$$

Thus the maximum output SNR depends only on the signal energy and input noise power. It is also observed that the maximum SNR is not dependent on the modulation type or form of transmitted signal. Consequently the signal bandwidth, type, and total energy can all be independently selected when a matched filter is used. In fact, most of the modern radar uses a matched filter in its receiver.

The equality in (8.9) holds when

$$H(\omega) = KS^*(\omega)\exp(j\omega t_0) \leftrightarrow h(t) = Ks(t_0 - t). \tag{8.12}$$

Thus the matched filter derives its name from the fact that its transfer function is proportional to the complex conjugate of the transmitted signal's Fourier transform. The filter must change with the change of the input signal at the filter input. Also the negative of time t indicates that the impulse response is proportional to the input signal "running backward." In other words, the impulse response is the delayed mirror image of the conjugate of the signal.

Example 8.1

The signal to a matched filter is represented by $s(t) = A\exp(-t/2T)$. Find the maximum SNR at the output of the matched filter.

134   Fundamental Principles of Radar

Solution:

Using (8.10) we can find the maximum SNR:

$$SNR = \frac{2E}{N_0}$$

where

$$E = \int_{-\infty}^{\infty} |s(t)|^2 \, dt = \int_{-\infty}^{\infty} e^{-(t^2/T)} \, dt = \sqrt{\pi T}$$

This leads to

$$SNR = \frac{2\sqrt{\pi T}}{N_0}$$

where $(N_0 / 2)$ is the input noise power spectral density.

Example 8.2

Find the causal transfer function of the matched filter when the input signal is $s(t) = A \exp(-t^2 / 2T)$.

Solution:

Using (8.12) for the causal transfer function given by

$$H(\omega) = KS^*(\omega) \exp(j\omega t_0) \leftrightarrow h(t) = Ks(t_0 - t)$$
$$H(\omega) = [S(\omega)]^* \exp(-j\omega t_0) = \sqrt{2\pi T} \exp(-(\omega^2 / 2)T) \exp(-j\omega t_0)$$

A Classic Example of a Matched Filter

Consider that an input signal to causal matched filter is given by

$$s(t) = \frac{A}{T}(T - t), \qquad 0 \le t \le T$$
$$= 0, \qquad \text{elsewhere.} \tag{8.13}$$

The impulse response of the matched filter as defined in (8.12) will become

$$h(t) = \frac{A}{T}[T - (\tau - t)], \qquad (\tau - T) \le t \le \tau. \tag{8.14}$$

The output signal is obtained by convolving the input signal and the impulse response

$$s_0(t) = \int_a^b \left(\frac{A}{T}(T - u) \right) \left(\frac{A}{T}(T - \tau + t - u) \right) du. \tag{8.15}$$

Pulse Compression Radar   135

The integral boundaries are determined by the signal and impulse response boundaries. Assuming $t \le \tau = T$, we can write output signal as

$$s_0(t) = \int_0^t \left(\frac{A}{T}(T-u) \right)\left(\frac{A}{T}(T-\tau+t-u) \right) du, \qquad t \le T$$

$$= \frac{A^2 t^2}{2T^2}\left(T - \frac{t}{3} \right). \tag{8.16}$$

The maximum output signal occurs at $t = \tau = T$, that is,

$$s_0(\tau) = s_0(T) = \frac{A^2 T}{3} \tag{8.17}$$

And the signal energy content can be obtained by integrating the square of the signal, as follows:

$$E = \int_0^T \left[\frac{A}{T}(T-\xi) \right]^2 d\xi = \frac{A^2 T}{3} \tag{8.18}$$

The output noise power is given by

$$R_{n_0}(t) = \frac{N_0}{2}\int_{-\infty}^{\infty} |h(u)|^2 \, du = \frac{N_0}{2}\int_0^T \left(\frac{At}{T} \right)^2 dt = \frac{N_0 A^2 T}{6} \tag{8.19}$$

Thus

$$SNR(T) = \frac{s_0^2(T)}{R_{n_0}(0)} = \frac{(A^2 T/3)^2}{N_0 A^2 T/6} = \frac{A^2 T/3}{N_0/2} = \frac{2E}{N_0} \tag{8.20}$$

This is consistent with (8.10) as expected.

8.3 THE RADAR AMBIGUITY FUNCTION

The ambiguity function is defined as the absolute value of the complex envelope of the output signal obtained when the matched filter is fed by a Doppler-shifted version of the original signal, to which the filter is matched. The mathematical functions called *time-frequency autocorrelation functions* or *ambiguity functions* were first introduced by Woodward[2] to allow representation of the time response of a signal processor when a target has significant radial velocity. The radar ambiguity function is used by radar designers to obtain a better insight about how different radar waveforms may be suitable for the various radar applications. It is also used to determine the range and Doppler resolution for a specific radar waveform.

Let $g(t)$ be the envelope of the signal, then we define a special function of the complex envelope, denoted by $\chi(\tau, \omega_d)$, at time τ in response to a transmit waveform that has been Doppler shifted by ω_d Hz is given by

$$\chi(\tau, \omega_d) = \int_{-\infty}^{\infty} g(t)g*(t-\tau)\exp(j\omega_d t)\,dt. \tag{8.21}$$

The amplitude $|\chi(\tau, \omega_d)|$ is sometimes called the *uncertainty function*, and the square of the amplitude $|\chi(\tau, \omega_d)|^2$ is called the *ambiguity function*. The filter is matched to the signal expected at a nominal center frequency and a nominal delay. The two parameters of the ambiguity function are the additional delay τ and additional Doppler shift ω_d. Therefore, $|\chi(0,0)|$ becomes the output when the input signal is returned from a point target at the nominal delay and Doppler shift for a matched filter. At any other values of τ and ω_d other than zero indicate a return from a target at some other range and velocity.

Now denote E as the energy of the complex signal envelope $g(t)$ given by

$$E = \int_{-\infty}^{\infty} |g(t)|^2 \, dt. \tag{8.22}$$

In the following discussion we will assume that the complex signal envelope $g(t)$ has a unit energy. For such normalized signals the following properties of the ambiguity function can be listed:

1. $|\chi(\tau, \omega_d)| \leq |\chi(0,0)| = 1$ \hfill (8.23)

2. $\dfrac{1}{2\pi} \int_{-\infty}^{\infty} \int_{-\infty}^{\infty} |\chi(\tau, \omega_d)|^2 \, d\tau \, d\omega_d = 1$ \hfill (8.24)

3. $|\chi(-\tau, -\omega_d)| = |\chi(\tau, \omega_d)|$ \hfill (8.25)

4. if $g(t) \leftrightarrow |\chi(\tau, \omega_d)|$

   then $g(t)e^{j\pi k t^2} \leftrightarrow |\chi(\tau, \omega_d + k\tau)|$ \hfill (8.26)

The property (1) indicates that the maximum value of the ambiguity function occurs at $(\tau, \omega_d) = (0,0)$ and for the normalized signal; the maximum value of the ambiguity function is 1. The property (2) indicates that the total volume under the ambiguity function is a constant. The property (3) implies that the ambiguity function is symmetric. The property (4) indicates that multiplying the envelope of any signal by a quadratic phase will shear the shape of the ambiguity function. This is applied to a pulse compression technique called linear FM or "chirp" in the subsequent article dealing with linear FM. Proofs of the previous properties of the ambiguity function can be found in Papoulis.[3] For proofs of the previous properties, see Problems 8.3 through 8.6.

The ideal radar ambiguity function is represented by a spike of infinitesimal width that peaks at the origin and is zero everywhere else, as illustrated in Figure 8.2. An ideal ambiguity function provides perfect resolution between neighboring targets regardless of how close they may be with respect to each other.

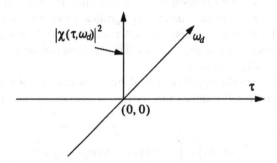

FIGURE 8.2 The ideal ambiguity function.

Pulse Compression Radar

A Classic Example of the Ambiguity and Uncertainty Functions

Ambiguity Function: Consider a normalized rectangular pulse $g(t)$ defined by

$$g(t) = \frac{1}{\sqrt{\tau_p}} \text{rect}\left(\frac{t}{\tau_p}\right) \tag{8.27}$$

Applying (8.21) for the ambiguity function, we have

$$\chi(\tau, \omega_d) = \int_{-\infty}^{\infty} g(t) g^*(t-\tau) \exp(j\omega_d t) \, dt \tag{8.28}$$

Substituting (8.27) into (8.28) and performing integration, we get

$$|\chi(\tau, \omega_d)|^2 = \left| \left(1 - \frac{|\tau|}{\tau_p}\right) \frac{\sin[(\omega_d/2)(\tau_p - |\tau|)]}{(\omega_d/2)(\tau_p - |\tau|)} \right|^2 \quad |\tau| \leq \tau_p \tag{8.29}$$

The Uncertainty Function: Consider a Gaussian pulse defined by

$$g(t) = A \exp[-t^2/(2\sigma)] \tag{8.30}$$

By using (8.31) in (8.21) and simplifying we obtain

$$\chi(\tau, \omega_d) = A^2 \sqrt{\pi \sigma^2} \exp\left[-\frac{\tau^2}{4\sigma^2} + j\frac{\omega_d \tau}{2} - \left(\frac{\omega_d \sigma}{2}\right)^2\right] \tag{8.31}$$

Figure 8.3 shows a 3-dimensionsl plot[4] of a single pulse ambiguity function with pulse width $\tau_p = 2$ seconds.

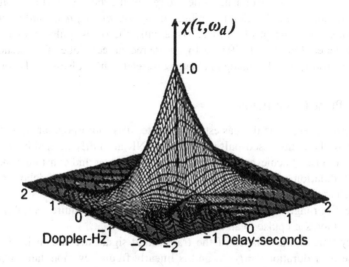

FIGURE 8.3 Single pulse 3-D ambiguity function with pulse width of 2 seconds (From Mahafza, 2005).

138   Fundamental Principles of Radar

8.4 PULSE COMPRESSION IN RADARS

As mentioned earlier, pulse compression allows a radar to transmit a pulse of a relatively long duration and low peak power to achieve large radiated power, but simultaneously to obtain the range resolution of a short pulse. This is accomplished by utilizing frequency or phase modulation to increase the signal bandwidth and then compressing the received echo waveform in a matched filter. The range resolution achievable with a radar system is

$$\Delta R = \frac{c\tau}{2}. \tag{8.32}$$

Equation (2.5) is reproduced here for consistency while retaining the symbols as before. In a pulse compression system, transmitted waveform is modulated either in frequency or phase, so that the bandwidth $B \gg 1/\tau$. If τ_c represents the effective pulse width of the resultant compressed pulse implying $\tau_c = 1/B$, (8.32) becomes

$$\Delta R = \frac{c\tau_c}{2}. \tag{8.33}$$

Thus the pulse compression radar can utilize a transmit pulse of duration τ and yet achieve a range resolution corresponding to the compressed pulse of duration τ_c. Define a term called *dispersion factor* or *compression ratio (CR)* as

$$CR = \frac{\tau}{\tau_c} = B\tau \tag{8.34}$$

which indicates that the compression ratio also equals the time-bandwidth[5] product of the system. Generally $B\tau \gg 1$; values from 100 to 300 might be considered as more typical. There are many types of modulations used for pulse compression, but two that have wide applications are the *frequency modulation technique* and the *phase modulation technique*.

8.5 FREQUENCY MODULATION IN PULSE COMPRESSION

In the frequency modulation technique, the transmitter frequency is modulated to increase the bandwidth of the transmitted waveform, while the received echo waveform is compressed. There are two types of frequency modulation techniques using: *linear frequency modulation* (LFM) and *frequency stepping*. The concept of LFM is historically the oldest pulse compression technique developed by Kluder et al. in the late 1940s.[6] In a more recent technique of frequency stepping,[7] the frequency of the transmit signal is made to change discretely on a pulse-to-pulse or subpulse basis.

8.5.1 LFM IN PULSE COMPRESSION

The LFM or *chirp* waveform is the easiest to generate. It is more popular than any other coded waveforms because the compressed-pulse shape and SNR are fairly insensitive to Doppler shifts. Pulse compression in LFM achieves increased transmitter power and constant bandwidth by a linear frequency modulation of the transmitted waveform and a receiver delay network utilizing frequency-time characteristics of the transmitted waveform. But there are some major disadvantages: it has excessive range-Doppler cross coupling errors, and it requires weighting to reduce the sidelobes of the compressed pulse to an acceptable level.

The frequency-time characteristics of the transmitted signal are shown in Figure 8.4(a). The transmitted pulse is of duration $\tau = t_2 - t_1$ and is linearly frequency modulated from f_1 to f_2 over the pulse length by an amount Δf as depicted in Figure 8.4(b). The effect of frequency modulation

Pulse Compression Radar

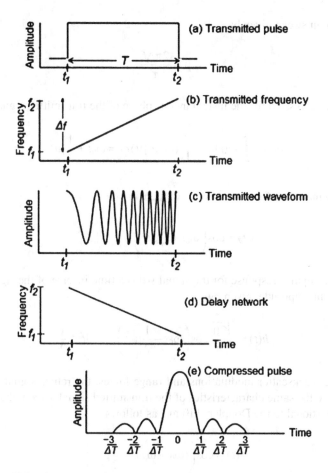

FIGURE 8.4 Linear FM pulse compression. (a) transmitted pulse, (b) transmitted frequency, (c) transmitted waveform, (d) delay network, and (e) compressed pulse.

on the transmitted signal is shown in Figure 8.4(c). The target return signal will be similar to the transmitted signal. The pulse compression filter in the radar receiver is matched to the transmitted waveform so that the received signal experiences a frequency-dependent time delay as given in Figure 8.4(d). It is observed that the lowest frequency f_1 is delayed the longest, while the highest received frequency f_2 is not delayed at all. The final output is a short pulse of duration $\tau_c = 2/\Delta f$ and with a large amplitude of $\sqrt{CR}$, as given in Figure 8.4(e). The compressed pulse of Figure 8.4(e) displays the amplitude characteristics after passage through the matched filter or dispersive delay line. The sidelobes of the amplitude–time characteristics are often undesirable since they may result in false detections. These sidelobes can be reduced by amplitude weighting of the received signal, and will be discussed in a subsequent section. The effect of weighting the received signal to lower the sidelobes, unfortunately, widens the main lobe and reduces the peak SNR compared to the unweighted LFM pulse compression.

Matched Filter Analysis of LFM Pulse Compression

Mathematical derivations to the equations that represent the general LFM waveforms will be given to justify the previously mentioned facts. Consider the LFM waveform of Figure 8.4, and designate the transmitted angular frequency of the carrier ω_0, up-chirped over the transmitted pulse length τ, by

$$\omega(t) = \omega_0 + \mu t \tag{8.35}$$

where μ is the radian sweep given by

$$\mu = \frac{2\pi\Delta f}{\tau}. \tag{8.36}$$

We can obtain the expression for the instantaneous phase of the transmitted signal as

$$\theta(t) = \int_0^t \omega(t)\,dt = \int_0^t (\omega_0 + \mu t)\,dt = \omega_0 t + \frac{1}{2}\mu t^2. \tag{8.37}$$

The transmitted signal, therefore, can now be written

$$s(t) = \cos\left(\omega_0 t + \frac{1}{2}\mu t^2\right), \qquad |t| \le \frac{\tau}{2}. \tag{8.38}$$

The matched filter impulse response for the signal $s(t)$ is a time inverse of the signal at the receiver input, normalized in amplitude, given by

$$h(t) = \sqrt{\frac{2\mu}{\pi}} \cos\left(\omega_0 t - \frac{1}{2}\mu t^2\right), \qquad |t| \le \frac{\tau}{2} \tag{8.39}$$

Ignoring the target backscatter modulations and range losses, the return signal from the target at any time t will have the same characteristics of the transmitted signal, except that it will be shifted in frequency proportional to the Doppler shift ω_d, as follows

$$s(t) = \cos\left((\omega_0 + \omega_d)t + \frac{1}{2}\mu t^2\right). \tag{8.40}$$

The matched filter output $\psi(t, \omega_d)$ is obtained by the convolution of $s(t)$, given by (8.40) and $h(t)$ given by (8.39). Accordingly, it follows

$$\psi(t, \omega_d) = \sqrt{\frac{2\mu}{\pi}} \int_{-\tau/2}^{+\tau/2} \cos\left[(\omega_0 + \omega_d)\tau' + \frac{1}{2}\mu\tau'^2\right] \cos\left[\omega_0(t - \tau') - \frac{1}{2}\mu(t - \tau')^2\right] d\tau' \tag{8.41}$$

where τ' is a dummy time variable. The closed form solution of $\psi(t, \omega_d)$ can be obtained after a considerable amount of trigonometric and algebraic manipulations,[8] and is given by

$$\psi(t, \omega_d) = \sqrt{\frac{\mu}{2\pi}} \frac{\sin\left[\dfrac{\omega_d + \mu t}{2}(\tau - |t|)\right]}{\dfrac{\omega_d + \mu t}{2}} \cos\left[\left(\omega_0 + \frac{\omega_d}{2}\right)t\right], \qquad -\frac{\tau}{2} \le t \le \frac{\tau}{2}. \tag{8.42}$$

Equation (8.42) represents the output of the matched filter as a function of time and the Doppler shift of the received signal. A good approximation of the maximum value of $\psi(t, \omega_d)$ occurs at $t \ll \tau$ and $\omega_d + \mu t = 0$, that is

$$t = -\frac{\omega_d}{\mu} = -\frac{\tau f_d}{\Delta f}. \tag{8.43}$$

Pulse Compression Radar

Since the values of τ and Δf are known, the time in which the maximum amplitude occurs will depend on the Doppler shift of the returned signal. For the case of no Doppler shift ($\omega_d = 0$), this becomes

$$\psi(t) = \sqrt{\frac{\mu}{2\pi}} \; \frac{\sin\frac{\mu t}{2}(\tau - |t|)}{\frac{\mu t}{2}} \cos\omega_0 t, \qquad -\frac{\tau}{2} \le t \le \frac{\tau}{2}. \tag{8.44}$$

Neglecting $|t|$ compared to τ, we get

$$\psi(t) \approx \sqrt{\frac{\mu}{2\pi}} \; \frac{\sin\frac{\mu t \tau}{2}}{\frac{\mu t}{2}} \cos\omega_0 t = \sqrt{\frac{\mu \tau^2}{2\pi}} \; \frac{\sin\frac{\mu t \tau}{2}}{\frac{\mu t \tau}{2}} \cos\omega_0 t. \tag{8.45}$$

The first null beamwidth of the compressed pulse is the distance between the first zeros of $\psi(t)$, and these occur at $\frac{\mu t \tau}{2} = \pm\pi$, which gives

$$t = \pm\frac{2\pi}{\mu\tau} = \frac{2\pi}{(2\pi\Delta f / \tau)\tau} \pm \frac{1}{\Delta f}. \tag{8.46}$$

Thus the null-to-null width τ_c of the compressed pulse is

$$\tau_c = \frac{2}{\Delta f} \tag{8.47}$$

which corresponds to the points 4 dB down from the peak of output power $\psi^2(t)$. The amplitude of the compressed pulse is written from (8.45) as

$$A = \sqrt{\frac{\mu\tau^2}{2\pi}} = \sqrt{\frac{(2\pi\Delta f / \tau)\tau^2}{2\pi}}\sqrt{\tau\Delta f} \Rightarrow A = \sqrt{CR} \tag{8.48}$$

Example 8.3

Consider that an up-chirp LFM pulse compression radar centered at $f_0 = 150$ kHz is used in a sonar system with a range resolution $\Delta R = 3$ cm. It is desired to increase the transmitted energy by a factor of 15, which can be achieved by increasing the width of the pulse. Assume that the velocity of propagation of the electromagnetic energy in water is 350 m/s.

    a. Find the bandwidth of the radar and the expression of the transmitted pulse.
    b. Find the impulse response of the causal matched to the LFM transmitted pulse.

Solution:

    a. The resolution of the sonar radar system is

$$\Delta R = \frac{v\tau_c}{2} \Rightarrow \tau_c = \frac{2\Delta R}{v} = \frac{(2)(0.03)}{300} = 0.2 \text{ ms}$$

So the receiver bandwidth of the radar is

$$B = (1/\tau_c) = (1/0.2\times10^{-3}) = 5 \text{ kHz}$$

An LFM up-chirp transmitted signal has the form

$$s(t) = \cos\left(\omega_0 t + \frac{1}{2}\mu t^2\right), \qquad |t| \le \frac{\tau}{2}$$

where μ is called the radian sweep and given by $\mu = (2\pi\Delta f / \tau)$. Since Δf is the bandwidth B of this signal, we get $\Delta f = 5$ kHz, which yields

$$\mu = \frac{2\pi\Delta f}{\tau} = \frac{(2\pi)(5\times 10^3)}{3\times 10^{-3}} = 10.47\times 10^6 \text{ radian}$$

where we have used the transmitted pulse width $\tau = 15\tau_c = (15)(0.2\times 10^{-3}) = 3\times 10^{-3}$ s. It follows then

$$s(t) = \cos\left((2\pi)(150\times 10^3)t + \frac{15.7\times 10^6}{2}t^2\right), \qquad |t| \le \frac{3\times 10^3}{2}$$

b. The causal matched filter impulse response for the signal $s(t)$ is a time inversion of $s(t)$ (normalized in amplitude) given by

$$h(t) = \left(\frac{(2)(15.7\times 10^6)}{\pi}\right)^{1/2} \cos\left((2\pi)(150\times 10^3)t - \frac{15.7\times 10^6}{2}t^2\right), \qquad |t| \le \frac{3\times 10^3}{2}$$

Example 8.4

Consider an X-band pulse compression radar operating at a frequency of 10 GHz is designed to achieve a range resolution of 15 m. A target is an aircraft approaching the radar at a speed of 300 m/s with a squint angle of 60°.

a. Find the Doppler frequency shift and the maximum frequency excursion of the transmitted frequency modulated signal.
b. Find the time at which the maximum amplitude occurs at the output of the matched filter when the transmitted pulse length is 3 ms.

Solution:

a. The Doppler frequency shift is computed as

$$f_d = \frac{2v\cos\theta_s}{\lambda} = \frac{(2)(300)\cos 60°}{3\times 10^8 / (10\times 10^9)} = 10 \text{ kHz}$$

and the frequency excursion is found from

$$\Delta R = \frac{c}{2\Delta f} \Rightarrow \Delta f = \frac{c}{2\Delta R} = \frac{3\times 10^8}{(2)(15)} = 10 \text{ MHz}$$

b. Using (8.43) we get

$$t = -\frac{\tau f_d}{\Delta f} = -\frac{(3\times 10^{-3})(10\times 10^3)}{(10\times 10^6)} = -3.0 \text{ μs}$$

8.5.2 Frequency Stepping in Pulse Compression

In frequency stepping, the frequency of the transmit signal is made to change discretely on a pulse-to-pulse or subpulse basis. As discussed previously, LFM uses continuous modulation of the high frequency carrier, while frequency stepping uses discrete modulation to encode the high frequency carrier. Three different versions of frequency stepping (Nathanson 1991) are used: (a) discrete linear stepped FM, (b) scrambled frequency stepping, and (c) interpulse frequency stepping. In this section only a detailed analysis of the discrete linear frequency stepping in pulse compression is presented.

Consider a linearly stepped frequency waveform containing N equal-amplitude rectangular pulses of length τ_c each on a different frequency f_n. The envelope of this waveform is illustrated in Figure 8.5. The symbols Δf and τ represent the spacing between frequencies and the subpulse or segment length, respectively. It is assumed that the frequency spacing Δf is equal to the inverse of the subpulse length τ, and the subpulse envelopes are rectangular. It can be shown that when $\tau \Delta f = 1$, the nulls of the autocorrelation function of the subpulse envelope tend to suppress the range ambiguities that could have otherwise resulted because of a slightly smaller or larger value of $\tau \Delta f$ than unity. The time and phase origin is taken to be the leading edge of the first pulse. The transmitted signal for such a linearly stepped frequency waveform can be written as

$$s(t) = \sum_{n=0}^{N-1} \left[u(t - n\tau) - u(t - [n+1]\tau) \right] \cos(\omega_0 + n\Delta\omega)t \tag{8.49}$$

where $\omega_0 = 2\pi f_0$, is the lowest frequency in the transmission comb, $\Delta\omega = 2\pi\Delta f$ is the angular frequency spacing, and $u(t)$ is the unit step function. It can be shown that the matched filter has an impulse response, which can be written[9] as

$$h(t) = s(-t) = \sum_{n=0}^{N-1} \left[u(-t - n\tau) - u(-t - [n+1]\tau) \right] \cos(\omega_0 + n\Delta\omega)t. \tag{8.50}$$

The matched filter output, with the target delay neglected, is then

$$s_0(t) = \sum_{n=0}^{N-1} \exp j(\omega_0 + n\Delta\omega)t, \quad (N-1)\tau < t < N\tau, \tag{8.51}$$

$$s_0(t) = \exp(j\omega_0 t) \sum_{n=0}^{N-1} \exp(jn\Delta\omega)t = \exp(j\omega_0 t) \frac{1 - \exp(jN\Delta\omega t)}{1 - \exp(.j\Delta\omega t)} \tag{8.52}$$

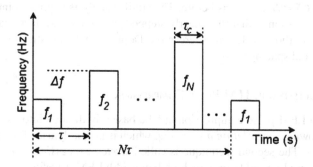

FIGURE 8.5 Linear stepped FM in pulse compression.

144 Fundamental Principles of Radar

This can be finally simplified to

$$s_0(t) = \left(\frac{\sin N(\Delta\omega / 2)t}{\sin(\Delta\omega / 2)t} \right) \exp j[\omega_0 + (N-1)\Delta\omega / 2]t. \qquad (8.53)$$

Nulls in the compressed-pulse envelope occur when $\sin[N(\Delta\omega / 2)t = 0$ or when

$$N\frac{\Delta\omega}{2}t = \pm m\pi, \quad m = 0,1,2,3,\cdots. \qquad (8.54)$$

The first sidelobe is reduced from the main lobe by 13.5 dB when $N \geq 50$, and is 13.06 dB down for $N = 8$. From (8.54) we get

$$t = \pm\frac{2m\pi}{N2\pi\Delta f} = \pm\frac{m}{N\Delta f}. \qquad (8.55)$$

The first null beamwidth of the compressed pulse is the distance between the first zeros of $s_0(t)$, and these occur at

$$t = \pm\frac{1}{N\Delta f}. \qquad (8.56)$$

At one half the distance between the nulls, the linear envelope response is down by 4 dB from the peak of the envelope output power of $s_0^2(t)$. Hence, the compressed pulse width is

$$\tau_c = \frac{1}{N\Delta f} = \frac{1}{N(N / \tau)} = \frac{\tau}{N^2}. \qquad (8.57)$$

The pulse compression ratio CR is then written as

$$CR = \frac{\tau}{\tau_c} = N^2. \qquad (8.58)$$

It indicates that the pulse compression ratio is equal to the square of the number of subpulses with the customary definition of 3-dB pulse width. Thus the compressed matched filter output for a stepped linear FM waveform has almost the same shape as that for a continuous LFM waveform for large compression ratios.

In a scrambled frequency stepping scheme, the component frequencies are the same as those of the linearly stepped FM scheme, but they occur randomly in time. It can be shown that the pulse compression ratio is $CR = N^2$, same as before. The similar analysis applies to interpulse frequency stepping. It must be mentioned that the linearly stepped-frequency waveform has Doppler characteristics approximating those of LFM, whereas the Doppler characteristics of other two schemes depend on the order and spacing used.

8.5.3 ACTIVE PROCESSING IN LFM PULSE COMPRESSION

Active processing in LFM pulse compression can be basically divided into two techniques.[10] The first technique is known as *correlation processing*, which is normally used for narrow and medium band radar operation. The second technique is called *stretch processing* or *deramp compression processing*, and is normally used to process high bandwidth LFM waveforms. The stretch processing technique is normally employed since it is much easier to implement. Other advantages include

Pulse Compression Radar

reduced signal bandwidth, dynamic range increase due to signal compression, and the baseband frequency offset, which is directly proportional to the target range.

Correlation Processing

In correlation processing, the pulse is compressed by adding frequency modulation to a long pulse at transmission, and by employing a matched filter in the receiver in order to compress the received signal as used in LFM. In practice the correlation processor is implemented by using the *fast Fourier transform* (FFT). When using LFM, returns from neighboring targets are resolved as long as they are separated by the compressed pulse width. For the LMF case, the first sidelobe is approximately 13.5 dB below the peak; and for most radar applications this may not be sufficient. Unfortunately, high sidelobe levels are not desirable because noise located at the sidelobes may interfere with target returns in the main lobe. Weighting functions can be employed on the compressed pulse spectrum in order to reduce the sidelobe levels. This approach results in a loss in the main lobe resolution, and a reduction in the SNR.

All radar operations are usually performed in a *receive window*, which is defined as the difference between the radar maximum and minimum range. Echoes from all targets within the receive window are collected and then passed through the matched filter to perform pulse compression using FFT. The process is illustrated in Figure 8.6. The receiver window in correlation processing is defined by

$$R_{rec} = R_{max} - R_{min} \tag{8.59}$$

where R_{max} and R_{min} are the maximum and minimum ranges, respectively, over which the radar performs detection. The normalized complex transmitted signal is expressed as

$$s_t(t) = \exp\left[j\left(\omega_0 t + \frac{1}{2}\mu t^2\right)\right], \quad |t| \leq \frac{\tau}{2} \tag{8.60}$$

where the symbols have their usual meanings, and $\mu = 2\pi B / \tau$. The received signal from a target at R_1 that returns after time delay $\tau_1 = 2R_1 / c$ is

$$s_r(t) = b_1 \exp\left[j\left(\omega_0(t-\tau_1) + \frac{1}{2}\mu(t-\tau_1)^2\right)\right] \tag{8.61}$$

where b_1 is proportional to RCS of the target located at R_1. The phase of the mixer output is extracted by multiplying $s_r(t)$ with a reference signal $s_{ref}(t) = \exp(j\omega_0 t)$, and then by low-pass filtering. This results in an expression given by

$$\theta(t) = -\omega_0 \tau_1 + \frac{\mu}{2}(t - \tau_1)^2. \tag{8.62}$$

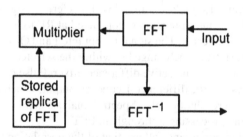

FIGURE 8.6 The convolution process in the correlation process.

It follows that the instantaneous frequency is

$$f_i(t) = \frac{1}{2\pi}\frac{d\theta(t)}{dt} = \frac{1}{2\pi}\mu(t-\tau_1) = \frac{1}{2\pi}\left(\frac{2\pi B}{\tau}\right) = \frac{B}{\tau}\left(t - \frac{2R_1}{c}\right). \tag{8.63}$$

Assuming n targets at ranges $R_1, R_2, \cdots, R_n$ within the receive window R_{rec}, we can write the phase of the mixer output from (8.62) as

$$\theta(t) = \sum_{i=1}^{n}\left[-\omega_0\tau_i + \frac{\mu}{2}(t-\tau_i)^2\right] \tag{8.64}$$

where $\tau_i = 2R_i/c$, $i = 1, 2, \cdots, n$ represent the two-way delay times. To avoid the ambiguity in the spectrum, sampling frequency f_s is chosen according to the Nyquist rate as

$$f_s \geq 2B \Rightarrow \Delta T = \frac{1}{f_s} \leq \frac{1}{2B} \tag{8.65}$$

where ΔT is the sampling interval. It can be shown, using (8.63), that the frequency resolution of FFT is

$$\delta f = 1/\tau. \tag{8.66}$$

Then the minimum required number of samples is

$$N = \frac{1}{\delta f \Delta T} = \frac{\tau}{\Delta T}. \tag{8.67}$$

Using (8.66) and (8.67) in (8.65) gives

$$N \geq 2B\tau. \tag{8.68}$$

Equation (8.68) implies that the total number of samples N is sufficient enough to completely describe an LFM waveform of duration τ and bandwidth B.

Stretch Processing

Stretch processing is normally used to process extremely high bandwidth LFM waveforms. The functional block diagram for a stretch processing receiver is shown in Figure 8.7(a). It consists of a mixer, an LFM generator, a timing circuitry, and a spectrum analyzer. In this diagram, τ_M is the range delay to which the stretch processor is matched and is usually close to τ_R. Details of the functional block diagram are shown in Figure 8.7(b). The stretch processing consists of the following steps: first of all the received signals are mixed with a reference signal from the local oscillator (LO). Hence, active-correlation multiplication is conducted at RF followed by low pass filtering to extract the difference frequency terms. The signal is split into I and Q components and digitized for further processing. The return from each range bin within the selected range gate thus corresponds to a pulsed sinusoid at the output of the active difference mixer. Pulse compression is completed by performing a spectral analysis of the difference-frequency output, which transforms the pulse tones into corresponding frequency resolution cells. Spectral analysis is performed by digitizing the difference frequency output and processing it through an FFT.

The analysis of the stretch processing follows that of the correlation processing. The frequency of the reference signal is linearly swept over an RF bandwidth B. The received signal is a replica of

Pulse Compression Radar

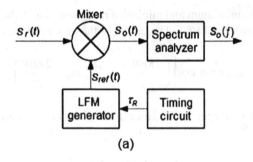

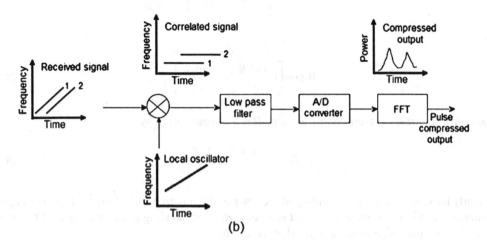

FIGURE 8.7 Stretch processing in pulse compression.

the transmitted signal with a time delay τ_R due to range R. The normalized LFM transmitted signal is reproduced here for convenience:

$$s_t(t) = \cos\left(\omega_0 t + \frac{1}{2}\mu t^2\right), \quad |t| \leq \frac{\tau}{2} \tag{8.69}$$

where the symbols have their usual meanings, and $\mu = 2\pi B / \tau$. The received signal that returns after the time delay of $\tau_R = 2R/c$ is

$$s_r(t) = b\cos\left[\omega_0(t-\tau_R) + \frac{\mu(t-\tau_R)^2}{2}\right]. \tag{8.70}$$

The reference signal is

$$s_{ref}(t) = \cos\left(\omega_0 t + \frac{1}{2}\mu t^2\right), \quad |t| \leq \frac{T_{rec}}{2} \tag{8.71}$$

where the receive window in seconds is

$$T_{rec} = \frac{2(R_{max} - R_{min})}{c} = \frac{2R_{rec}}{c} \tag{8.72}$$

148   Fundamental Principles of Radar

where R_{max} and R_{min} are the maximum and minimum range over which the radar performs detection. It follows that the bandwidth $B = 1 / T_{rec}$. The output of the mixer followed by low pass filtering is

$$s_0(t) = b \, \cos\left[\left(\frac{4\pi BR}{c\tau}\right)t + \frac{2R}{c}\left(\omega_0 - \frac{2\pi BR}{c\tau}\right)\right]$$ (8.73)

where b is proportional to target RCS. Since $\tau \gg \tau_R = 2R/c$, we can approximate the above relation to

$$s_0(t) = b \, \cos\left[\left(\frac{4\pi BR}{c\tau}\right)t + \frac{2\omega_0 R}{c\tau}\right] = b \, \cos\theta(t)$$ (8.74)

where

$$\theta(t) = \left[\left(\frac{4\pi BR}{c\tau}\right)t + \frac{2\omega_0 R}{c\tau}\right].$$ (8.75)

The return frequency corresponding to range R is a constant given by

$$f_R = \frac{1}{2\pi}\frac{d\theta(t)}{dt} = \frac{2BR}{c\tau}.$$ (8.76)

It clearly indicates that proper sampling of the low pass filter output and taking FFT of the sampling sequence lead to the presence of a target at a range R_n corresponding to the peak of the FFT output occurring at a particular frequency f_n. More precisely,

$$R_n = \left(\frac{c\tau}{2B}\right)f_n.$$ (8.77)

For close targets at ranges $R_1, R_2, \cdots (R_1 < R_2 < \cdots < R_n)$, we can extend the previous result from (8.73), by superposition, to arrive at an expression for the composite signal given by

$$s_0(t) = \sum_{i=1}^{n} b_n \, \cos\left[\left(\frac{4\pi BR_i}{c\tau}\right)t + \frac{2R_i}{c}\left(\omega_0 - \frac{2\pi BR_i}{c\tau}\right)\right].$$ (8.78)

Thus the output of the stretch processor depends on the critical determination of the sampling rate and the size of FFT. If we define the resultant bandwidth after stretch processing as the frequency resolution δf, then we can use (8.76) to represent this by

$$\delta f = f_2 - f_1 = \frac{2B}{c\tau}(R_2 - R_1) = \frac{2B}{c\tau}R_{rec} = \left(\frac{2B}{c\tau}\right)\left(\frac{c}{2B}\right) = \frac{1}{\tau}.$$ (8.79)

Since $B > 1/\tau$ in LFM, it follows from (8.79) that, through stretch processing, the bandwidth at the receiver output is less than the original signal bandwidth, thereby facilitating the implementation of a digital signal processing (DSP) system in an LFM radar system. The maximum frequency resolvable by the FFT, limited in the region $\mp N\delta f / 2$, is given by

$$\frac{N\delta f}{2} > \frac{2BR_{rec}}{c\tau}.$$ (8.80)

Pulse Compression Radar 149

Combining (8.79) and (8.80) and denoting $T_{rec} = 2R_{rec} / c$, we can obtain the size of the FFT for implementation as follows:

$$N_{FFT} \geq 2BT_{rec} \tag{8.81}$$

Example 8.5

Consider a certain radar that uses a stretch processor to process an extremely high bandwidth of 450 MHz in an LFM pulse compression radar. It transmits a pulse of length 30 μs to perform detection of two resolvable targets with a range receive window of 750 m. Assume that one of the targets is located at a range of 15 km.

  a. Find the frequency tones corresponding to the above targets.
  b. Find the minimum FFT size.

Solution:

$R_2 = R_1 + R_{rec} = 15,000 + 750 = 15,750$ m. The receive window in seconds is then

$$T_{rec} = \frac{2(R_2 - R_1)}{c} = \frac{2R_{rec}}{c} = \frac{(2)(750)}{3 \times 10^8} = 5 \ \mu s$$

  a. The frequency tones corresponding to the targets are obtained from (8.77):

$$R_n = \left(\frac{c\tau}{2B} \right) f_n \Rightarrow f_n = \left(\frac{2B}{c\tau} \right) R_n$$

It follows then

$$f_1 = \left(\frac{2B}{c\tau} \right) R_1 = \left(\frac{(2)(450 \times 10^6)}{(3 \times 10^8)(30 \times 10^{-6})} \right)(15 \times 10^3) = (10^5)(15 \times 10^3) = 1.500 \ \text{GHz}$$

$$f_2 = \left(\frac{2B}{c\tau} \right) R_2 = \left(\frac{(2)(450 \times 10^6)}{(3 \times 10^8)(30 \times 10^{-6})} \right)(15.75 \times 10^3) = (10^5)(15.75 \times 10^3) = 1.575 \ \text{GHz}$$

  b. Using (8.81) we get

$$N_{FFT} = 2BT_{rec} = (2)(450 \times 10^6)(5 \times 10^{-6}) = 4,500$$

8.6 PHASE-CODED MODULATION IN PULSE COMPRESSION

In the phase-coded modulation technique, the long pulse is subdivided into an N number of shorter subpulses of equal time duration T. The modulation, however, is not of the frequency of each phase, but of the phase of each pulse selected in accordance with a phase code. Thus the complex envelope of the phase-modulated signals is expressed as

$$g(t) = \frac{1}{(NT)^{1/2}} \sum_{n=0}^{N-1} g_n(1 - nT) \tag{8.82}$$

where

$$g_n(t) = \begin{cases} \exp(j\phi_n), & 0 \leq t \leq T \\ 0, & \text{elsewhere.} \end{cases} \tag{8.83}$$

TABLE 8.1
Known Barker Codes

Code Symbols	Code Lengths	Code Elements	Sidelobe Reductions (dB)
B_2	2	10	6.0
		11	
B_3	3	110	9.5
B_4	4	1101	12.0
		1110	
B_5	5	11101	14.0
B_7	7	1110010	16.9
B_{11}	11	11100010010	20.8
B_{13}	13	1111100110101	22.3

There are many phase-coding schemes of the sequence $\{\phi_n\}$. Here we will discuss two major families: *Barker codes*, which are binary sequences, and *Frank codes,* which are polyphase sequences.

8.6.1 Barker Coding for Pulse Compression

Barker codes[11] are binary phase coding sequences with the property that the peak sidelobes of the resulting ambiguity functions are all less than or equal to $1/N$ in magnitude where the output signal voltage is normalized to 1. There are only seven known Barker codes that share this unique property, which are listed in Table 8.1, where $\exp(j0) = +1$ is represented by "1" and $\exp(j\pi) = -1$ is represented by "0."

Extensive searches for Barker codes of lengths greater than 13 have been conducted by various researchers without any success. Barker codes have some unique features that their sidelobe structures contain minimum energy, which is uniformly distributed among the sidelobes. Because of these features, Barker codes are sometimes called *perfect codes.*

In general, the autocorrelation function for a B_N Barker code is $2NT$ wide. The main lobe is $2T$ wide and the peak value is equal to N. There are $(N-1)/2$ sidelobes on either side of the main lobe. This is illustrated in Figure 8.8 for B_{13}. It is observed that the peak value of the main lobe is 13, while the sidelobes are all unity. A Barker code can offer sidelobe reduction -22.3 dB at most, which may not be sufficient for many radar applications. However, one scheme to generate codes longer than 13 is the method of forming combined codes using known Barker codes. As an example, a B_m code can be used within a B_n code to generate a code of length mn. The compression ratio for the combined B_{mn} code is equal to $1/mn$. Combined codes consisting of any number of individual codes can be analogously defined. The filter associated with a combined code is a combination of filters

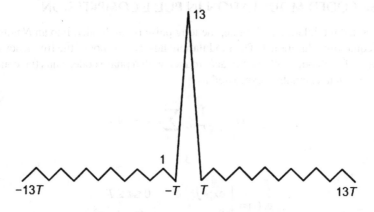

FIGURE 8.8 Autocorrelation function of the Barker code of length 13.

Pulse Compression Radar 151

matched to the individual codes. The individual codes are called the subcodes of the full code. The matched filter for a combined code may be implemented directly as a tapped delay line whose impulse response is the time inverse of the code.

As an example, in a combined matched filter for the 5×4 combined Barker code, the first stage of the filter is simply the matched filter to the inner 4 bit code, while the second stage represents a filter matched to the 5-bit code. The resulting Barker code is a 20-bit code of 1101 1101 11101 0010 1101. The filter is equivalent to the 20-bit tapped delay line matched filter.

Example 8.6

Construct a Barker code by combining into a longer sequence of length 35-bit using both B_{75} and B_{57}.

Solution:

B_{75} can be constructed of seven sequences, phasing the individual sequences as 5-bit Barker, and setting the phase of each group according to the 7-bit Barker code. Accordingly, the resulting sequence is

$$B_{75} = \{11101, 11101, 11101, 00010, 00010, 11101, 00010\}$$

Likewise B_{57} can be constructed of five sequences, phasing the individual sequences as 7-bit Barker code, and setting the phase of each group according to the 5-bit Barker code. Accordingly, the resulting sequence is

$$B_{57} = \{1110010, 1110010, 1110010, 0001101, 1110010\}$$

8.6.2 Frank Coding for Pulse Compression

Polyphase coding is similar to binary coding in that a wave of duration T is subdivided into N equal length subpulses, and each subpulse has one of M possible phases. The Frank code[12] is a polyphase code used for pulse compression. Codes in the Frank code are harmonically related phases based on certain fundamental phase increments. This form of coding was developed by Robert L. Frank. A Frank code must have a length $N = M^2$, where M is an integer. The phase of the qth element in pth sequence is given by

$$\varphi_{p,q} = \frac{2\pi k}{M}(p-1)(q-1), \qquad p = 1, 2, \dots, M, \quad q = 1, 2, \dots, M \tag{8.84}$$

where k is any integer that is relatively prime to M, and usually is chosen as $k = 1$. When the sequence is listed one underneath the other, they form an $M \times M$ matrix of phases, whose element in the pth row and qth column is given in (8.84). Here, the phase within each subphase is kept constant with reference to the CW signal. A Frank code with $N = M^2$ subpulses is known as M phase Frank code. The first step in computing a Frank code is to define the fundamental phase increment as $\Delta\varphi = 2\pi / M$. For M phase Frank code, phase shift of each of the subpulses are computed by establishing the following matrix equation:

$$\begin{bmatrix} 0 & 0 & 0 & 0 & \cdots & 0 \\ 0 & 1 & 2 & 3 & \cdots & M-1 \\ 0 & 2 & 4 & 6 & \cdots & 2(M-1) \\ 0 & 3 & 6 & 9 & \cdots & 3(M-1) \\ \vdots & \vdots & \vdots & \vdots & \vdots & \vdots \\ 0 & (M-1) & 2(M-1) & 3(M-1) & \cdots & (M-1)^2 \end{bmatrix} \frac{2\pi}{M} \tag{8.85}$$

152   Fundamental Principles of Radar

where each row represents a group, and a column represents the subpulses for that group.

We illustrate the procedure by an example. Consider a four-phase Frank code with $M = 4$, so the fundamental phase increment is $\Delta\varphi = 2\pi / 4 = \pi / 2$. Then it follows that

$$\begin{bmatrix} 0 & 0 & 0 & 0 \\ 0 & 1 & 2 & 3 \\ 0 & 2 & 4 & 6 \\ 0 & 3 & 6 & 9 \end{bmatrix} \frac{\pi}{2} = \begin{bmatrix} 0 & 0 & 0 & 0 \\ 0 & 1 & 2 & 3 \\ 0 & 2 & 0 & 2 \\ 0 & 3 & 2 & 1 \end{bmatrix} \frac{\pi}{2}. \qquad (8.86)$$

The second form replaces elements where the phase exceeds 2π by their equivalent values. An example is the element in row 4, column 3, where $6\pi / 2 = 2\pi + 2(\pi / 2)$ has the equivalent value of $2(\pi / 2)$ leading to a value of 2. Thus, the Frank code of 16 elements is given by

$$F_{16} = \{ 0000;0123;0202;0321 \}. \qquad (8.87)$$

Note that the first four code word elements are from row 1 where the semicolon represents the end of row 1. With this convention the next twelve elements follow from rows 2, 3, and 4 in order. The phases of these elements are obtained by multiplying each by $\pi / 2$.

Example 8.7

Generate a Frank code of length $N = 25$.

Solution:

We have $M = \sqrt{25} = 5$, which gives the phase increment of $2\pi / 5$. Using (8.85) to generate the following matrix:

$$\begin{bmatrix} 0 & 0 & 0 & 0 & 0 \\ 0 & 1 & 2 & 3 & 4 \\ 0 & 2 & 4 & 6 & 8 \\ 0 & 3 & 6 & 9 & 12 \\ 0 & 4 & 8 & 12 & 16 \end{bmatrix} \frac{2\pi}{5} = \begin{bmatrix} 0 & 0 & 0 & 0 & 0 \\ 0 & 2 & 4 & 6 & 8 \\ 0 & 4 & 8 & 12 & 16 \\ 0 & 6 & 12 & 18 & 24 \\ 0 & 8 & 16 & 24 & 32 \end{bmatrix} \frac{\pi}{5} \Rightarrow \begin{bmatrix} 0 & 0 & 0 & 0 & 0 \\ 0 & 2 & 4 & 6 & 8 \\ 0 & 4 & 8 & 2 & 6 \\ 0 & 6 & 2 & 8 & 4 \\ 0 & 8 & 6 & 4 & 2 \end{bmatrix} \frac{\pi}{5}$$

The resulting matrix can be written by replacing the elements where phase exceeds by 2π by their equivalent values of modulo 2π. The phases of these elements are obtained by multiplying each by $\pi / 5$ and then using their equivalent values of modulo 2π. Then it follows that the Frank code of length 25 is written as

$$F_{25} = \{00000;02468;04826;06284;08642\}$$

Note that the first five code word elements are from row 1 where the semicolon represents the end of row 1. With this convention the next twenty elements follow from rows 2, 3, 4, and 5 in order.

There are two basic modes of use for Frank codes: (a) the *Frank periodic code*, which is generated if the sequence of length M^2, derived from sequencing through all elements of the matrix of (8.85), is continually repeated; and (b) the *Frank aperiodic code*, which is generated by cycling through the matrix of (8.85) once during the transmitted pulse, and then stopping until the next pulse is to be produced. The Frank periodic code has the sidelobes completely suppressed, while

Pulse Compression Radar 153

sidelobes of the Frank aperiodic code are no longer zero. Frank code is the discrete approximations to linear FM resulting in a better Doppler tolerance than the binary codes, and relatively better sidelobe characteristics.

PROBLEMS

8.1 Show that the following property given by (8.23) is true:

$$|\chi(\tau,\omega_d)| \le |\chi(0,0)| = 1$$

8.2 Show that the following property given by (8.24) is true:

$$\frac{1}{2\pi} \int_{-\infty}^{\infty} \int_{-\infty}^{\infty} |\chi(\tau,\omega_d)|^2 \, d\tau d\omega_d = 1$$

8.3 Show that the following property given by (8.25) is true:

$$|\chi(-\tau,-\omega_d)| = |\chi(\tau,\omega_d)|$$

8.4 Show that the following property given by (8.26) is true:

$$\text{if } g(t) \leftrightarrow |\chi(\tau,\omega_d)|$$
$$\text{then } g(t)\exp(j\pi kt^2) \leftrightarrow |\chi(\tau,\omega_d + k\tau)|$$

8.5 Consider an LFM pulse compression radar operating at K_u-band using a 200 ns pulse that is compressed with a compression ratio of 100. Find the chirp bandwidth and the range resolutions due to uncompressed and compressed pulses.

8.6 A certain LFM pulse compression radar with a very large bandwidth of 500 MHz performs stretch processing over 1500 m. Find the minimum FFT size of the processor.

8.7 Consider a certain radar that uses a stretch processor to process a bandwidth of 600 MHz and transmits a pulse of length $40\,\mu s$ with a range receive window of 600 m.
(a) Find the frequency tones corresponding to two targets if the minimum range is 10 km.
(b) Find the minimum FFT size.

8.8 Generate a Barker of length $N = mn = 39$.

8.9 Work Problem 8.8 except for $N = mn = 55$.

8.10 Generate a Frank code of length $N = 9$.

REFERENCES

1. D. O. North, "An Analysis of the Factors Which Determine Signal/Noise Discrimination of Pulsed Carrier Systems," *Proceedings of the IEEE,* 51, no. 7 (1963): 1015–1027.
2. P. M. Woodward, *Probability and Information Theory with Applications to Radar* (New York: McGraw-Hill Book Company, 1963).
3. A. Papoulis, *Signal Analysis* (New York: McGraw-Hill Book Company, 1977).
4. B. R. Mahafza, *Radar System Analysis and Design Using MatLab,* Second Edition (Boca Raton, FL: Chapman & Hall/CRC, 2005).
5. A. W. Rihaczek, *Principle of High Resolution Radar* (New York: McGraw-Hill Book Company, 1959).
6. J. R. Kluder, A. C. Price, S. Darlington, and W. J. Albersheim, "The Theory and Design of Chirp Radar," *Bell System Technical Journal,* 39, no. 4 (1960): 745–808.
7. F. E. Nathanson, *Radar Design and Principles* (New York: McGraw-Hill Book Company, 1969).

8. R. W. Burkowitz, ed., *Modern Radar* (New York: John Wiley, 1966).
9. D. M. White, "Synthesis of Pulse Compression Waveform with Weighted Finite Frequency Combs," *Applied Physics Laboratory Report, TG-934* (1967).
10. B. F. Mahafza, *Introduction to Radar Analysis* (Boca Raton, FL: CRC Press, 1998).
11. R. H. Barker, "Group Synchronizing of Binary Digital Systems," in *Communication Theory* , ed. W. Jackson, ed. (New York: Academic Press, 1953), 273–287.
12. R. L. Frank, "Polyphase Codes with Good Nonperiodic Correlation Properties," *IEEE Transaction on Information Theory,* IT-9 (1963): 43–45.

9 Synthetic Aperture Radars

"We live on the leash of our senses. There is no way in which to understand the world without first detecting it through the radar-net of our senses."

—Diane Ackerman

9.1 INTRODUCTION

Two major advances in radar and signal processing techniques have made it possible, in the last 30 years, to obtain very high-resolution radar maps. These advances occurred in the development of *synthetic aperture radar* (SAR)[1] and *pulse compression techniques*. High resolution can be obtained using a conventional noncoherent radar system employing a large antenna or operating at a short wavelength. Airborne platforms, however, impose severe physical limitations on the size of the antenna. The antenna that is used on airborne platforms must be a reasonable size but still maintain acceptable flight characteristics of the aircraft. A practical SAR is implemented by taking advantage of the motion of the vehicle carrying the radar to synthesize the effect of a large antenna aperture. The radar motion is a prerequisite of the SAR technique. SAR achieves high resolution in the cross range dimension.

SAR is mainly applied for a high-resolution two-dimensional (three-dimensional in some cases) radar imagery of the ground surface. The imaging of terrain on the earth's surface replaces optical or infrared imagery in applications requiring all weather operation. Applications of SAR technology include high-resolution surveillance, planetary mapping, and rotating object imagery. Specifically, the imaging of the earth's surface by SAR to provide a map-like display can be applied to military reconnaissance, geological and mineral explorations, GPS, and other remote sensing applications.[2] The use of SAR for remote sensing is particularly suited for tropical countries. By proper selection of operating frequency, the microwave signal can penetrate clouds, haze, rain and fog, and precipitation with very little attenuation, thus allowing operation in unfavorable weather conditions that preclude the use of a visible/infrared system.[3] Since SAR is an active sensor, which provides its own source of illumination, it can therefore operate day or night, be able to illuminate with a variable look angle, and can select a wide area coverage. In addition, the topography change can be derived from the phase difference between measurements using radar interferometry. SAR has been shown to be very useful over a wide range of applications, including sea and ice monitoring, geological and mineral explorations, mining, oil pollution monitoring, oceanography, snow monitoring, and many more. The potential of SAR in a diverse range of applications led to the development of a number of airborne and spaceborne SAR systems

9.2 SAR HISTORY[4]

SAR designs and associated applications have grown exponentially since the 1950s when Carl Wiley, of the Goodyear Aircraft Corporation, observed that a one-to-one correspondence exists between the along-track coordinate of a reflecting object (being linearly traversed by a radar beam) and the instantaneous Doppler shift of the signal reflected to the radar by that object. He concluded that a frequency analysis of the reflected signals could enable finer along-track resolution than that permitted by the along-track width of the physical beam itself, which governed the performance of the *real aperture radar* (RAR) designs of that era. In 1952 the Doppler beam-sharpening system was developed by Wiley of Goodyear Corporation. This system was not a side-looking radar. It operated in squint mode with the beam point around 45° ahead. The radar group at the Goodyear research facility in Litchfield, Arizona, pursued Wiley's concept and built the first airborne SAR system, flown aboard a DC-3 in 1953. The radar system operated at 930 MHz used a Yagi antenna with a real aperture beamwidth of 100°. During the late 1950s and early 1960s, the classified development of SAR systems took place at the University of Michigan and at some companies. At the same time, similar developments were conducted in other countries such as Russia, France, and the United Kingdom.

Industrial and military developments, using airborne platforms, continued at Goodyear, Hughes, and Westinghouse. The Jet Propulsion Laboratory (JPL), University of Michigan, Environmental Research Institute of Michigan (ERIM), Sandia Laboratories, and others also began to explore this new technology. In 1974, engineers at JPL formed an alliance with a group of international ocean scientists led by the National Oceanic and Atmospheric Administration (NOAA) to determine if an ocean application satellite featuring a space-based SAR could be achieved. Up until this time, the major emphasis of space-based remote sensing had been on land applications using visible and infrared sensors. The resulting NASA/NOAA alliance assembled a multiagency, interdisciplinary group of engineers and scientists that focused on ocean and ice applications using active and passive microwave sensors that could collect data, day or night, with a general disregard for cloud obscuration.

SEASAT was the first earth-orbiting satellite designed for remote sensing of the earth's oceans and had on board the first spaceborne SAR. The mission was designed to demonstrate the feasibility of global satellite monitoring of oceanographic phenomena and to help determine the requirements for an operational ocean remote-sensing satellite system. Specific objectives were to collect data on sea-surface winds, sea-surface temperatures, wave heights, internal waves, atmospheric water, sea ice features, and ocean topography. SEASAT was managed by NASA's JPL and was launched on June 27, 1978, into a nearly circular 800 km orbit with an inclination of 108°. SEASAT operated for 106 days until October 10, 1978, when a massive short circuit in the satellite's electrical system ended the mission. SEASAT was followed by the Shuttle Imaging Radar-A (SIR-A) and Shuttle Imaging Radar-B (SIR-B) flown in 1981 and 1984, respectively. Both the SIR-A and SIR-B radars were variations on the SEASAT radar operating at L-band and horizontal transmit, horizontal receive (HH) polarization. SIR-B had the added capability of operating at different incident angles (the angle of incidence is defined as the angle between the radar LOS and the local vertical at the point where the radar intersects the earth or ocean surface).

With the exception of the Soviet 1870 SAR, the 1980s saw only the space shuttle–based SEASAT derivative, space-borne SAR activity. The 1990s witnessed a significant expansion of SAR missions with the launch of five earth-oriented SAR satellites along with two more Shuttle Imaging Radar missions, as well as the pioneering interplanetary use of the Magellan SAR to map Venus. The satellite systems ALMAZ, European Remote Sensing, the Japanese Earth Resources Satellite (JERS)-1, ERS-2, and RADARSAT-1, each operated at a single frequency and single polarization, like SEASAT. ALMAZ and RADARSAT-1 had the added ability to operate at different incident angles. RADARSAT-1 also has a frequently used ScanSAR mode, where the coverage swath extends up to 500 km.

One of the most advanced SAR systems, the Shuttle Imaging Radar-C/X-band SAR (SIR-C/XSAR), was a joint NASA/German Space Agency/Italian Space Agency mission, flown in April and October 1994 on the *Endeavor*. The system could be operated simultaneously at three frequencies (L, C, and X) with the C- and L-band having the ability to alternately transmit and receive at both horizontal and vertical polarization. By collecting a near-simultaneous and mutually coherent version of the scattered field in a minimum basis set of polarizations, this quadrature polarimetry or "fully polarimetric" capability allows for a more complete characterization of the target's scattering characteristics within the illuminated resolution cell area. The C- and X-band portions of the SIR-C radar were again flown in 2002 for the Shuttle Radar Topography

Synthetic Aperture Radars

Mission (SRTM). During this flight, a second receiving antenna was placed at the end of a 60-m mast, extended perpendicular to the main radar antenna. The purpose of the mast antenna was to provide a second receiving point in space for each radar pulse. The slight variations in phase, between the receipts of the radar pulses at each of the antennas, will be processed into a height measurement of the reflecting point on earth's land surface.

Future SAR missions are expected to provide enhanced capabilities, where the radar can be operated in several collection modes. SEASAT began the evolution of space-based SAR that continues to this day. The active international participation of the SEASAT User Working Group led to international cooperation in the form of data collection and processing facilities and experiments with ground truth. In retrospect, the most important result of these cooperative interactions was the zealous expansion of the SAR technology in space. As a result, more than two decades of data collection have provided a rich data source, with each system adding unique characteristics in terms of applications, radar design, and mission data collection parameters.

9.3 SAR GENERAL DESCRIPTION

A synthetic aperture radar operates from a moving vehicle, offset from the mapping area. A simplified block diagram of such a radar is shown in Figure 9.1. The transmitter emits RF electromagnetic energy, which is coherent. This energy is the result of sampling and amplifying the output of a stable oscillator, which is also used as the coherent reference for phase detection of the returned signals.

In SAR, forward motion of the actual antenna carried by the aircraft is used to "synthesize" a very long antenna. At each position a pulse is transmitted, the return echoes pass through the receiver, and recorded in an "echo store." The Doppler frequency variation for each point on the ground is a unique signature. SAR processing involves matching the Doppler frequency variations and demodulating by adjusting the frequency variation in the return echoes from each point on the ground. Result of this matched filter is a two-dimensional high-resolution image. Here, one dimension is the *cross-range resolution* across the line of sight (XLOS) in the direction of the flight path and the other is *range resolution* along the LOS perpendicular to the flight path.

9.3.1 Resolution along the LOS Axis

Ground resolution is defined as the ability of the system to distinguish between two targets on the ground. Ground range resolution is shown in Figure 9.2. In this figure, W_g represents the ground area illuminated by the radar antenna. This resolution is achieved by time discrimination of the transmitted pulse and the received pulse. The time delay ΔT between the transmitted pulse and the received pulse at a range R_0 will be

$$\Delta T = \frac{2R_0}{c}. \tag{9.1}$$

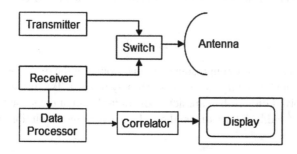

FIGURE 9.1 Simplified block diagram of an SAR.

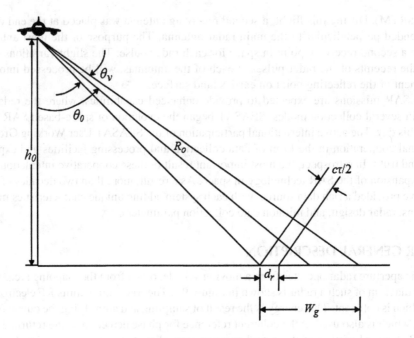

FIGURE 9.2 Range resolution of a real aperture radar.

The return signals from two distinct targets are nonoverlapping in time if these targets are separated by a slant distance ΔR_s given by

$$\Delta R_s = \frac{c\tau}{2} \tag{9.2}$$

where τ is the duration of the transmitted pulse. We can now define the ground range resolution d_R to be the projection onto the ground of the slant range resolution ΔR_s:

$$d_R = \frac{c\tau}{2\cos\theta_0} \tag{9.3}$$

where θ_0 is the angle of incidence (or sometimes known as depression angle). Assuming the transmission bandwidth B to be inversely proportional to width transmitted pulse, we have

$$B \cong \frac{1}{\tau}. \tag{9.4}$$

Using (9.4) in (9.3) we get a relation between the range resolution and the bandwidth,

$$d_R = \frac{c}{2B\cos\theta_0}. \tag{9.5}$$

Cross-range resolution is the minimum distance on the ground in the direction parallel to the flight path of the aircraft at which two targets can be separately imaged. Two targets located at the same slant range can be resolved only if they are not in the radar beam at the same time. For an antenna with a half-power beamwidth of θ_B radians, the radar cross-range resolution d_X in a real antenna radar at a range R_0 will be

$$d_X = \theta_B R_0. \tag{9.6}$$

Synthetic Aperture Radars

At a given range R_0, the improvement in resolution can be obtained by reducing the half-power beamwidth. The value of θ_B, in terms of antenna aperture ℓ, is usually given by

$$\theta_B \cong \lambda / \ell \tag{9.7}$$

which results in a cross-range resolution given by

$$d_X = \frac{\lambda R_0}{\ell}. \tag{9.8}$$

Considering an X-band radar with $\lambda = 3$ cm, and aperture of 1 m, the XLOS resolution at a range of 10 km will result in $d_X = 300$ m, an absolutely unacceptable resolution compared with the available resolution along the LOS axis.

9.3.2 Resolution along the XLOS Axis

SAR can provide significant improvement in the resolution along the XLOS axis over that obtained from the *side-looking airborne radar* (SLAR) system. The SAR concept employs a coherent radar system and a single moving antenna to simulate the function of all the antennas in a real linear antenna array. This single antenna sequentially occupies the spatial positions of a synthesized array. All received signals, both amplitude and phase, are stored for each antenna position in the synthetic array. These stored data are then processed to re-create the image of the illuminated area seen by the radar.

Typical SAR configuration is side-looking, in which the radar antenna is oriented normal to the aircraft flight path and downward at some appropriate depression angle. Figure 9.3 illustrates this scenario[5] and ground area illuminated by the radar. Cross-range resolution, as defined earlier, is in the direction of the flight path, and is also referred to in the literature as the XLOS resolution.

The phase history of a ground return can be obtained from the range history to a specific point on the ground during the aircraft flight through the observation section. The aircraft is mapping the point P located at $(x_0, y_0, 0)$. The aircraft flies a synthetic array of length L, centered at $y = 0$ with speed v, along the y-axis and at a constant altitude h_0 above the ground. The range from the aircraft to point P at time $t = 0$ is

$$R_0 = \sqrt{x_0^2 + y_0^2 + h_0^2}. \tag{9.9}$$

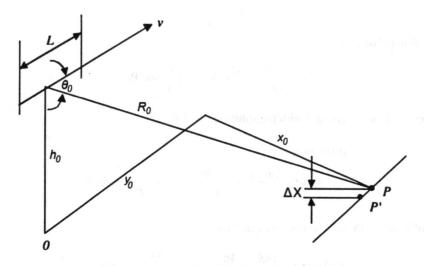

FIGURE 9.3 A geometry of synthetic aperture radar.

At time $t = 0$, the angle between the aircraft velocity vector and the radar range vector to point P is θ_0 given by

$$\cos\theta_0 = \frac{y_0}{R_0}. \qquad (9.10)$$

The angle θ_0 is called the *squint angle* or *the depression angle*. The observation section L is the section of the aircraft path during which the return from point P is processed. It is required that all along the path L the point P will be within the beam of the real antenna, and is located at the *far-field pattern* of the antenna defined by $(2D^2 / \lambda)$, where D is the size of the antenna. The time to fly the array is $T = L / v$. The range from the aircraft to point P at any time during the observation time T is given by

$$R(t) = \sqrt{x_0^2 + (y_0 - vt)^2 + h_0^2}, \qquad -T/2 \leq t \leq T/2. \qquad (9.11)$$

Using (9.9) and (9.10) in (9.11) yields

$$R(t) = [R_0^2 - 2R_0 vt \cos\theta_0 + v^2 t^2]^{1/2}, \qquad -T/2 \leq t \leq T/2. \qquad (9.12)$$

This range equation can be expanded[6] about $t = 0$ in a Taylor series as follows:

$$R(t) \approx R_0 - vt \cos\theta_0 + \frac{v^2 t^2}{2R_0} \sin^2\theta_0. \qquad (9.13)$$

A signal transmitted toward a point scatterer at P as

$$e_1(t) = A_1 \cos\omega_0 t \qquad (9.14)$$

where A_1 is the amplitude, and ω_0 is the angular frequency. This signal is then reflected and received at the aircraft with a time delay τ' so that the received signal is given by

$$e_2(t) = A_2 \cos\omega_0(t - \tau') \qquad (9.15)$$

where A_2 is the amplitude of the received signal. The time delay τ' is written as

$$\tau' = \frac{2R(t)}{c}. \qquad (9.16)$$

Using (9.13) in (9.16) gives

$$\tau' = \frac{2R_0}{c} - \frac{2vt}{c} \cos\theta_0 + \frac{v^2 t^2}{R_0 c} \sin^2\theta_0 \qquad (9.17)$$

The argument of the signal $e_2(t)$ with the substitution of τ' is

$$\theta(t) = \omega_0(t - \tau')$$
$$= \omega_0 t - \frac{2\omega_0 R_0}{c} + \frac{2\omega_0 vt}{c} \cos\theta_0 - \frac{\omega_0 v^2 t^2}{R_0 c} \sin^2\theta_0. \qquad (9.18)$$

In terms of wavelength $\lambda = 2\pi c / \omega_0$, we can write $\theta(t)$ as

$$\theta(t) = \omega_0 t - \frac{4\pi R_0}{\lambda} + \frac{4\pi}{\lambda} vt \cos\theta_0 - \frac{2\pi v^2 t^2}{R_0 \lambda} \sin^2\theta_0. \qquad (9.19)$$

Synthetic Aperture Radars

161

The instantaneous frequency can now be obtained:

$$f_i(t) = \frac{1}{2\pi} \frac{d\theta(t)}{dt}. \tag{9.20}$$

Using (9.19) in (9.20) we get

$$f_i(t) = f_0 + \frac{2v}{\lambda} \cos\theta_0 - \frac{2v^2 t}{R_0 \lambda} \sin^2\theta_0. \tag{9.21}$$

The second term in (9.21) is the Doppler shift associated with the squint angle and the third term represents the change in Doppler shift due to the forward motion of the aircraft. The instantaneous frequency for the two targets at P and P', both at a distance of R_0 but separated in azimuth by a distance ΔX will be $f_0 + \frac{2v}{\lambda} \cos\theta_0$ and $f_0 + \frac{2v}{\lambda} \cos\theta_0 - \frac{vD}{R_0 \lambda} \sin\theta_0$, respectively, since the time to fly the distance between them is $t = (\Delta X \sin\theta_0 / v)$. Then the observed differential frequency shift Δf_i will be

$$\Delta f_i = \frac{2v\Delta X}{R_0 \lambda} \sin\theta_0. \tag{9.22}$$

Thus to resolve two targets separated by a distance ΔX having a frequency difference of Δf_i, data must be collected for a time $T \cong 1 / \Delta f_i$. This results in the synthetic array length $L = vT$, which is written as

$$L = \frac{R_0 \lambda}{2\Delta X \sin\theta_0}. \tag{9.23}$$

Thus the azimuth resolution d_X is finally obtained from (9.23) given by

$$d_X = \Delta X = \frac{\lambda R_0}{2L \sin\theta_0}. \tag{9.24}$$

Maximum resolution in the y direction can be obtained by considering the fact that the antenna illumination coverage on the ground should be greater than the equivalent antenna array length. Using the antenna beamwidth relationships, we get

$$\frac{R_0 \lambda}{\ell} \geq L = \frac{\lambda R_0}{2\Delta x \sin\theta_0} \tag{9.25}$$

where ℓ is the actual antenna length and L is the equivalent array length on the ground. For $\theta_0 = 90^\circ$, (9.25) yields $2\Delta x \geq \ell$. It follows then

$$d_x = \Delta X \geq \frac{\ell}{2}. \tag{9.26}$$

The best cross-range resolution in the y obtainable is equal to one half of the actual antenna length ℓ, which is independent of range, wavelength, and incidence angle. This implies that the smaller antenna can result in a better resolution.

Example 9.1

A synthetic aperture radar operating at 10.09 GHz is designed to achieve a cross-range resolution of 1.0 m at a range of 2 km. Find (a) the length of the synthetic antenna necessary to achieve this resolution, and (b) the maximum size of the antenna and the distance of the far field of the synthetic antenna. Assume that the antenna depression angle is 60°.

Solution:

a. The synthetic antenna length can be obtained from (9.23):

$$L = \frac{R_0 \lambda}{2\Delta X \sin\theta_0} = \frac{(2\times 10^3)(.03)}{(2)(1)\sin 60°} = 17.32 \text{ m}$$

where $\lambda = \frac{3\times 10^8}{10\times 10^9} = 0.03$ m has been used.

b. The maximum length of the antenna, using (9.26), is

$$d_x = \Delta X \geq \frac{\ell}{2} \Rightarrow \ell \leq 2d_x = (2)(1.0) = 2.0 \text{ m}$$

and the distance to the far field of the synthetic antenna of length 17.32 m is

$$R_{FF} = \frac{2L^2}{\lambda} = \frac{(2)(173.2)^2}{0.03} \approx 20 \text{ km}$$

9.4 SAR SIGNAL PROCESSING

The main goal of SAR data processing is the determination of the range and azimuth coordinates of the targets lying in the strip map. The SAR data space is a conceptual collection of SAR data arranged by range line $(1, 2, 3, \cdots)$ with the first data point from each range line at the top and the last data from each range line at the bottom. A range line is a recording of all the reflections from a single transmitted chirp pulse. This data space is shown in Figure 9.4.

Generally there are two approaches for SAR processing, namely *two-dimension algorithm* and *range-Doppler processing algorithm*. The two-dimension algorithm processes the range and azimuth data simultaneously, whereas the range-Doppler processing algorithm implements range compression processing followed by azimuth compression processing. However, the two-dimensional algorithm proposed by Alan Di Cenzo[7] required larger memory and computational power.

The most common algorithm employed in most SAR processing systems is the range-Doppler processing algorithm. It is a two-dimensional correlating procedure. The two dimensions of the correlation processing are realized as two one-dimensional matched filter operations, namely range

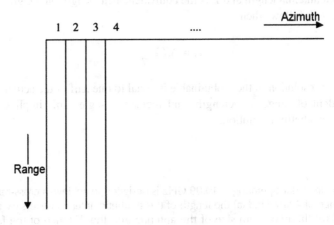

FIGURE 9.4 A two-dimensional SAR data space.

Synthetic Aperture Radars

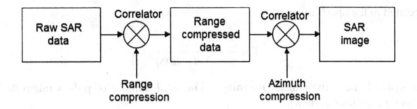

FIGURE 9.5 Range Doppler processing.

compression and azimuth compression. The first matched filtering operates on the single pulse radar returns, and the second matched filtering operates on the Doppler signal. Figure 9.5 shows the basic concept of the range-Doppler processing.

9.5 RADAR EQUATION OF THE SAR SYSTEM

The power received at the antenna of a monostatic radar is given by the familiar radar range equation:

$$P_r = \frac{P_t G^2 \lambda^2 \sigma}{(4\pi)^3 R^4} \qquad (9.27)$$

where
  P_r = power received at the antenna
  P_t = peak power transmitted by the antenna
  G = gain of the antenna
  R = distance of the target from the radar
  λ = wavelength of the operation
  σ = radar cross section of the target

Atmospheric attenuation effects discussed earlier in Chapter 4 have been ignored. An expression will now be derived for the radar equation pertinent to the SAR system. Signals received by radar are usually contaminated by noise due to random modulations of the radar pulse during atmospheric propagation, or due to fluctuations in the receiving circuits. The SNR is defined as:

$$SNR = \frac{S}{N} = \frac{P_t G^2 \lambda^2 \sigma}{(4\pi)^3 R^4 F K T_0 B} \qquad (9.28)$$

where K is the Boltzmann's constant and is equal to 1.38×10^{-23} in Joules/Kelvin, T_0 is the effective absolute noise temperature in Kelvin, B is the effective noise bandwidth of the receiver in Hz, and F represents the noise figure.

The improvement in SNR by a factor of n is achieved by coherent integration of synthetic aperture length, and is given by

$$SNR = \frac{P_t G^2 \lambda^2 \sigma n}{(4\pi)^3 R^4 F K T_0 B} \qquad (9.29)$$

where n is the number pulses integrated coherently, and can be expressed in terms of the pulse repetition frequency f_r and the time T over which the synthetic aperture length L is formed. This is given by

$$n = T f_r. \qquad (9.30)$$

164   Fundamental Principles of Radar

But T is related to the aperture length L as

$$T = \frac{L}{v} = \frac{\lambda R}{2vd_X \sin\theta_0} \tag{9.31}$$

where the symbols have their usual meanings. The total number of pulses integrated over the coherent integration time is then

$$n = \frac{f_r \lambda R}{2vd_X \sin\theta_0}. \tag{9.32}$$

Thus the *SNR* equation for a point target is then modified to

$$SNR = \frac{P_t G^2 \lambda^3 \sigma f_r}{2(4\pi)^3 R^3 FKT_0 Bv d_X \sin\theta_0} = \frac{P_{av} G^2 \lambda^3 \sigma}{2(4\pi)^3 R^3 FKT_0 v d_X \sin\theta_0}. \tag{9.33}$$

We have used the relation $P_{av} = P_t \tau f_r = P_t f_r / B$ in (9.33).

For a distributed clutter, the RCS is described by the following expression:

$$\sigma = \sigma^\circ d_R d_X \sec\theta_0 \tag{9.34}$$

where the symbols have their usual meanings and σ° is called the normalized RCS of the ground per unit area illuminated. The corresponding expression for the SNR for the distributed clutter is then

$$SNR = \frac{P_t G^2 \lambda^3 f_r \sigma^\circ d_R}{2(4\pi)^3 R^3 FKT_0 Bv \sin^2\theta_0}. \tag{9.35}$$

P_t used above can be expressed in terms of the average power P_{av} as

$$P_t = \frac{P_{av}}{\tau f_r} = \frac{P_{av} B}{f_r}. \tag{9.36}$$

Using (9.36) in (9.35), we get the final form of the radar equation for the SAR system as

$$SNR = \frac{P_{av} G^2 \lambda^3 \sigma^\circ d_R}{2(4\pi)^3 R^3 FKT_0 v \sin^2\theta_0}. \tag{9.37}$$

It can be observed that the SNR in an SAR system is (1) inversely proportional to the third power of range, (2) independent of azimuth resolution, (3) a function of the ground range resolution, (4) inversely proportional to the velocity, and (5) proportional to the third power of wavelength. Also it can be seen that for a fixed SNR, the average power must be increased if the resolution in range or azimuth are decreased or the range is increased.

The explicit expression for P_{av} can be readily written as

$$P_{av} = \left(\frac{(4\pi)^3 R^3 FKT_0}{G^2 \lambda^2 \sigma^\circ d_R}\right)\left(\frac{2v}{\lambda}\right)(SNR)\sin^2\theta_0. \tag{9.38}$$

For a strip-mode SAR, where the squint angle $\theta_0 = 90°$, the previous equation modifies to

$$P_{av} = \left(\frac{(4\pi)^3 R^3 FKT_0}{G^2 \lambda^2 \sigma^\circ d_R}\right)\left(\frac{2v}{\lambda}\right)(SNR). \tag{9.39}$$

Synthetic Aperture Radars 165

The result provides the average power requirements of synthetic aperture radar. It is observed that the average power is independent of XLOS resolution. However, for a given cross-range resolution, the time to fly the aperture increases at the same rate so that the power-time product (energy) is independent of incidence angle. To increase the effective peak power for small values of d_R, a number of synthetic aperture radars employ pulse compression techniques.

Example 9.2

An airborne SAR system operating at 8.4 GHz is designed to achieve a clutter SNR of 30 dB and an LOS range resolution of $d_R = 0.3$ m in order to map a distributed ground clutter with $\sigma° = 0.015$ located at a distance of 15 km. For a strip-mode SAR flown at a speed of 300 m/s, where the depression angle $\theta_0 = 90°$ and antenna gain $G = 35$ dB, find the P_{av} necessary. Assume that the noise figure of the radar receiver is 3 dB.

Solution:

The P_{av} necessary to achieve the desired SNR can be found by using (9.39) as

$$P_{av} = \left(\frac{2(4\pi)^3 R^3 F K T_0 v}{G^2 \lambda^3 \sigma° d_R} \right) (SNR)$$

where $\lambda = 3 \times 10^8 / (8.5 \times 10^9) = 0.035$ m, $SNR = 30$ dB $= 10^3$, $F = 3$ dB $= 2$, $G = 35$ dB $= 3162.2$. Thus

$$P_{av} = \frac{(2)(4\pi)^3 (15 \times 10^3)^3 (2)(1.38 \times 10^{-23})(290)(300)(10^3)}{(3162.2)^2 (0.036)^3 (0.015)(0.25)} = 18.38 \text{ W}.$$

9.6 SAR SYSTEM DESIGN CONSIDERATION

To obtain high-quality imagery, careful attention must be paid to the SAR system's design. Different system-level objectives will lead to different radar configurations. Some of the parameters are selected by the radar engineers in a logical manner. When the system is too complex for logical selection of the radar parameters, system engineers must make decisions based on experience and judgment. Some of the SAR system design considerations include the following.

Mode of operation: Available data acquisition mode includes *strip mapping, squint mode, spotlight mode, ScanSAR, interferometry,* and *polarimetric SAR.* Spotlight-mode SAR provides a high-resolution image but involves complex hardware and processing algorithm. Squint-mode SAR is used to image while maneuvering. It is usually employed in military aircraft. ScanSAR is normally used in space-borne SAR to increase the swath width. It requires powerful computation hardware. The polarimetric system is capable of measuring a scattering matrix of targets whereas interferometry SAR is the latest technology developed with the capability to measure terrain height and construct a three-dimensional image. Hardware design of interferometry and polarimetric SAR is more complicated.

Polarization: Conventional SAR system that employs a single polarization such as VV or HH to acquire information from the earth. For remote sensing of earth terrain such as oil palm plantation or tropical forest, single polarization is sufficient. A device that measures the full polarization response of the scattered wave is called a *polarimeter.* Polarimetric systems differ from conventional systems in that they are capable of measuring the complete scattering matrix of the remotely sensed media. By having the knowledge of the complete scattering matrix, it is possible to calculate the backscattered signal for any given combination of the transmitting and receiving antennas. This process is called the *polarization synthesis*, which is an important technique used in terrain classification.[8]

Operating platform: Basically SAR systems can be separated into two groups: an airborne system operating on variety of aircraft, and a space-borne system that operates on satellite or space shuttle platforms. Space-borne SARs have a smaller range of incidence angle and a larger swath width, but a high data rate is a problem, whereas, airborne SARs illuminate a smaller footprint on the earth and the data rate is much lower.

Dynamic range of back-scattering coefficient $\sigma°$: The required system sensitivity is determined based on the various categories of the earth terrain to be mapped such as man-made target, ocean, sea ice, forest, natural vegetation and agriculture, geological targets, mountain, and land and sea boundary. From the open literature, the typical value of $\sigma°$ falls in the range of +20 dB to −40 dB.[9] For vegetation the typical value of $\sigma°$ varies from +0 dB to −20 dB.

Operating frequency: For remote-sensing applications, a frequency range from 1 to 30 GHz is normally used. In the 1–10 GHz range, the transmissivity through air approaches 100%. Thus an SAR operating in this frequency range is always able to image the earth's surface independent of the cloud cover or precipitation. As the radar frequency increases, the transmission attenuation increases. At 22 GHz there is a water vapor absorption band that reduces transmission to about 85%.

Modulations: In an SAR system there are basically three types of widely used modulation schemes: pulse, LFM or chirp, and phase coded. A pulse system is used in older generation radar. Modern radar uses LFM waveform to increase range resolution when long pulses are required to get a reasonable SNR. The same average transmitting power as in a pulse system can be achieved with lower peak amplitude. The LFM configuration is employed in this project, since it gives better sensitivity without sacrificing range resolution and ease of implementation. The lower peak power allows for the use of commercially available microwave components that have moderate peak-power handling capability. Phase-coded modulation is not preferable due to its difficulty to generate. Phase-coded modulation is normally used for long duration waveforms and when jamming may be a problem.

Pulse repetition frequency (PRF): The upper limit of PRF is attained from a consideration of the maximum mapping range, and the fact that the return pulse from this range should come within the interpulse period. To adequately sample the Doppler bandwidth, the radar must have a pulse train with a PRF greater or equal to this bandwidth.

Antenna: Yagi, slotted-waveguide, horn, dish, and microstrip antennas have found some applications in SAR at one time or another. Yagi is suited for lower frequency applications, while a dish is suited for a very high-frequency application. Modern civilian SAR systems generally operate in L-, C-, and X-band, where slotted-waveguide and microstrip antennas provide the best performance.

Swath width and range of incidence angle: From the open literature, the swath width for an airborne SAR ranges from 5 km to 60 km depending on the altitude of the operating platform and incidence angle. An incidence angle from 0° to 80° is usually utilized by present airborne SARs. The backscattering coefficient of nature targets such as soil, grass, and vegetable are maintained almost constant over the incidence angle of 40° to 60°.

PROBLEMS

9.1 An airborne coherent SAR operating at 10 GHz is flown with a depression angle of 90° at a constant speed of 300 m/s. The Doppler frequency change measured is 30 Hz from two separate targets at a constant range of 2 km. Determine the distance of separation between the two targets and the corresponding synthetic aperture length.

Synthetic Aperture Radars

9.2 An airborne SAR moving at a constant speed of 300 m/s operates at 10 GHz and has the following parameters as depicted in Figure 9.4: $y_0 = 0$, $R_0 = 2$ km, $L = 30$ m. Find the cross-range resolution and the maximum length of the actual antenna.

9.3 For the SAR depicted in Figure 9.4 with $y_0 = 0$, derive the expression of the range difference at the two edges of synthetic aperture length given by

$$\delta R \approx \frac{L^2}{8R_0}.$$

9.4 Consider an airborne SAR flown in an aircraft with the ground point geometry, as depicted in Figure 9.4, has the following parameters: $L = 400$, $R_0 = 20$ km, and $y_0 = 0$. Determine the range difference at the two edges of the synthetic aperture.

9.5 Work Problem 9.4 in the case of a space-borne SAR with the following parameters: $L = 12$ km, $R_0 = 850$ km, and $y_0 = 0$.

9.6 Consider an airborne SAR system transmitting an average power of 10 W at 10.0 GHz operating frequency, which has the following parameters: $G = 35$ dB, $d_R = 0.2$ m, $d_X = 2$ m, $\sigma = 1$ m², $F = 3$ dB, $v = 210$ m/s, $R = 12$ km. Find the SNR for this radar.

9.7 Work Problem 9.6 to calculate the clutter-to-noise ratio for the same radar but the target is distributed, which is described by the following parameters: the depression angle $\theta_0 = 30°$, the normalized RCS of the ground per unit area illuminated $\sigma° = -20$ dB.

REFERENCES

1. W. M. Brown and L. J. Porcello, "An Introduction to Synthetic Aperture Radar," *IEEE Spectrum,* 6 (1969): 52–62.
2. M. I. Skolnik, "A Perspective of Synthetic Aperture Radar for Remote Sensing," *Naval Research Laboratory Memorandum Report 3783*, Washington, D.C.
3. F. T. Ulaby, R. K. Moore, and A. K. Fung, *Microwave Remote Sensing: Active and Passive*, vol. 1 (Norwood, MA: Artech House, 1981).
4. S. W. McCandless, "1989: SAR in Space—The Theory, Design, Engineering and Application of a Space-based SAR System," in *Space Based Radar Handbook*, ed. L. Cantafio (Norwood, MA: Artech House), 121–166.
5. S. A. Hovanessian, *Introduction to Synthetic Array and Imaging Radars* (Dedham, MA: Artech House, 1980).
6. R. C. Heimiller, "Theory and Evaluation of Gain Patterns of Synthetic Arrays," *IRE Transaction* on *Military Electronics,* MIL-6 (1962): 122–129.
7. A. Di Cenzo, "A New Look at Nonseparable Synthetic Aperture Radar Processing," *IEEE Transactions on Aerospace and Electronic Systems,* 24, no. 3 (1988): 218–223.
8. D. L. Evans and J. J. Van Zyl, "Polarimetric Imaging Radar: Analysis Tools and Applications," *Progress in Electromagnetics Research,* 3 (1990): 371–389.
9. J. T. Pulliainen, K. J. Hyyppa, and M. T. Hallikainen, "Backscattering Properties of Boreal Forests at the C- and X-Bands," *IEEE Transactions on Geoscience and Remote Sensing,* 32, no. 5 (1994): 1041–1050.

10 Tracking Radars

"I have a radar built inside me to avoid punches."

—**Muhammad Ali**

10.1 INTRODUCTION

Tracking *radars* is the term applied to radars that measure the spatial position and provide data that may be used to determine the target path and predict the future position, in range, elevation angle, azimuth angle, and sometimes Doppler frequency shift. Tracking radars can be either continuous or discrete as in the case of track-while-scan (TWS) radars. Continuous tracking supplies continuous tracking data on a particular target. TWS radars, on the other hand, scan a given search volume and illuminate one or more targets, periodically extracting spatial position and range rate information and then processing it in a digital computer using sophisticated smoothing and prediction filters.

The tracking radar utilizes a pencil beam to find its target first before it can track. It is for this reason that a separate search radar is needed to facilitate target acquisition by the tracker. The search radar or the acquisition radar designates targets to the tracking radar by providing the coordinates where the targets are to be found. The tracking radar acquires a target by performing a limited search in the area of the designated target coordinates.

The major applications of tracking radars are weapon control and missile-range instrumentation, air-traffic control, antiaircraft artillery control, and satellite control. In all applications a high degree of precision and an accurate prediction of the future position of the target are generally required. The earliest use of tracking radar was in gunfire control. In missile-range instrumentation, the tracking radar output is used to measure the trajectory of the missile and to predict future position. This is also used with a beacon to provide a point source target with high SNR.

This chapter addresses three areas of target tracking: *range tracking, angle tracking,* and *TWS tracking.* Usually range tracking is inherent in both angle tracking and TWS tracking systems.

10.2 RANGE TRACKING

Range tracking provides continuous estimation of slant range from the radar to a specific target. It is accomplished by measuring the time required for each transmitted signal to travel from the radar to the target of interest and back to the radar. The time of the echo delay from a target at range R_0 at $t = 0$ of a monostatic radar is related by (1.1), as described in Chapter 1:

$$t_d = \frac{2R_0}{c} \tag{10.1}$$

where the symbols have their usual meanings. Measuring t_d actually measures R_0. If $\hat{t}_d$ denotes the radar's measurement of t_d, the corresponding measurement R_0, denoted by $\hat{R}_0$ is

$$\hat{R}_0 = \frac{c}{2}\hat{t}_d. \tag{10.2}$$

169

Now if we assume that the measurement is unbiased, the variance of measurement error and the mean-squared error are the same. Thus if the variance of the error is denoted by $\sigma_{\hat{t}_d}^2$ in the measurement of $\hat{t}_d$, the variance of error in the measurement of $\hat{R}_0$, denoted by $\sigma_{\hat{R}_0}^2$, is given by

$$\sigma_{\hat{R}_0}^2 = \left(\frac{c}{2}\right)^2 \sigma_{\hat{t}_d}^2. \tag{10.3}$$

Thus the accuracy (variance) of the errors in the measurement of range is related directly to the variance of the errors in the measurement of delay.

Example 10.1

A monostatic radar measures the target range with an rms error of 7.5 m. Find the corresponding rms error in measuring delay due to noise.

Solution:

The rms error in measuring delay time calculated from (10.3):

$$\sigma_{\hat{R}_0}^2 = \left(\frac{c}{2}\right)^2 \sigma_{\hat{t}_d}^2 \Rightarrow \sigma_{\hat{t}_d}^2 = \left(\frac{2}{c}\right)^2 \sigma_{\hat{R}_0}^2$$

$$\sigma_{\hat{t}_d} = \frac{2}{c}\sigma_{\hat{R}_0} = \frac{(2)(7.5)}{3 \times 10^8} = 0.05 \ \mu s$$

In early applications of radar, range tracking was accomplished by manual operation through observing the desired target signal on the radar A-scope. Continuous range and range rate information was obtained by mechanically translating the position and by moving the cursor. Corrections are necessary only when the radial range rate of the target with respect to the radar changes. Aided range tracking, producing both range position and range rate, is sometimes employed in modern tracking radars when they are operated in a manual range tracking. Automatic determination of the target range requires the use of a range tracker to obtain target range and range rate for application to a display. Applications of the range tracker include the supply of target range and range rate information to weapon system computers. Instrumentation radars record target range data as well as angle data to provide a record of target position versus time.

A typical search scenario in a tracking radar requires scanning the antenna and presenting the data from the sector of interest on a PPI display as shown in Figure 10.1.

When the target is acquired by the acquisition antenna, the radar is switched either manually or automatically to the range track mode where the tracking is continuously executed as long as it remains in the radar view.

Split-Gate Range Tracking

Split-gate tracking has gained the greatest prominence in range tracker implementations. The most common configuration of such an implementation is that of a split-gate tracker. The major component of the split-gate tracker is the time discriminator. Since the range to a moving target is continuously changing with time, the range tracker must be constantly adjusted to keep the target locked in range using two range gates: the *early gate* and the *late gate*. This technique for automatically tracking range is thus based on a split range gate. Two range gates are generated as shown in Figure 10.2. The echo pulse is shown in Figure 10.2(a), the relative position of the gates at a particular time in Figures 10.2(b) and 10.2(c), and the responses of early and late gates in Figures 10.2(d) and 10.2(e), respectively.

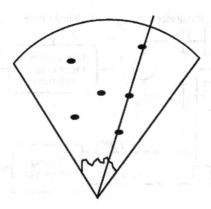

FIGURE 10.1 PPI display of range tracking.

The early gate passes only the first half of the tracked target video signal. The late gate starts at the end of the early gate, and passes only the second half of the target video signal. The on-target gate passes the complete tracked target video. The output of the on-target video gate is not required in the range tracker itself. The early gate generates positive voltage output, while the late gate generates negative voltage output.

The voltage outputs of the early and late are then applied to a difference amplifier to generate an error signal, which may be used to reposition the center of the gates.[1] If the target video energy during the early-gate enabled period is equal to that during the late-gate enabled period, the difference amplifier input voltages are equal, and, consequently, the output will be zero implying range gates are centered on the echo pulse. This condition exists when the range tracker is correctly following the centroid of the target return. On the other hand, if the partition between the early and late gates is not centered on the target video, the difference amplifier output will be nonzero. This gives an indication that the gates must be repositioned in time, left or right, depending on the distance of the gates' partition from the centroid of the target video, and on the polarity of the amplifier output. This is accomplished by using a feedback-control system.

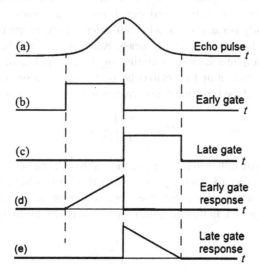

FIGURE 10.2 Split-range gate tracking. (a) Echo pulse, (b) early range gate, (c) late range gate, (d) early gate response, and (e) late gate response (From Mahafza, 1998).

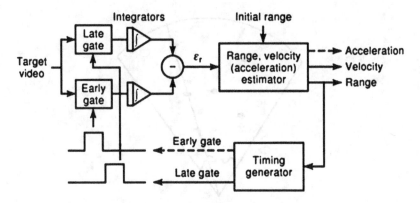

FIGURE 10.3 Analog split-gate trackers.

A split gate tracker consists of two *sample-and-hold* (S&H) circuits separated in time (range) by about one pulse width. They are known as the early and late gates. In early analog systems, these were in fast FET switches that allowed the charge to flow into an integrator for the periods of early and late gates as shown in Figure 10.3.

The split-gate range tracking may be associated with noise in the range-tracking circuits in the case of a finite length target. This range tracking noise depends on the length of the target and its shape. It has been reported[2] that the rms value of the range noise is approximately 0.8 of the target length when tracking is accomplished with a video split-gate range error detector.

Range Tracking Accuracy

Range accuracy is the ability of a radar to determine, within a specified range error, the actual range of the target from the radar. In other words, range accuracy is the measure of how accurately the radar determines the time delay of the received signal. The accuracy of the signal echo time delay is, in turn, determined primarily by how precisely the radar can locate an echo.

Range tracking accuracy depends on a number of factors[3] including radar systematic errors, target dependent errors, and propagation effects. The radar-induced range errors are primarily considered as bias errors that include the improper setting of antenna pedestal encoders, inadequate compensation in range for internal cable lengths and signal transmission lines. Target-induced errors are caused by the target mostly because of target scintillation. However, propagation effects for most radar have little effect on the range tracking accuracy. Range noise is another source of errors caused by thermal noise, PRF jitter, interference contaminating the echo pulse, and other instabilities in the radar system. The main contributing factors to delay time determination are the pulse width and the SNR. According to Barton,[4] the rms time delay error, denoted by $\sigma_{\hat{t}_{d1}}$, is given by

$$\sigma_{\hat{t}_{d1}} = \frac{1}{2B\sqrt{SNR}} \tag{10.4}$$

where B is the receiver bandwidth matched with τ_0, the width of the pulse. This result assumes that the approximate range to the target is known, a priori. Also, it determines the range noise on a single pulse basis. For the target range over several pulses, the range noise is reduced by a factor $1/\sqrt{n}$, where n is the number of pulses integrated. In this case, the range noise is also given by Barton (1988) as

$$\sigma_{\hat{t}_d} = \frac{1}{2B\sqrt{f_r t_0 SNR}} \tag{10.5}$$

where f_r is the pulse repetition frequency of radar and t_0 is the dwell time or time of observation over which data is gathered for measurement.

Tracking Radars

Example 10.2

Assuming that the primary source of range noise is mostly due to thermal noise, determine the rms time delay error and the corresponding rms range error for a radar in air with an SNR of 30 dB and receiver noise bandwidth of 2 MHz on a single pulse basis. Assume that the pulse repetition frequency of the radar is 12 kHz.

Solution:

On a single pulse basis, the rms time delay error can be found from (10.4) given by

$$\sigma_{\hat{t}_{d1}} = \frac{1}{2B\sqrt{SNR}} = \frac{1}{(2)(2 \times 10^6)\sqrt{10^3}} \approx 0.008 \ \mu s$$

The rms range error is therefore

$$\sigma_{\hat{R}_1} = \frac{c}{2}\sigma_{\hat{t}_{d1}} = \left(\frac{3 \times 10^8}{2}\right)(0.008 \times 10^{-6}) = 1.2 \ m.$$

10.3 ANGLE TRACKING

The accurate determination of target position is one of the important functions of radar systems. However, when the accurate detection of target angular position is desired, several methods are used. Among them are *sequential lobing, conical scan,* and *monopulse.* Angle tracking is concerned with generating continuous measurements of angular position in the azimuth and elevation coordinates. Each of these methods involves the use of information obtained from offset antennas to develop signals related to angular errors between the target position and the boresight axis of the tracking antenna. The resultant error signal indicates how much the target has deviated from the axis of the main beam. The antenna beam in the angle tracking radar is continuously positioned in an angle by a servomechanism, actuated by the error signal, in an attempt to generate a zero error signal. The error signal needs to be a linear function of the deviation angle.

10.3.1 SEQUENTIAL LOBING

One of the earliest tracking radars methods was sequential lobing. Sequential lobing is often referred to as *lobe switching* or *sequential switching.* The antenna pattern commonly employed with sequential lobing is the symmetrical pencil beam in which the elevation and azimuth beamwidths are approximately equal. However, a simple pencil beam antenna is not suitable for tracking radars unless means are provided to determine the magnitude and direction of the target's angular position with respect to some reference direction, usually the antenna axis. The difference in the target position and the reference direction is the angular error. The tracking radar attempts to position the antenna continuously to make the angular error zero. When the angular error becomes zero, the target is located along the reference direction implying that the target is tracked.

One method of obtaining the direction and magnitude of the angular error includes alternately switching the antenna beam between two predetermined symmetrical positions around the reference direction (tracking or switching axis) as illustrated in Figure 10.4. In each position, target strength is measured and converted into a voltage. The difference in amplitude between the voltages obtained in the two switched positions is a measure of angular displacement of the target from the switching axis. The polarity of the voltage difference determines the direction in which the antenna beam must be moved in order to align the switching axis with the direction of the target. When the

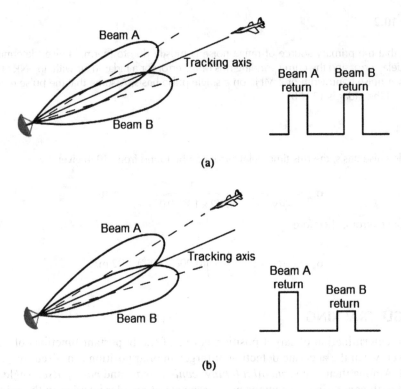

FIGURE 10.4 Sequential lobing: (a) target is on the tacking axis, (b) target is off the axis (From Mahafza, 1998).

voltages in the two switched positions are equal, the target is on the axis and its position may be determined from the direction of the antenna axis.

If both azimuth and elevation data are needed, two additional switching positions are required. Thus, two-dimensional sequential lobing radar consists of a cluster of four feed horns (two for each coordinate) illuminating a single antenna, arranged so that the right-left and up-down sectors are covered by successive antenna positions. Both transmission and reception are accomplished at each position. A cluster of five feeds might also be employed, with the central feed used for transmission while the outer four feeds are for reception.

An important feature of sequential lobing is the accuracy of the target position, which can be improved by carefully determining the equality of the signals in the switched positions, and limiting the system noise, caused either by electrical and mechanical fluctuations, to a minimum.

10.3.2 Conical-Scan Tracking

Conical-scan tracking is conceptually a logical extension of the sequential lobing technique described in the previous section where, in this case, the offset antenna beam is continuously rotated about the antenna axis, shown in Figure 10.5. The angle between the axis of rotation and the axis of the antenna beam (LOS of the antenna beam) is called the *squint angle*, denoted by a symbol θ_q. The echo signal will be amplitude modulated at a frequency equal to the frequency of rotation of the antenna beam. The amplitude of the echo signal depends on the shape of the antenna beam pattern, the squint angle, and the angle between the target LOS and the rotation axis. The phase of the modulation is a function of the angle between the target and the rotation axis. The conical-scan modulation is extracted from the echo signal, and applied to a servo-control system, which continually positions the antenna on the target. When the antenna is on the target, the LOS to the target and the rotation axis coincide, and the modulation is zero.

Tracking Radars

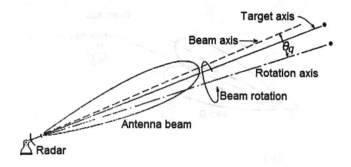

FIGURE 10.5 Conical-scan tracking antenna.

Figure 10.6 illustrates how the tracking is achieved by the conical-scan tracking. In this case, as the antenna rotates about the rotation axis, the echo signal will have zero modulation indicating that the target is tracked and no further action is needed. Next, consider the case, as depicted in Figure 10.7, where the amplitude of the echo signal is maximum for the target lying along the beam's axis at position B, and is minimum for the beam at position A. Between these two positions, the amplitude of the target return will vary between the maximum and minimum values. Thus the extracted amplitude modulated signal can be fed to the servo-control system in order to position the target on the desired tracking axis.

Derivation of the Error Signal

The relationships between the scanning beam, the target, and the error voltage in a conical-scan system are shown in Figure 10.8, which illustrates the top view of the beam axis location. Assume that the beam starts at $t = 0$. The beam positions for maximum and minimum voltage amplitudes, and the symbols used are illustrated in the figure. The quantity ε defines the distance between the target location and the tracking axis of the antenna. The azimuth and elevation errors are, respectively, given by

$$\varepsilon_a = \varepsilon \sin \theta_q \tag{10.6}$$

$$\varepsilon_e = \varepsilon \cos \theta_q \tag{10.7}$$

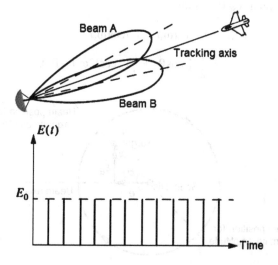

FIGURE 10.6 Modulated error signal when the target is on the tracking axis (Mahafza, 1998).

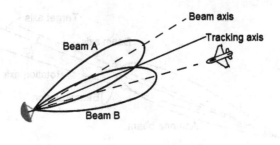

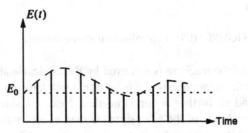

FIGURE 10.7 Modulated error signal when the target is off the tracking axis (From Mahafza, 1998).

and the envelope of signal history voltage $v_{SE}(t)$ can then be written as

$$v_{SE}(t) = V_0 \cos(\omega_s t + \theta_q)$$
$$= V_0 \cos\omega_s t \cos\theta_q - V_0 \sin\omega_s t \sin\theta_q \quad (10.8)$$
$$= V_0 \varepsilon_e \cos\omega_s t - V_0 \varepsilon_a \sin\omega_s t$$

with the scan reference signal $v_{SC}(t)$ with unity amplitude (henceforth normalized to unity) written in terms of the scan frequency as

$$v_{SC}(t) = \cos\omega_s t \quad (10.9)$$

where V_0 is a constant called the error slope, ω_s is the scan frequency in radians per second, and θ_q is the squint angle as already defined. The elevation error is the product of the signal history and the scan reference, both of which vary sinusoidally at the same frequency. Thus heterodyning $v_{SE}(t)$ with the scan reference signal $v_{SC}(t)$ results

$$v_{el}(t) = V_0 \cos(\omega_s t + \theta_q)\cos\omega_s t$$
$$= \frac{1}{2}V_0 \cos\theta_q + \frac{1}{2}V_0 \cos(2\omega_s t - \theta_q). \quad (10.10)$$

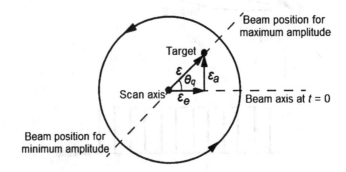

FIGURE 10.8 Error detection in a conical-scan radar.

Tracking Radars 177

After low-pass filtering we get the d-c term, which is applied to the tracking circuits to reposition the antenna beam. This target position results in a large positive average value in its elevation error given by

$$V_{el} = \frac{1}{2} V_0 \cos \theta_q.$$ (10.11)

This implies that the elevation error drives the antenna beam upward due to the positive sign because in this tracking scheme, a positive elevation error drives the beam upward and the negative error drives it downward. Similarly, the azimuth error signal is obtained by heterodyning $v_{SE}(t)$ with the quadrature component $\sin \omega_s(t)$ of the scan frequency signal. Thus

$$v_{az}(t) = V_0 \cos(\omega_s t + \theta_q) \sin \omega_s t$$
$$= \frac{1}{2} V_0 \sin(2\omega_s t - \theta_q) - \frac{1}{2} V_0 \sin \theta_q.$$ (10.12)

After low-pass filtering, the azimuth error becomes

$$V_{az} = -\frac{1}{2} V_0 \sin \theta_q.$$ (10.13)

Similar conclusions can be made with respect to positioning the beam depending on the sign of V_{az}. The *automatic gain control* (AGC) holds the average level constant, while preserving the rapid variations caused by the scan. It makes the tracking errors a function of only the amount of target angular displacement. The error signal outputs are sometimes filtered in a bandpass filter, centered at the scan frequency with a bandwidth equal to twice the servo bandwidth in order to reduce the effects of target scintillation and *amplitude modulated* (AM) jamming.

The scan rate of the conical scan is dependent on sampling and the mechanism used to generate the scan. The conical-scan radar needs at least four target returns to be able to determine the target azimuth and elevation coordinates (two returns per coordinate). Thus the maximum conical-scan rate is equal to one fourth of the PRF. Beyond the Nyquist sampling criterion, the scan rate is set by the scan mechanism. The squint angle needs to be large enough so that a good error signal can be measured. However, due to the squint angle, the antenna gain in the direction of the tracking axis is less than maximum value. Thus when the target is in track (located on the tracking axis), the SNR suffers a loss equal to the drop in the antenna gain. This loss is known as the *squint* or *crossover loss*. The squint angle is normally chosen such that the two-way (transmit and receive) crossover loss is less than a few decibels.

Example 10.3

For a conical-scan tracking radar, the antenna makes one complete rotation around the rotation axis in 3 seconds. The radar emits 15,000 pulses during one rotation to be able to track the target. Calculate the radar PRF and the unambiguous range.

Solution:

The conical-scan radar needs at least four target returns to be able to track the target in azimuth and elevation. Therefore, the maximum conical-scan rate is equal to $f_r / 4$. During conical scanning the radar sends emits $(15,000 / 3) = 5,000$ pulses per second. Hence the scan rate is

$$\omega_s = \frac{\text{maximum scan rate}}{5,000} = \frac{f_r / 4}{5,000} = \frac{f_r}{20,000} \text{ rad/second}$$

It follows then

$$\omega_s = \frac{1}{2} = \frac{f_r}{20{,}000} \Rightarrow f_r = 10 \text{ kHz}$$

The unambiguous range is

$$R_u = \frac{c}{2f_r} = \frac{3 \times 10^8}{(2)(10 \times 10^3)} = 15 \text{ km}.$$

Block Diagram of a Conical-Scan Radar System

A typical conical-scan pulsed radar system may be represented by the block diagram in Figure 10.9. As shown in the schematic diagram of the tracker, the AM signal out of the range gate is demodulated by the azimuth and elevation reference signals to produce the two angle error signals. These angle errors drive the angle servos, which in turn control the position of the antenna, and drive it to minimize the error (a null tracker). Since the conical-scan system utilizes amplitude changes to sense position, amplitude fluctuations at or near the conical-scan frequency will adversely affect the operation of the conical-scan radar system by inducing tracking errors. Three major causes of amplitude fluctuations include the inverse-fourth-power relationship between the echo signal and range, conical-scan modulation, and amplitude fluctuations in the target cross section. The function of the AGC is to maintain a constant level of the receiver output and to smooth amplitude fluctuations as much as possible without disturbing the extraction of the desired error signal. The d-c level of the receiver output must be maintained constant by preventing saturation caused by large signals. Thus the AGC that is required to normalize the pulse amplitude must be carefully designed. The required dynamic range of the AGC depends on the variation in range over which the targets are tracked, and should be in the order of 90 dB or more. Two/three stages of IF amplifiers are normally used to stabilize the dynamic range of the system. An alternative AGC filter design would maintain the AGC loop gain up to frequencies much higher than the conical-scan frequency resulting in the measurement of range in the normal manner. In such a case the error signal can be extracted from the AGC voltage.

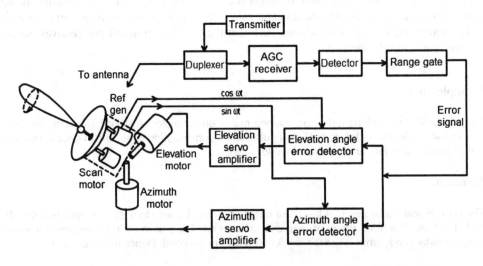

FIGURE 10.9 Block diagram of a conical-scan radar system (From Skolnik, 1980).

10.3.3 Monopulse Tracking Radar

Fluctuations in echo signal can have serious effects in some applications to severely limit the accuracy of the conical scan and sequential lobing tracking radars requiring many pulses to be processed in extracting the error signal. These effects on tracking accuracy can be eliminated if the angular measurement is made on the basis of one pulse rather than many. More than one antenna beam is used simultaneously in these methods, in contrast to the conical scan or sequential lobing tracking radar, which utilizes one antenna beam on a time-shared basis. The angle of arrival of the echo signal may be determined in a single-pulse system by measuring the relative amplitude of the echo signal received in each beam. The tracking systems that use a single pulse to extract all the information necessary to determine the angular errors are called *monopulse* systems. Angular errors in these systems are obtained by using either *amplitude comparison monopulse* or *phase comparison monopulse*. The advantages are the greater efficiency of the measurement, higher data rate, and reduced vulnerability to gain inversion and AM jamming. Also, monopulse tracking is more accurate, and is not susceptible to lobing anomalies.

Amplitude Comparison Monopulse

The generation of angular track errors in an amplitude comparison monopulse angle tracking is similar to lobing in the sense that multiple squinted antenna beams and the relative amplitude of the echoes in each beam are required to determine the angular error. The difference is that the beams are produced simultaneously rather than sequentially. There are various methods of monopulse tracking, only some of which employ four beams. We will restrict our discussions to four beams only. Monopulse tracking radars can employ both reflector antennas as well as phased array antennas to generate four partially overlapping antenna beams. In the case of reflector antennas, a compound feed of four horn antennas is placed at the parabolic focus. The pattern, in the case of signals arriving from the off-axis, is determined by the amount and direction of the signal angular offset from the reflector axis. The distances between horns are small and the phases of the four signals A, B, C, and D are within a few degrees of one another. It is assumed that the phases are identical for all practical purposes. Amplitude comparison monopulse tracking with phased array antennas is more complex than with reflectors.

A typical antenna pattern of a monopulse tracking radar with four beams representing four beam positions is shown in Figure 10.10. When viewed from the front, the beam patterns are the cross section that appears in Figure 10.11. All four feeds generate the sum pattern. The difference pattern in one plane is formed by taking the sum of two adjacent feeds and subtracting this from the sum of the other adjacent feeds. The difference pattern in the orthogonal planes is obtained by adding the differences of the orthogonal adjacent pairs.

A total of four hybrid junctions generate the sum channel, the azimuth difference channel, and the elevation difference channel, as illustrated in Figure 10.11. Monopulse processing consists of

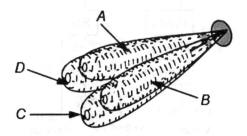

FIGURE 10.10 Monopulse antenna pattern.

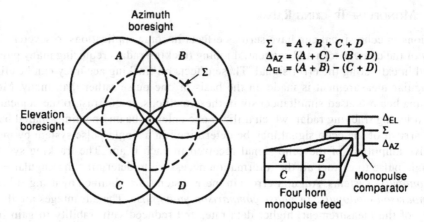

FIGURE 10.11 A Monopulse comparator to produce sum and difference signals.

computing a sum Σ and two difference Δ (one for azimuth and the other for elevation) antenna patterns. The difference patterns provide the magnitude of the angular error, while the sum pattern provides the range measurement, and is also used as a reference to extract the sign of the error signal. The difference patterns Δ_{AZ} and Δ_{EL} are produced on reception using a microwave hybrid circuit called a *monopulse comparator*.

The hybrids perform phasor additions and subtractions of the RF signal to produce output signals, as shown in Figure 10.12. If a target is on boresight, then the amplitudes of the signals received in the four channels (A, B, C, D) will be equal, and so the difference signals will be zero.

However, as the target moves off boresight, the amplitude of the signals received will differ, and the difference signal will take on the sign and magnitude proportional to the error that increases in amplitude with increasing displacement of the target from the antenna axis. The difference signals also change 180° in phase from one side of center to the other. The sum of all four horn outputs provides the video input to the range tracking system and establishes the AGC voltage level for automatic gain control.

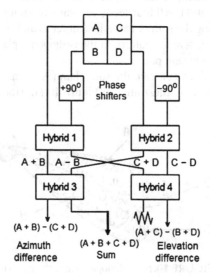

FIGURE 10.12 Hybrid configurations to produce sum and difference channel signals.

Tracking Radars

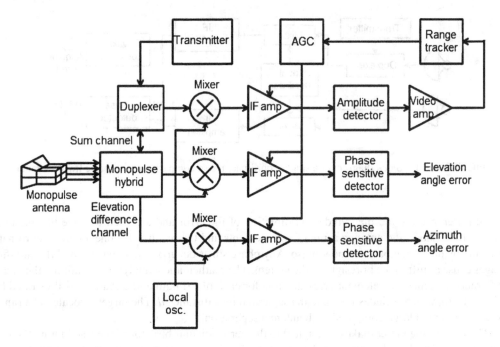

FIGURE 10.13 Block diagram of a two-coordinate amplitude comparison monopole radar.

For small angles, the ratio of relative amplitudes of target returns from sum and difference channels will result in a voltage proportional to the angular error. This ratio can be obtained using an AGC circuit that operates on the two difference channels and is driven by the detected sum-channel output of the tracking gate, or by division in a digital tracker. Phase detectors demodulate the azimuth and elevation error signals using the sum channel IF signal as a reference to produce the two error voltages. These must also be range-gated so that their magnitudes represent the error signals from the correct target

A block diagram of an amplitude comparison monopulse radar with provision for extracting error signals in both azimuth and elevation is shown in Figure 10.13. Hybrid junctions generate the sum channel, the azimuth difference channel, and the elevation difference channel. Two phase-sensitive detectors are included to extract the angular error information, one for azimuth and the other for elevation, while the range information is extracted from the output of the sum channel. According to Page,[5] the phase difference between channels must be maintained to within 25° or better. AGC is needed to maintain a stable closed-loop servo-control system for angle tracking. The AGC in a monopulse radar is achieved by employing a voltage proportional to the sum channel IF output to control the gain of all three receiver channels.

Phase Comparison Monopulse

Phase comparison monopulse differs from amplitude comparison monopulse in that the phase of the signal received in different antenna elements determines the angular errors. The major difference is that the four signals produced in amplitude comparison monopulse have similar phases but different amplitudes, however, in phase comparison monopulse; the signals have the same amplitudes but different phases. Phase comparison monopulse tracking radar uses an array of at least two antennas separated by some distance from one another. Separate arrays are required for azimuth and elevation, with a complete phase comparison monopulse tracking radar needing at least four antennas. The phases of the signals received by elements are compared. If the antenna axis is pointed at the target, the phases are equal; if not, they differ. The amount and the direction of the

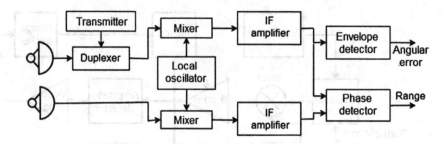

FIGURE 10.14 Block diagram of a single-coordinate phase comparison monopulse tracking radar.

phase difference are the magnitude and direction of the error and are used to drive the antenna. Figure 10.14 shows the antenna array of two elements and the receiver for one angular coordinate tracking by phase comparison monopulse. Any phase shifts occurring in the mixer and IF amplifier stages cause a shift in the boresight of the system. The mathematical analysis in obtaining the phase difference assumes two-element array antenna for each of azimuth and elevation, as illustrated by Figure 10.15, which includes two antenna separated by a distance d. The target is located at a range R and is assumed large compared with antenna separation.

The LOS to the target makes an angle θ to the perpendicular bisector of the line joining the two antennas. It, therefore, follows from Figure 10.16 that the distance from antenna 1 to the target can be obtained by applying cosine law,

$$R_1^2 = R^2 + \left(\frac{d}{2}\right)^2 - 2R\left(\frac{d}{2}\right)\cos\left(\theta + \frac{\pi}{2}\right)$$

$$= R^2 + \frac{d^2}{4} + dR\sin\theta.$$

(10.14)

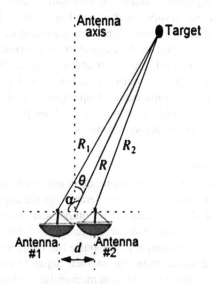

FIGURE 10.15 Single-coordinate phase comparison monopulse.

Tracking Radars

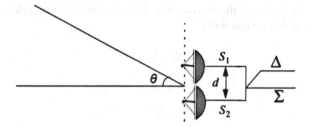

FIGURE 10.16 Phase comparison half-angle monopulse tracker.

Since $d \ll R$, (10.14) can be simplified to

$$R_1 = R\left[1 + \left(\frac{d}{2R}\right)^2 + \frac{d}{R}\sin\theta\right]^{1/2}$$
$$\approx R\left[1 + \frac{d}{R}\sin\theta\right]^{1/2}.$$

(10.15)

Applying binomial series expansion to (10.15) gives

$$R_1 \approx R\left(1 + \frac{d}{2R}\sin\theta\right) = R + \frac{d}{2}\sin\theta.$$

(10.16)

Similarly, the distance from the antenna 2 to the target is

$$R_2 = R - \frac{d}{2}\sin\theta.$$

(10.17)

The phase difference between the two antenna elements is then given by

$$\Delta\phi = \frac{2\pi}{\lambda}(R_1 - R_2) = \frac{2\pi}{\lambda}d\sin\theta$$

(10.18)

where λ is the wavelength of operation. The phase difference $\Delta\phi$ is used to determine the angular location of the target. For $\theta = 0$, the target will be on the axis of the antennas. For small angles, $\sin\theta \approx \theta$, the phase difference $\Delta\phi$ is a linear function of the angular error and may be used to position the antenna employing the servo-control system.

The disadvantages of phase comparison monopulse compared with amplitude comparison monopulse are the relative difficulty in maintaining a highly stable boresight, and the difficulty in providing the desired antenna illumination taper for both sum and difference signals. The longer paths from the antenna outputs to the comparator circuitry make the phase comparison system more susceptible to boresight change due to mechanical loading or sag, differential heating, etc. A technique providing greater boresight stability combines the two antenna outputs at RF with passive circuitry to yield sum and difference signals, as shown in Figure 10.16. These signals may then be processed as in a conventional amplitude comparison receiver. The single coordinate sum and difference signals are, respectively, given by

$$\Sigma(\theta) = S_1 + S_2$$

(10.19)

$$\Delta(\theta) = S_1 + S_2$$

(10.20)

184   Fundamental Principles of Radar

where S_1 and S_2 are the signals in the two antenna elements. Since S_1 and S_2 have the same amplitude but differ in phases by $\Delta\phi$, we can write

$$S_1 = S_2 e^{-j\Delta\phi}. \tag{10.21}$$

Thus

$$\Sigma(\theta) = S_2(1 + e^{-j\Delta\phi}), \tag{10.22}$$

$$\Delta(\theta) = S_2(1 - e^{-j\Delta\phi}). \tag{10.23}$$

We can write the ratio as

$$\frac{\Delta}{\Sigma} = \frac{1 - e^{-j\Delta\phi}}{1 + e^{-j\Delta\phi}} = j\tan\left(\frac{\Delta\phi}{2}\right). \tag{10.24}$$

which is purely imaginary. The modulus of the error signal is given by

$$\left|\frac{\Delta}{\Sigma}\right| = \tan\left(\frac{\Delta\phi}{2}\right). \tag{10.25}$$

This kind of phase comparison monopulse tracker is sometimes called a *half-angle tracker.*

Example 10.4

A phase comparison monopulse tracking radar using a two-element array antenna operating at 600 MHz measures a phase difference of 25° between the signal outputs of the antenna elements. Assume that the antenna elements are separated by 1.5 m. Determine the angular error of the target it makes with the antenna axis.

Solution:

The angular error is obtained directly using (10.18):

$$\Delta\phi = \frac{2\pi}{\lambda} d\sin\theta$$

where

$$\lambda = \frac{3 \times 10^8}{600 \times 10^6} = 0.5 \text{ m.}$$

It follows then

$$\frac{25}{57.3} = \frac{2\pi}{0.5}(1.5)\sin\theta = 6\pi\sin\theta \Rightarrow \theta = \sin^{-1}\left[\frac{25}{(57.3)(6\pi)}\right] = 1.33°$$

Example 10.5

The phase comparison monopulse tracking radar in Example 10.4 is now used as a half-angle tracker that employs a technique providing greater boresight stability by combining the two antenna outputs' yield sum and difference signals. The radar measurement shows that the amplitude of the sum signal is four times that of the difference signal. Find the angular error of the target it makes with the antenna axis.

Solution:

Using (10.25) gives

$$\left|\frac{\Delta}{\Sigma}\right| = \tan\left(\frac{\Delta\phi}{2}\right)$$

which then with this *half-angle tracker* becomes

$$\frac{1}{4} = \tan\left(\frac{\Delta\phi}{2}\right) \Rightarrow \Delta\phi = (2)\tan^{-1}\left[\frac{1}{4}\right] = 28.07°$$

Then following the steps in Problem 10.12, we can directly obtain the angular error as

$$\theta = \sin^{-1}\left[\frac{28.07}{(57.3)(6\pi)}\right] = 1.49°$$

Comparison between Conical Scan and Monopulse

The greater the SNR and the steeper the slope of the error signal in the vicinity of zero angular error, the more accurate the measurement of the angle. The error signal slope as a function of the squint angle or beam crossover is illustrated in Figure 10.17. The maximum slope occurs at a beam crossover of about 1.1 dB. The monopulse option gives greater SNR for the same-size target due to the higher on-boresight antenna gain; only the gains of the difference channel signals are reduced by the beam squint angle. In addition, the steeper error slope near the origin results in superior tracking accuracy, and because new tracking information is generated with each new pulse, tracking is not degraded by fluctuations in echo amplitude.

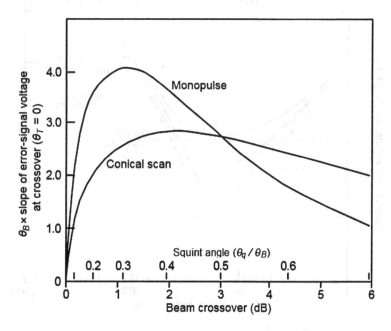

FIGURE 10.17 Slope of the angular error signal at crossover for a monopulse and conical-scan tracking radar. θ_B = half power beamwidth, θ_q = squint angle (From Skolnik, 1980).

Accuracy of Angle Tracking Techniques

The choice of angle tracking techniques for use in a given tracking system depends on the system requirements. Conical-scan systems in general are less expensive and less complex than either monopulse or sequential lobing systems. The gain, resolution, data rate, bandwidth, and overall accuracy of the monopulse systems are superior to any other types. One advantage of sequential lobing over the conical scan is the higher scanning rates.

Several factors that influence the accuracy to which a radar can track a target in angular dimensions are the antenna beamwidth, SNR of the receiver, target amplitude fluctuations (scintillation), target phase fluctuations (glint), the servo-system noise in the case of tracking radars, the receiver noise, atmospheric fluctuations, the stability of the electronic circuits, and the reflection characteristics of the targets. All these factors can degrade the tracking accuracy, which may cause random fluctuations of the antenna beam about the true target.

The contributions of the various factors affecting the tracking error are illustrated in Figure 10.18. As noted, angle fluctuations (glint) vary inversely as the range; receiver noise varies directly as the square of the range; amplitude fluctuations (scintillation) and servo-control system noise are independent of the range. Two different resultant curves are shown. Curve (A) is the sum of all effects and is representative of conical-scan and sequential lobing tracking radars, while Curve (B), on the other hand, does not include amplitude fluctuations, and is, therefore, representative of monopulse radars. The line labeled "receiver noise" represents the relative random noise generated in the receiver as a function of range. Also note that, whereas conical-scan and sequential lobing tracking systems are affected by target scintillation, monopulse is not. In the mid-range region, the contribution due to angular noise and receiver noise is small. Since the monopulse system is unaffected by amplitude noise, the limit for monopulse systems is primarily due to servo noise. As observed in Figure 10.19, the tracking accuracy, in general, deteriorates at both short and long target range, with the best tracking occurring at some intermediate range.

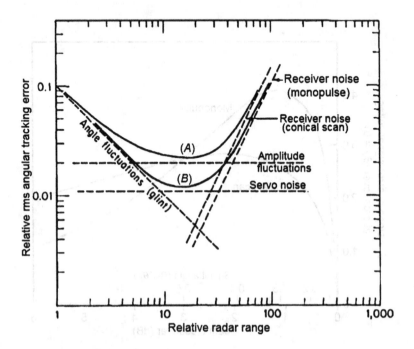

FIGURE 10.18 Relative angular tracking errors. Curve (A) composite error for a conical-scan radar; curve (B) composite error for a monopulse radar (From Skolnik, 1980).

Tracking Radars

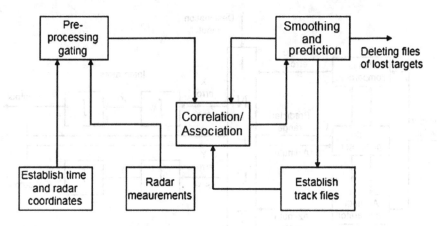

FIGURE 10.19 TWS data processing.

10.4 TRACK-WHILE-SCAN (TWS) RADAR

The *track-while-scan* (TWS) is a mode of radar operation in which the radar allocates part of its power to tracking the target or targets while part of its power is allocated to scanning, unlike the straight-tracking mode, when the radar directs all its power to tracking the acquired targets. In the TWS mode the radar has a possibility to acquire additional targets as well as providing an overall view of the airspace and helping maintain better situational awareness.

Modern scanning radar are designed for handling several modes of operation such as simultaneous tracking of multiple targets, prediction of future target location, and in the case of airborne radars, ground mapping, weather detection, and aircraft surveillance. Depending on the configuration, the TWS radar can either provide full hemispherical coverage or cover a limited angular segment. Because of the complexity of the TWS process and the necessity for storing both present and past target positions and velocities for multiple targets, digital computers or phased-array radars are generally required to provide TWS processing. TWS radars became possible with the introduction of two new technologies: *phased-array radars* and *computer memory devices*. Phased-array antennas became practical with the introduction of tunable high-power coherent radio frequency oscillators in the 1960s. By shifting the phase slightly between a series of antennas, the resulting additive signal can be steered and focused electronically. Much more important to the development of TWS was the development of digital computers and their associated memories, which allowed the radar data to be remembered from scan to scan.

Figure 10.19 represents the general sequence for TWS processing. The basic operations are the computation of the target's initial coordinates and measurements, correlating and associating target observations with existing target tracks to avoid redundant tracks, the computation of the information for displays or other system inputs.

Target positions inherently performed in polar coordinates are converted to the direction cosines (N, E, and V) of the inertial coordinate systems, which are more convenient for computer processing of target tracks. The inertial angular position of each target specifies the inertial target position. To convert the radar measurements to the inertial coordinate system, the measured range to the target must be computed by the following expressions:

$$R_N = \hat{N}R$$
$$R_E = \hat{E}R \qquad (10.26)$$
$$R_V = \hat{V}R$$

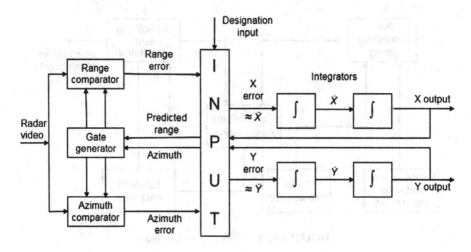

FIGURE 10.20 A typical converter from polar to inertial coordinates.

where R_N, R_E and R_V are in the northerly, easterly, and vertical components of the target positions respectively, R is the target range, and $\hat{N}$, $\hat{E}$ and $\hat{V}$ are the unit directional cosines in the respective inertial coordinate system. The method of converting polar measurements to the inertial coordinate system is shown in a typical block diagram shown in Figure 10.20. After the coordinate transformation has been performed, the observed target position must be correlated with the established target tracks stored in the computer. If the target position is near the predicted target position for one of the previously established tracks and the difference between the observed and predicted position is within the preset error bound, a positive correlation is obtained.

If the observed target does not correlate with any of the existing tracks, then a new track is established for the target. If the observed target correlates with two or more of the established tracks, then an established procedure such as that described by Hovanessian[6] must be followed in assigning the observation to a particular track. As a note, the process of assigning observations to the proper track is referred to as *association*. After the observed targets are associated with established or new tracks, estimated target positions must be computed for each target along with predictions of the target positions for the next radar scan. The current estimated target positions are computed by digital filtering of the current observed target position along with a weighted estimate of previous target observations associated with the target track. The predicted target positions for each track are then computed based on the current target position estimate, the time between scans, velocity components along each of the directional cosines. The predicted target positions are then used in the correlation process for each target observation on the next radar scan. For a newly established target track, if Doppler information is available from the radar, the computer can determine the radial velocity of the moving target.

The target velocity components in three inertial coordinate directions can be obtained in terms of $\dot{R}_N$, $\dot{R}_E$ and $\dot{R}_V$. The target velocity V_t can then be computed using the following equation:

$$V_t = \left[(\dot{R}_N)^2 + (\dot{R}_E)^2 + (\dot{R}_V)^2 \right]^{1/2}. \tag{10.27}$$

10.4.1 Target Prediction and Smoothing

The tracking radar system has a wide application in both the military and civilian fields. In the military, tracking is essential for fire control and missile guidance, whereas in civilian applications it is useful for controlling traffic of manned maneuverable vehicles such as ships, submarines, and aircrafts, which require accurate tracking. Tracking filters play the key role of target state estimation from which the tracking system is updated continuously. One of the tracking filters in use today in many applications is the α-β-γ *filter,* which is a development of the α-β *filter* aimed in tracking an accelerating

Tracking Radars

target since the α-β filter is only effective when input of the target model is a constant velocity model. A filter developed in the mid-1950s, the α-β[7] is popular because of its simplicity and computational inexpensive requirements. This allows its use in limited power capacity applications like passive *sonobuoys*. The α-β tracker[8] is now recognized as a simplified subset of the *Kalman*[9] *filter*. Low-cost and high-speed digital computing capability has made Kalman filters practical for more applications.

Performance of the operation of *smoothing* and *prediction* of target coordinates take place after the completion of *correlation* and *association*. Smoothing provides the best estimate of the present target position, velocity, and acceleration to predict future parameters of the target. Typical smoothing and prediction equations, for the direction cosines and range, are implemented using the α-β-γ filter, which is a simplified version of the Kalman filter. This α-β-γ filter can also provide a smoothed estimate of the present position used in guidance and fire control operation. The application of Kalman filtering to velocity tracking will be described by first considering a system with only velocity measurements and then employing the α-β tracker. Then the concept will be extended to α-β-γ tracker related to Kalman filtering and to an application where range, velocity, and acceleration are measured independently.

10.4.2 The α-β Tracker

The α-β tracker (also called α-β filter, f-g filter, or g-h filter) is a simplified form of observer for estimation, data smoothing, and control applications. It is closely related to Kalman filtering and to linear state observers used in control theory. Its principal advantage is that it does not require a detailed system model.

The α-β filter presumes that a system is adequately approximated by a model having two internal states, where the first state is obtained by integrating the value of the second state over time. Measured system output values correspond to observations of the first model state, plus disturbances. This very low order approximation is adequate for many simple systems, for example, mechanical systems where position is obtained as the time integral of velocity. Based on a mechanical system analogy, the two states can be called *position x* and *velocity v*. Assuming that velocity remains approximately constant over the small time interval T between measurements, smoothing is performed to reduce the errors in the predicted position through adding a weighted difference between the measured and predicted position, as follows:

Prediction:

$$x_p(k) = x_s(k-1) + T v_s(k-1) \tag{10.28}$$

$$v_p(k) = v_s(k-1), \tag{10.29}$$

Smoothing:

$$x_s(k) = x_p(k) + \alpha(x_0(k) - x_p(k)) \tag{10.30}$$

$$v_s(k) = v_p(k) + \frac{\beta}{T}(x_0(k) - x_p(k)) \tag{10.31}$$

where

x_s = smoothed values of position x
x_p = predicted values of position x
v_s = smoothed values of measured velocity v
v_p = predicted values of measured velocity v
k = scanning time at k
α, β = predetermined smoothing parameters
T = scan interval between samples

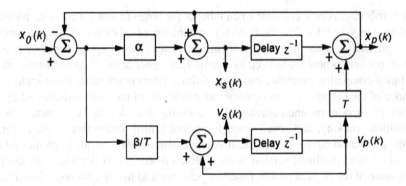

FIGURE 10.21 An implementation of an α-β tracker (From Mahafza, 1998).

An implementation of the α-β tracker is shown in Figure 10.21. The performance of the tracker depends on the choice of α and β, but choices are not independent. For stability and convergence, the values of α and β constant multipliers should be positive and small according to the following relations:

$$0 < \alpha < 1$$
$$0 < \beta \leq 2 \qquad (10.32)$$
$$0 < (4 - 2\alpha - \beta).$$

Noise is suppressed only if $0 < \beta < 1$, otherwise noise increases significantly. In general, larger α and β gains tend to produce a faster response for tracking transient changes, while smaller α and β gains reduce the level of noise in the estimate. If a reasonable balance between the accurate tracking and noise reduction is found, and the algorithm is effective, the filtered estimates are more accurate than the direct measurements. This motivates the α-β tracker a filter.

Prediction equations (10.28) and (10.29) can be rewritten in state space as follows:

$$X_p(k) = \Phi X_s(k-1) \qquad (10.33)$$

where the state vectors X_p and X_s are

$$X_p = \begin{bmatrix} x_p \\ v_p \end{bmatrix}, \qquad X_s = \begin{bmatrix} x_s \\ v_s \end{bmatrix}, \qquad (10.34)$$

respectively, and the corresponding transition matrix Φ is defined by

$$\Phi = \begin{bmatrix} 1 & T \\ 0 & 1 \end{bmatrix}. \qquad (10.35)$$

Similarly smoothing equations (10.30) and (10.31) can be rewritten in state space as follows:

$$X_s(k) = X_p(k) + K(x_0(k) - \Gamma X_p(k)) \qquad (10.36)$$

where the gain K is represented by

$$K = \begin{bmatrix} \alpha \\ \beta/T \end{bmatrix} \qquad (10.37)$$

and

$$\Gamma = [1 \quad 0]. \qquad (10.38)$$

Tracking Radars

Example 10.6

Consider an α-β filter used in a tracking radar with a scanning time interval of 1.2 ms between samples that assumes $\alpha = 0.75$, $\beta = 1.5$. Estimate the predicted values of position and velocity of a target corresponding to the desired estimated values of the target at 10 km moving with a velocity of 300 m/s.

Solution:

The specifications of the given filter satisfy the following relations for stability and convergence:

$$0 < \alpha < 1$$
$$0 < \beta \le 2$$
$$0 < (4 - 2\alpha - \beta)$$

Now using (10.33) gives

$$X_p(k) = \Phi X_s(k-1)$$

where

$$X_p = \begin{bmatrix} x_p \\ v_p \end{bmatrix}, \qquad X_s = \begin{bmatrix} 10^4 \\ 300 \end{bmatrix},$$

and

$$\Phi = \begin{bmatrix} 1 & T \\ 0 & 1 \end{bmatrix} = \begin{bmatrix} 1 & 1.2 \times 10^{-3} \\ 0 & 1 \end{bmatrix}$$

Solving the above equations gives $X_p = \begin{bmatrix} 10000.36 & 300 \end{bmatrix}^T$

10.4.3 THE α-β-γ TRACKER (KALMAN FILTERING)

The α-β-γ tracker estimates the values of state variables and corrects them in a manner similar to α-β filter or a state observer. However, an α-β-γ tracker does this in a much more formal and rigorous manner. The α-β-γ tracker is a steady-state Kalman filter, which assumes that the input model of the target dynamics is a constant acceleration model. The model has a low computational load, since the two steps are involved, that is the estimation and updating of position, velocity, and acceleration. In addition, smoothing coefficients of the filter are constants for a given sensor, which further contributes to its design simplicity. The selection of the weighting coefficients is an important design consideration because it directly affects the error-reduction capability.

The α-β-γ tracker is a one-step forward position predictor that uses the current error, called the *innovation*, to predict the next position. The innovation is weighted by the smoothing parameters α, β and γ. These parameters influence the behavior of the system in terms of stability and ability to track the target. Therefore, it is important to analyze the system using control theoretic aspects to gauge its stability and performance. Based on these weighting parameters, the α-β-γ equations applied in estimating predicted and smoothed values of position x, velocity v, and acceleration a are expressed as

Prediction:

$$x_p(k) = x_s(k-1) + Tv_s(k-1) + \frac{T^2}{2} a_s(k-1) \tag{10.39}$$

$$v_p(k) = v_s(k-1) + Ta_s(k-1) \tag{10.40}$$

$$a_p(k) = a_s(k-1), \tag{10.41}$$

Smoothing:

$$x_s(k) = x_p(k) + \alpha(x_0(k) - x_p(k)) \qquad (10.42)$$

$$v_s(k) = v_p(k) + \frac{\beta}{T}(x_0(k) - x_p(k)) \qquad (10.43)$$

$$a_s(k) = a_p(k) + \frac{2\gamma}{T^2}(x_0(k) - x_p(k)) \qquad (10.44)$$

where the subscripts $0, p$, and s denote the observed, predicted, and smoothed state parameters, respectively; x, v, and a are the target position, velocity, and acceleration, respectively; T is the simulation time interval; k is the sample number as used in the analysis of the α-β tracker.

An implementation of the α-β-γ tracker is shown in Figure 10.22. The parameter constraints of α, β, and γ selected to obtain the region of stability are:

$$\begin{aligned} 0 &< \alpha < 2 \\ 0 &< \beta \le 4 - 2\alpha \\ 0 &< \gamma < \frac{4\alpha\beta}{2-\alpha}. \end{aligned} \qquad (10.45)$$

Prediction equations (10.39)–(10.41) can be rewritten in state space as follows:

$$\mathbf{X}_p(k) = \Phi \mathbf{X}_s(k-1) \qquad (10.46)$$

where the state vectors X_p and X_s are

$$\mathbf{X}_p = \begin{bmatrix} x_p \\ v_p \\ a_p \end{bmatrix}, \quad \mathbf{X}_s = \begin{bmatrix} x_s \\ v_s \\ a_s \end{bmatrix}, \qquad (10.47)$$

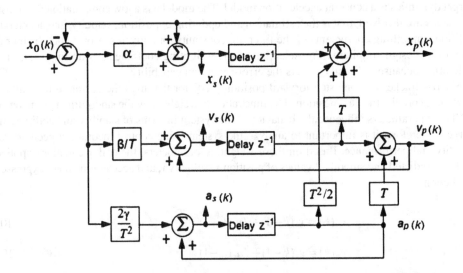

FIGURE 10.22 An implementation of an α-β-γ tracker (From Mahafza, 1998).

Tracking Radars 193

respectively, and the corresponding transition matrix Φ is defined by

$$\Phi = \begin{bmatrix} 1 & T & T^2/2 \\ 0 & 1 & T \\ 0 & 0 & 1 \end{bmatrix}. \tag{10.48}$$

Example 10.7

Consider an α-β-γ tracker with a scanning time interval of 2 ms between samples that assumes $\alpha = 1.7$, $\beta = 0.75$, and $\gamma = 5$. Estimate the predicted values of position, velocity, and acceleration of the target corresponding to the desired estimated values of the target at 10 km having a velocity of 300 m/s and an acceleration of 18 m/s².

Solution:

It can be easily shown that the given specifications of α, β, γ satisfy the following relations for proper stability of the filter:

$$0 < \alpha < 2$$
$$0 < \beta \le 4 - 2\alpha$$
$$0 < \gamma < \frac{4\alpha\beta}{2 - \alpha}$$

Now using (10.46) yields

$$X_p(k) = \Phi X_s(k-1)$$

where the state vectors X_p and X_s are

$$X_p = \begin{bmatrix} x_p \\ v_p \\ a_p \end{bmatrix}, \qquad X_s = \begin{bmatrix} 10^4 \\ 300 \\ 18 \end{bmatrix},$$

respectively, and the corresponding transition matrix Φ is defined by

$$\Phi = \begin{bmatrix} 1 & 2 \times 10^{-3} & 2 \times 10^{-6} \\ 0 & 1 & 2 \times 10^{-3} \\ 0 & 0 & 1 \end{bmatrix}.$$

Solving the above equations gives $X_p = \begin{bmatrix} 10000.6 & 300.036 & 18 \end{bmatrix}^T$

Similarly, smoothing equations (10.42)–(10.44) can be rewritten in state space as follows:

$$X_s(k) = X_p(k) + K(x_0(k) - \Gamma X_p(k)) \tag{10.49}$$

where the gain K is represented by

$$K = \begin{bmatrix} \alpha \\ \beta / T \\ 2\gamma / T^2 \end{bmatrix} \tag{10.50}$$

and the output matrix Γ is represented by

$$\Gamma = \begin{bmatrix} 1 & 0 & 0 \end{bmatrix}. \tag{10.51}$$

The predicted and smoothed positions are the first element of the vector X_s and X_p, respectively, which can be computed as:

$$x_p(k) = \Gamma X_p(k) \tag{10.52}$$

$$x_s(k) = \Gamma X_s(k). \tag{10.53}$$

If only the predicted estimates are considered, we can combine (10.46) and (10.49) as

$$X_p(k) = H X_p(k-1) + P x_{0(k)} \tag{10.54}$$

where

$$H = \Phi(I - KP) = \begin{bmatrix} (1-\alpha-\beta-\gamma) & T & T^2/2 \\ -(\beta+2\gamma)/T & 1 & T \\ -2\gamma/T^2 & 0 & 1 \end{bmatrix} \tag{10.55}$$

$$P = \Phi K = \begin{bmatrix} \alpha+\beta+\gamma \\ (\beta+2\gamma)/T \\ 2\gamma/T^2 \end{bmatrix} \tag{10.56}$$

where I is the identity matrix of order 3 given by

$$I = \begin{bmatrix} 1 & 0 & 0 \\ 0 & 1 & 0 \\ 0 & 0 & 1 \end{bmatrix} \tag{10.57}$$

Similarly, If only the smoothed estimates are considered, we can combine (10.46) and (10.49) as

$$X_s(k) = H' X_s(k-1) + K x_{0(k)} \tag{10.58}$$

where

$$H' = (I - K\Gamma)\Phi = \begin{bmatrix} 1-\alpha & (1-\alpha)T & (1-\alpha)T^2/2 \\ -\beta/T & -\beta+1 & (1-\beta/2)T \\ -2\gamma/T^2 & -2\gamma/T & (1-\gamma) \end{bmatrix}. \tag{10.59}$$

Tracking Radars

Equations (10.54) and (10.58) are the true representations of linear shift time-invariant systems, and can be expressed in the frequency domain using z-transform given by

$$z[X_p] = z^{-1}Hz[X_p] + Pz[x_0], \text{ and} \tag{10.60}$$

$$z[X_s] = z^{-1}H'z[X_s] + Kz[x_0], \tag{10.61}$$

respectively. Finally, we can determine the transfer function for the predicted and smoothed state variables by simply substituting the proper values of H, H', P, and K in (10.60) and (10.61), respectively.

PROBLEMS

10.1 A monostatic radar system measures the delay time of 0.12 ms caused by a return echo from a target. Find the distance of the target.

10.2 A target range measured from a monostatic radar is 30 km. Find the corresponding delay time.

10.3 A radar measures the target range with a mean-squared error of 100 m^2. Find the corresponding mean-squared error in measuring delay due to noise.

10.4 A monostatic radar measures a target's delay with mean-squared error due to noise of 0.01×10^{-12} s^2. What is the corresponding rms error in measuring range?

10.5 Consider a monostatic radar that measures the angular position of a target at a distance of 60 km to an rms error of 0.6 mrad. Find the distance in an off-axis direction from the boresight axis corresponding to this error.

10.6 For the monostatic radar of Problem 10.6, if the delay measurement system provides an rms range error due to noise of half the off-axis error, calculate the target's rms delay error.

10.7 Work **Example 10.2** when the range is determined by observing the target range for 2 ms.

10.8 A conical-scan tracking radar uses a pulse train with a PRF of 100 kHz. Calculate the maximum scan rate so that at least 10 pulses are emitted in one scan to be able to track the target.

10.9 For the α-β filter with the same specifications as in **Example 10.6**, estimate the desired smoothed values corresponding to the predicted values of a target at 12 km with a velocity of 340 m/s.

10.10 For the α-β-γ filter with the same specifications as in **Example 10.7**, estimate the desired smoothed values corresponding to the predicted values of a target at 12 km with a velocity of 330 m/s and an acceleration of 36 m/s^2.

REFERENCES

1. A. S. Locke, *Guidance* (Princeton, NJ: D. Van Nostrand Company, Inc., 1955), 408–413.
2. J. H. Dunn, D. D. Howard, and A. M. King, "Phenomenon of Scintillation Noise in Radar Tracking Systems," *Proceedings of the IRE*, 47 (1959): 855–863.
3. R. S. Berkowitz, *Modern Radar* (New York: John Wiley and Sons, 1965).
4. D. K Barton, *Radar System Analysis* (Englewood Cliffs, NJ: Prentice-Hall, 1964).
5. R. M. Page, "Monopulse Radar," *IRE National Convention Record,* 3, pt. 8 (1955): 132–134.
6. S. A. Hovanessian, *Radar Detection and Tracking Systems* (Dedham, MA: Artech House, 1978)
7. J. K. Sklansky, "Optimizing the Dynamic Parameter of a Track-While Scan System." *RCA Laboratories Journal* (June 1957).
8. R. T. Benedict and G. W. Brodner, "Synthesis of an Optimal Set of Radar Track-While-Scan Smoothing Equations," *IRE Transaction on Automatic Control,* AC-7 (1962): 27–32.
9. R. E. Kalman, "A New Approach to Linear Filtering and Prediction Problems," *Journal of Basic Engineering,* 82, no. 1 (1960): 35–45.

11 Aperture and Phased Array Antennas

"Sometimes I feel a lot of things but I keep it in. I'm sure we all have this built-in radar of what we predict and when it happens, we feel 'I knew it.'"

—**Muhammad Ali**

11.1 INTRODUCTION

The radar antenna acts as a transducer between the free-space propagation and transmission-line propagation. During transmission the radar antenna concentrates the radiated energy into a shaped beam, which points in the desired direction. On reception the antenna performs the inverse function; it collects the energy contained in the echo signal and delivers it to the radar receiver. There are many types of antennas, and the detailed design of any type is typically quite involved. Rather than detailed treatment of various antennas, this chapter addresses only two types, *aperture* (particularly *parabolic reflector*) and *array antennas*, which are widely used radar applications.

Many of the important radar antennas are aperture antennas, which are characterized by directive beams. Aperture antennas are most common at microwave frequencies. There are many different geometrical configurations of an aperture antenna. They may take the form of a waveguide or a horn whose aperture may be rectangular, square, elliptical, circular, or any other configuration. Aperture antennas can be flush-mounted on the surface of spacecraft or aircraft. The parabolic reflector has been extensively employed in radar. The vast majority of radar antennas use the parabolic reflector in one form or other.

The phased array is essentially the special case of an antenna composed of a number of small radiating elements—suitably spaced with respect to one another—acting together to generate a radiation pattern, whose shape and direction are determined by the relative amplitudes and phases of the currents at the individual elements. The radiating elements might be dipoles, open-ended waveguides, slots cut on the conducting surface, or any other type of antenna. The flexibility to control the aperture illumination offered by the phased array antenna in steering the beam is by means of electronic control. The steering of the beam may be controlled by the use of computer, which facilitates to rapidly change the phase and amplitude of signals to or from the individual array elements. This offers the full potential of a phased array antenna to operate in a multifunctional way in performing surveillance and tracking of a variety of targets.

Phased array radar systems emerged after the introduction of short-wave radio equipment, in the 1920s and during World War I, which made possible the use of reasonably sized antenna arrays to achieve a directive radiation pattern for radio communications. During World War II, UHF and microwave antennas were introduced for radar applications with fixed phased array antennas in which the beam was scanned by mechanically actuated phase shifters. The major advances in phased array technology were essentially made in the early 1950s with the introduction of the electronic scanning method. The phased arrays now have many applications, particularly in radar.

Many geometrical configurations of phased array antennas are used. The most elementary is that of a *linear array,* which consists of array elements arranged in a straight line in one dimension. The elements may be equally or unequally spaced. When the array elements are arranged in two-dimensional configurations to lie in a plane, it is said to be a *planar array.* An array in which the

197

array elements are distributed on nonplanar surface such as found on an aircraft or missile is called a *conformal array*. A *broadside array* is one in which the direction of maximum radiation is perpendicular to the line (or plane) of the array. On the other hand, an *endfire* array has its maximum radiation parallel to the array configuration.

11.2 FUNDAMENTAL ANTENNA PARAMETERS

A number of antenna parameters are important, some of these are general and apply to any antenna, not just parabolic reflectors and antenna arrays. Some of the parameters are interrelated and not all of them need be specified for a complete description of the antenna performance. In this section we discuss and define these parameters. Parameters defined with quotation marks have been taken from *IEEE Standard Definitions of Terms for Antennas*.[1]

11.2.1 RADIATION PATTERN

An antenna *radiation pattern* is defined as

> "a graphical representation of the radiation properties of the antenna as a function of space coordinate. In most cases, the radiation pattern is determined in the far-field region and is represented as a function of the directional coordinates. Radiation properties include radiation intensity, field strength or polarization."

A trace of the received power at a constant radius with the antenna at the center of an imaginary sphere is called the *power pattern*. On the other hand, a graph of the spatial variation of the electric field intensity along a constant radius is called a *field pattern*.

The antenna patterns (azimuth and elevation plane patterns) are frequently shown as plots in polar coordinates. This gives the viewer the ability to easily visualize how the antenna radiates in all directions as if the antenna were "aimed" or mounted already. Occasionally, it may be helpful to plot the antenna patterns in Cartesian (rectangular) coordinates, especially when there are several sidelobes in the patterns and where the levels of these sidelobes are important. Various parts of the radiation pattern are referred to as *lobes*, which may be categorized into *major, minor, side, and back lobes*. In general, a lobe is any part of the pattern that is surrounded by regions of relatively weaker radiation. So a lobe is any part of the pattern that "sticks out" and the names of the various types of lobes are somewhat self-explanatory. Figure 11.1 provides a view of a radiation pattern with the lobes labeled in each type of plot. A main lobe (also called a major lobe) is defined as "the radiation of lobe containing the direction of maximum radiation." A minor lobe is the lobe except the main lobe. In Figure 11.1 the main lobe is pointing in the $\theta = 0$ direction. A sidelobe is "a radiation lobe in any direction other than the intended lobe." Usually the sidelobe is adjacent to the main lobe. A back lobe usually refers to a minor lobe that occupies the hemisphere in a direction opposite to that of the main lobe. Sidelobes usually represent radiation in undesired directions. Sidelobes are the largest of the minor lobes. Sidelobe level of the order of −20 dB or smaller is normally acceptable in most radar applications.

11.2.2 BEAMWIDTHS

The *half-power beamwidth* is defined as: "In a plane containing the direction of maximum of a beam, the angle between the two directions in which the radiation intensity is one-half the maximum value of the beam." Often, the term *beamwidth* is used to describe the angle between any two points on the pattern. However, the term *beamwidth* by itself is usually referred to as the half-power (or 3-dB) beamwidth.

The beam solid angle[2] Ω_A is defined as the solid angle through which all the power of the antenna would flow if its radiation intensity is constant for all angles within Ω_A. For antennas with one

Aperture and Phased Array Antennas

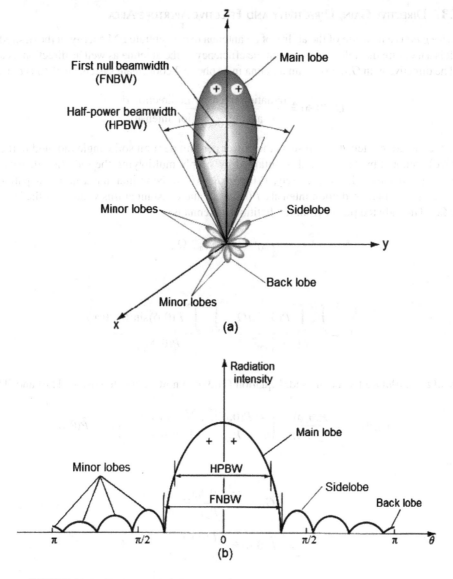

FIGURE 11.1 Typical radiation patterns in (a) polar and (b) Cartesian coordinates.

narrow major lobe and very negligible minor lobes, the beam solid angle is approximately related to the product of the half-power (3 dB) beamwidths θ_B and ϕ_B in two perpendicular principal planes. For a rotationally symmetric pattern, the half-power beamwidths in any two perpendicular planes are the same. With this rough approximation,

$$\Omega_A \approx \theta_B \phi_B \quad \text{(steradian).} \tag{11.1}$$

If the beamwidths are known in degrees, (11.1) can be written as

$$\Omega_A \approx \theta_B^\circ \phi_B^\circ \left(\frac{\pi}{180}\right)^2 = \frac{\theta_B^\circ \phi_B^\circ}{3283}. \tag{11.2}$$

Accurate determination of solid angle and beamwidths must be obtained from the actual radiation pattern.

11.2.3 Directive Gain, Directivity and Effective Aperture Area

Directive gain is a measure of the ability of an antenna to concentrate EM energy in the desired direction. It is a measure that takes into account the efficiency of the antenna as well as directional capabilities. The directive gain $G_D(\theta, \phi)$ of an antenna in a spherical coordinate system (R, θ, ϕ) is defined as

$$G_D(\theta, \phi) \triangleq \frac{\text{radiation intensity in direction } (\theta, \phi)}{\text{average radiation intensity}} \tag{11.3}$$

where the radiation intensity represents the power per unit per unit solid angle radiated in the direction (θ, ϕ), denoted by $P(\theta, \phi)$, and is obtained by simply multiplying the radiation density by the square of the distance. More precisely, $P(\theta, \phi) = R^2 \hat{P}(\theta, \phi)$. Now imagine a fictitious pattern with the value of maximum radiation intensity $P(\theta, \phi)|_{max}$ and constant at this value over the beam solid angle Ω_A. The radiated power from the fictitious antenna is

$$P_{rad} = P(\theta, \phi)|_{max} \, \Omega_A \tag{11.4}$$

where

$$\Omega_A = \frac{\int_0^{2\pi} \int_0^{\pi} P(\theta, \phi) d\Omega}{P(\theta, \phi)|_{max}} = \frac{\int_0^{2\pi} \int_0^{\pi} P(\theta, \phi) \sin(\theta) d\theta d\phi}{P(\theta, \phi)|_{max}} \tag{11.5}$$

where $d\Omega = \sin \theta d\theta d\phi$ has been used. Equation (11.3) can now be written using (11.4) and (11.5) as

$$G_D(\theta, \phi) = \frac{P(\theta, \phi)}{P_{rad} / 4\pi} = \left(\frac{4\pi P(\theta, \phi)|_{max}}{P_{rad}} \right) \left(\frac{P(\theta, \phi)}{P(\theta, \phi)|_{max}} \right) = G_D \hat{P}(\theta, \phi) \tag{11.6}$$

where

$$G_D = \frac{4\pi P(\theta, \phi)|_{max}}{P_{rad}}, \tag{11.7}$$

$$\hat{P}(\theta, \phi) \triangleq \frac{P(\theta, \phi)}{P(\theta, \phi)|_{max}}. \tag{11.8}$$

Directivity is the directive gain in the direction of maximum gain. It is a measure that describes only the directional properties of the antenna, and it is therefore controlled by the radiation pattern. More precisely,

$$G_D = G_D(\theta, \phi)|_{max}. \tag{11.9}$$

In (11.6) we have defined

$$G_D = \frac{4\pi P(\theta, \phi)|_{max}}{P_{rad}} = \frac{4\pi P(\theta, \phi)|_{max}}{\int_0^{2\pi} \int_0^{\pi} P(\theta, \phi) \sin(\theta) d\theta d\phi} = \frac{4\pi}{\Omega_A} \tag{11.10}$$

where G_D is the maximum value of $G_D(\theta, \phi)$, and is called directivity. $\hat{P}(\theta, \phi)$ is a normalized version of $P(\theta, \phi)$ such that its maximum is unity.

Aperture and Phased Array Antennas

Effective aperture, denoted by $A_e(\theta,\phi)$, is another parameter of importance. It is defined by

$$A_e(\theta,\phi) \triangleq \frac{\lambda^2 P(\theta,\phi)}{P_{rad}}. \tag{11.11}$$

By using this definition in (11.4) and (11.5), we get in terms of $A_e(\theta,\phi)$,

$$G_D(\theta,\phi) = \frac{4\pi A_e(\theta,\phi)}{\lambda^2} \tag{11.12}$$

and

$$G_D = \frac{4\pi A_e(\theta,\phi)_{max}}{\lambda^2}. \tag{11.13}$$

11.2.4 Power Gain and Antenna Efficiency

Power gain $G(\theta,\phi)$ includes some dissipative losses in the antenna structure, but it does not involve system losses resulting from mismatch of impedance or from polarization. The power gain is defined by

$$G(\theta,\phi) = \frac{\text{radiation intensity in the direction } (\theta,\phi)}{\text{radiation intensity from lossless isotropic antenna with same power input}} \tag{11.14}$$

Mathematically,

$$G(\theta,\phi) = \frac{P(\theta,\phi)}{P_{ac}/4\pi} = \frac{P(\theta,\phi)}{P_{rad}/4\pi} \cdot \frac{P_{rad}}{P_{ac}} = G_D(\theta,\phi)\frac{P_{rad}}{P_{ac}}. \tag{11.15}$$

where P_{ac} is the average power accepted at the antenna input.

Antenna efficiency η_a is used to take into account for losses at the input terminals due to mismatch between the transmission line and the antenna, and dissipative losses (conduction and dielectric). In general, the antenna efficiency is defined by

$$\eta_a \triangleq \frac{P_{rad}}{P_{ac}} \leq 1 \tag{11.16}$$

so that

$$G(\theta,\phi) = \eta_a G_D(\theta,\phi). \tag{11.17}$$

This indicates that the power gain $G(\theta,\phi)$ of an antenna is never greater than its directive gain.

Example 11.1

For an infinitesimal linear dipole antenna of length $\ell \ll \lambda$ with the power density given by $\hat{P}(\theta) = \hat{R} A_0 \sin^2\theta / R^2$ (W/m^2), where A_0 is the peak value of the power density, and $\hat{R}$ is the unit vector in the direction R coordinate; find the radiation intensity and the directivity of the antenna.

Solution:

The radiation intensity is given by

$$P(\theta,\phi) = R^2 \hat{P}(\theta,\varphi) = R^2(A_0\sin^2\theta / R^2) = A_0\sin^2\theta$$

The maximum radiation is thus directed along $\theta = \pi/2$, and is given by

$$P_{max} = P(\theta,\phi)\,|_{max} = A_0$$

The total radiated power is given by

$$P_{rad} = \int_0^{2\pi}\int_0^{\pi} P(\theta,\phi)\,d\Omega = \int_0^{2\pi}\int_0^{\pi} P(\theta,\phi)\sin\theta\,d\theta\,d\phi$$

$$= \int_0^{2\pi}\int_0^{\pi} (A_0 \sin^2\theta)\sin\theta\,d\theta\,d\phi = 8\pi A_0/3$$

Alternatively, the radiated power could be obtained directly from the power density as

$$P_{rad} = \int_0^{2\pi}\int_0^{\pi} \hat{P}(\theta,\phi)\cdot\hat{R}\,ds = \int_0^{2\pi}\int_0^{\pi} (\hat{R}A_0 \sin^2\theta/R^2)\cdot\hat{R}R^2 \sin\theta\sin\phi\,d\theta\,d\phi = 8\pi A_0/3$$

Using (11.17) gives the directivity of the antenna equal to

$$G_D = \frac{4\pi P(\theta,\phi)\,|_{max}}{P_{rad}} = \frac{4\pi A_0}{(8\pi A_0/3)} = 1.5$$

11.2.5 POLARIZATION AND BANDWIDTH

Polarization of a radiated wave is defined as "that property of a radiated electromagnetic wave describing the time varying direction and relative magnitude of electric-field vector; specifically, the figure traced as function of time extremity of the vector at a fixed location in space, and the sense in which it is traced, as observer along the direction of propagation." Polarization may be categorized into linear, elliptical, or circular. If the electric field vector at a point in space as a function of time is always directed along a line, the field is called linearly polarized. Elliptical polarization may be considered as the combination of two linearly polarized waves of the same frequency, traveling in the same direction, which is perpendicular to each other in space. If the amplitudes of the two waves are equal, and if they are 90° out of phase, the polarization is circular. Circular polarization and linear polarization are special cases of elliptical polarization. Clockwise rotation of the electric field is called right-hand polarization and counterclockwise as left-hand polarization.

Bandwidth of an antenna is defined as "the range of frequencies within which the performance of the antenna, with respect to some characteristic, conforms to a specified standard." The bandwidth can be considered to be the range of frequencies, on either side of a center frequency (usually the resonance frequency), where the antenna characteristics are within the acceptable values of those at the center frequency.

11.3 APERTURE ANTENNAS

Aperture antennas are most common at microwave frequencies because of the ever increasing demand for sophisticated forms of antennas. Antennas of this type are very useful for aircraft and spacecraft applications, because they can be very conveniently flush mounted on the skin of the aircraft or spacecraft. Aperture antennas may include slits, slots, waveguides, horns, lenses, and reflectors. An antenna that has a physical aperture opening with a circular shape is called a circular aperture. Various forms of circular aperture antennas are encountered in practice. In this section we will discuss ideal aperture with uniform and tapered amplitudes. This is followed by a discussion of parabolic reflector antennas, which are the most popular circular aperture antennas.

Aperture and Phased Array Antennas

11.3.1 Uniform Circular Aperture

If the amplitude of aperture distribution is constant, it is called *uniform circular aperture*. It is a widely used microwave antenna. The attractive feature of this configuration is its simplicity in construction. Because of the circular profile of the aperture, it is often desirable and convenient to adopt cylindrical coordinates for the analysis of fields. The uniform circular aperture is approximated by a circular opening of radius a in a ground plane illuminated by a uniform plane wave normally incident from behind. To simplify the mathematical details, the field over the aperture will be assumed to be constant at E_0 and given by

$$\overline{E}_a = \hat{x}E_0 \quad r' \leq a. \tag{11.18}$$

The *radiation integral* is written as

$$\vec{\Re} = \hat{x}E_0 \iint_{S_a} \exp(j\beta\hat{r} \cdot \vec{r}')\,ds' \tag{11.19}$$

where r' is a distance of the source point on the circular aperture, $\hat{r}$ is a unit vector in the direction of $\vec{r}'$ from the origin to the field point, and $\beta = \omega\sqrt{\mu\varepsilon} = 2\pi/\lambda$. The radius vector $\vec{r}'$ is

$$\vec{r}' = \hat{x}r'\cos\phi' + \hat{y}r'\sin\phi' \tag{11.20}$$

and the unit vector $\hat{r}$ is represented by

$$\hat{r} = \hat{x}\sin\theta\cos\phi + \hat{y}\sin\theta\sin\phi. \tag{11.21}$$

Using (11.20) and (11.21) gives

$$\begin{aligned}
\hat{r} \cdot \vec{r}' &= r'\sin\theta(\cos\phi\cos\phi' + \sin\theta\sin\phi') \\
&= r'\sin\theta\cos(\phi - \phi').
\end{aligned} \tag{11.22}$$

Hence the radiation integral becomes

$$\vec{\Re} = \hat{x}E_0 \int_0^a \left[\int_0^{2\pi} \exp(j\beta r'\sin\theta\cos(\phi - \phi'))\,d\phi' \right] r'\,dr'. \tag{11.23}$$

The second integral in (11.23) is of the form

$$J_0(x) = \frac{1}{2\pi} \int_0^{2\pi} \exp(jx\cos\xi)\,d\xi \tag{11.24}$$

where $J_0(x)$ is the Bessel function of the first kind of order zero. Equation (11.23) can now be expressed by applying the Bessel function as

$$\vec{\Re} = \hat{x}2\pi E_0 \int_0^a r'J_0(\beta r'\sin\theta)\,dr'. \tag{11.25}$$

The previous integration can be modified using

$$\int xJ_0(x)\,dx = xJ_1(x). \tag{11.26}$$

Here, $J_1(x)$ is a Bessel function of the first kind of first order, which is zero at $x = 0$. This is a decaying oscillatory function as x increases. Using these conditions, we can express (11.25) as

$$\vec{\Re} = \hat{x}\frac{2\pi a E_0}{\beta \sin\theta}J_1(\beta a \sin\theta) = \hat{x}\Re_x \tag{11.27}$$

where

$$\Re_x = \frac{2\pi a E_0}{\beta \sin\theta}J_1(\beta a \sin\theta). \tag{11.28}$$

The equivalent magnetic current formulation, using the equivalence principle, renders the far-field radiation field

$$\bar{E}(r,\theta,\phi) = (\hat{\theta}\cos\phi - \hat{\phi}\sin\phi\cos\theta)j\beta\frac{e^{-j\beta r}}{2\pi r}\Re_x. \tag{11.29}$$

Substituting $\Re_x$ from (11.27) into (11.29) and properly rearranging, we get the expression of the radiated electric field for the uniform circular aperture with an x-directed aperture electric field

$$\bar{E}(r,\theta,\phi) = (\hat{\theta}\cos\phi - \hat{\phi}\sin\phi\cos\theta)\frac{j\beta E_0\,\pi a^2 e^{-j\beta r}}{2\pi r}\frac{2J_1(\beta a \sin\theta)}{\beta a \sin\theta}. \tag{11.30}$$

We can now write the expressions for the far-field $(r \to \infty)$ E-plane and H-plane patterns, obtained for $\phi = 0$, as

$$E_\theta(\theta) = \frac{2J_1(\beta a \sin\theta)}{\beta a \sin\theta} \tag{11.31}$$

$$H_\phi(\theta) = \cos\theta\frac{2J_1(\beta a \sin\theta)}{\beta a \sin\theta} \tag{11.32}$$

which are normalized for a maximum value of unity when $\theta = 0$. The larger the aperture, the less significant is the $\cos\theta$ factor in (11.32) because the main beam in the $\theta = 0$ direction is very narrow, and in this small solid angle $\cos\theta \approx 1$.

The half-power point occurs at $\beta a \sin\theta = 1.6$, so the half-power beamwidth (HPBW) is written as

$$\mathrm{HPBW} = 2\theta_{3-dB} = 2\sin^{-1}\left(\frac{1.6}{\beta a}\right) = 2\sin^{-1}\left(\frac{1.6\lambda}{2\pi a}\right)$$

$$\approx 1.02\frac{\lambda}{2a}\ \mathrm{rad.} = 58.45\frac{\lambda}{2a}\ \mathrm{degree} \tag{11.33}$$

The sidelobe level of any uniform circular aperture is −17.5 dB for $a = 5\lambda$. Any uniform aperture with unity taper aperture efficiency, the directivity can be found only in terms of its physical area:

$$G_D = \frac{4\pi}{\lambda^2}A_p = \frac{4\pi}{\lambda^2}(\pi a^2). \tag{11.34}$$

Example 11.2

For a circular aperture antenna of diameter 0.30 m operating at a frequency of 10 GHz, find the half-power beamwidth and the directivity.

Aperture and Phased Array Antennas

Solution:

The half-power beamwidth, denoted by HPBW, is obtained using (11.33) as

$$\text{HPBW} \approx 1.02 \frac{\lambda}{2a} \text{ rad.} = 58.45 \frac{\lambda}{2a} \text{ degree}$$

Thus

$$\text{HPBW} = (58.45) \frac{0.03}{0.3} = 5.845 \text{ degree}$$

The directivity is obtained from (11.34) as

$$G_D = \frac{4\pi}{\lambda^2} A_p = \frac{4\pi}{\lambda^2} (\pi a^2) = 4\pi^2 (a/\lambda)^2 = 4\pi^2 (0.15/0.03)^2 = 986.96 = 29.94 \text{ dB}$$

11.3.2 Tapered Circular Aperture

Circular aperture antennas are sometimes approximated as radially symmetric apertures with field amplitude distribution, which is tapered from the center toward the aperture edge. Then the radiation integral (11.25) has a more general form, called the *unnormalized radiation integral* $\Re_{un}$:

$$\Re_{un}(\theta) = 2\pi E_0 \int_0^a E_a(r') r' J_0(\beta r' \sin\theta) dr'. \tag{11.35}$$

This integral can be performed for various aperture tapers and normalized to obtain $\Re(\theta)$. The parabolic taper of order n is given by

$$E_r(r') = \left[1 - \left(\frac{r'}{a} \right)^2 \right]^n. \tag{11.36}$$

Eliminating $E_a(r')$ from (11.35) by using (11.36), we obtain

$$\Re_{un}(\theta) = 2\pi E_0 \int_0^a \left[1 - \left(\frac{r'}{a} \right)^2 \right]^n r' J_0(\beta r' \sin\theta) dr'. \tag{11.37}$$

We can use the following Bessel function relation to simplify (11.37):

$$\int_0^a (1 - x^2) x J_0(bx) dx = \frac{2^n n!}{b^{n+1}} J_{n+1}(b). \tag{11.38}$$

Comparing (11.37) with (11.38) we find, in such a case, that $x = r'/a$ and $b = \beta a \sin\theta$. Accordingly,

$$\Re_{un}(\theta) = E_0 \frac{\pi a^2}{n+1} \Re(\theta, n) \tag{11.39}$$

where the normalized pattern function

$$\Re(\theta, n) = \frac{2^{n+1}(n+1)! J_{n+1}(\beta a \sin\theta)}{(\beta a \sin\theta)^{n+1}}. \tag{11.40}$$

206   Fundamental Principles of Radar

The *parabolic taper* for $n = 1$ is a smooth taper from the aperture center to zero at the aperture edge. The parabolic squared taper for $n = 2$ gives an even more severe taper. Parabolic tapers on a pedestal provide a nonzero illumination. In fact, as n increases, the taper becomes more severe, the sidelobe levels decrease, and beamwidth increases.

11.3.3 PARABOLIC REFLECTOR

The distinct advantages of mechanical simplicity, lightness, and flexible radiation patterns with high gain make reflector antennas the most widely used radar antennas. Radar antennas are generally required to produce a *pencil beam*, a *fan beam*, or a shaped beam determined by the type and function of the radar system. All these pattern shapes may be obtained from reflector antennas. The parabolic reflector, or popularly known as the *parabolic dish antenna*, is one of the most commonly used antennas for radar applications. The parabolic reflector, when properly fed, produces a pencil beam; an asymmetrical beam shape, called a fan beam, may also be obtained by using only a part of the paraboloid. Parabolic antennas are used for tracking, search, navigation, and for almost all conceivable radar applications.

In parabolic reflector antennas the parabola is illuminated by a source of energy called *feed* placed at the focus of the parabola and directed toward the reflector surface. The basic parabolic contour is used in a variety of configurations. Rotating the parabola about its axis produces what is called a *paraboloid*. It can be shown by geometrical optics considerations that a spherical wave emerging from the focal point and incident on the paraboloid reflector is transformed after reflection into a plane wave traveling on the direction of the positive axis of the paraboloid.

Front-Fed Parabolic Reflector Systems

The *parabolic reflector antenna* is geometrically represented in Figure 11.2 (a). In an ideal case the reflector surface is assumed to be perfectly conducting, and the point source feed is placed at the focus F of the reflector. The cross section in Figure 11.2(b), choosing a plane perpendicular to the axis of the reflector through the focus F, is typical. The different coordinate systems to be used in various expressions are shown in Figure 11.2(a). Referring to Figure 11.2(b), it follows that

$$FP + PA = \text{constant} = 2L_f \tag{11.41}$$

where L_f is the focal length. Equation (11.41) can be written as

$$r'(1 + \cos\theta') = 2L_f \tag{11.42}$$

or, at the reflector edge,

$$r' = \frac{2L_f}{(1 + \cos\theta')} = L_f \sec^2\left(\frac{\theta'}{2}\right), \quad \theta' \leq \theta_0. \tag{11.43}$$

If the feed antenna is not isotropic, the effect of its normalized radiation pattern $F_f(\theta', \phi')$, using the coordinate system of Figure 11.4 can be included as

$$\bar{E}_a(\theta', \phi') = E_0 \frac{F_f(\theta', \phi')}{r'} \hat{u}_r \tag{11.44}$$

Aperture and Phased Array Antennas

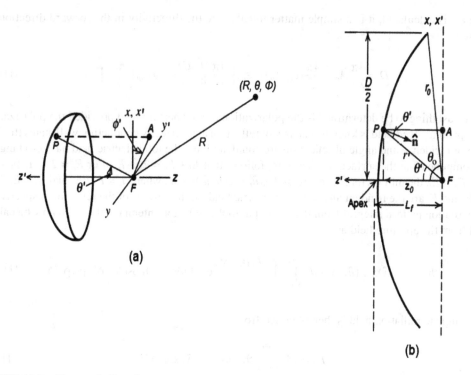

FIGURE 11.2 The parabolic reflector antenna: (a) parabolic reflector and coordinate system and (b) typical cross section (From Stutzman and Thiele, 1981).

where $\hat{u}_r$ is the unit vector of the aperture electric field, and the corresponding θ' and r' may be expressed, respectively, as

$$\theta' = 2\tan^{-1}\frac{r_0}{2L_f} \quad (11.45)$$

$$r' = \frac{4L_f^2 + r_0^2}{4L_f} \quad (11.46)$$

where r_0 denotes the distance of the rim from the focal point. There is another important expression relating the subtended angle θ_0 to the L_f/D ratio, where D is the diameter of the reflector. From the geometry of Figure 11.2(b), it can be shown that $\theta_0 = \tan^{-1}\left(\frac{D/2}{z_0}\right)$, where z_0 is the distance along z' axis from the focal point to the center of the paraboloid, which has the expression given by $z_0 = L_f - (D^2/16L_f)$. Then it follows that

$$\theta_0 = \tan^{-1}\left(\frac{D/2}{L_f - (D^2/16L_f)}\right) = \tan^{-1}\left(\frac{(1/2)(L_f/D)}{(L_f/D)^2 - (1/16)}\right). \quad (11.47)$$

Since directivity is perhaps the most important parameter for a reflector antenna, we shall start with the introduction of a parameter called the *aperture efficiency*, denoted by ε_{ap}. It is a measure of how efficiently the physical area A_p of the antenna is utilized. Balanis (1982) derived the expression for ε_{ap} as a function of the subtended angle θ_0 given by

$$\varepsilon_{ap} = 24\left\{\sin^2\left(\frac{\theta_0}{2}\right) + \ln\left[\cos\left(\frac{\theta_0}{2}\right)\right]\right\}^2 \cot^2\left(\frac{\theta_0}{2}\right). \quad (11.48)$$

208 Fundamental Principles of Radar

Once ε_{ap} is calculated, it is a simple matter to calculate the directivity in the forward direction as follows:

$$D = \frac{4\pi}{\lambda^2} A_e = \frac{4\pi}{\lambda^2}(\varepsilon_{ap} A_p) = \varepsilon_{ap}\left(\frac{4\pi}{\lambda^2}\right)\left(\frac{\pi D^2}{4}\right) = \varepsilon_{ap}\left(\frac{\pi D}{\lambda}\right)^2. \qquad (11.49)$$

The last thing to be determined is the polarization of the aperture field provided the polarization of the primary-feed field is known. The law of reflection at a perfectly conducting wall states that the angle of incidence and angle of reflection are equal, and that the total electric field has zero tangential components at the surface. It, therefore, follows that $\bar{E}_i + \bar{E}_r = 2(\hat{n} \cdot \bar{E}_i)\hat{n}$ or $\bar{E}_r = 2(\hat{n} \cdot \bar{E}_i)\hat{n} - \bar{E}_i$, since $E_r = E_i$, the unit vector $\hat{u}_r$ is expressed as $\hat{u}_r = 2(\hat{n} \cdot \hat{u}_i)\hat{n} - \hat{u}_i$, where $\hat{u}_i = \bar{E}_i / E_i$ and $\hat{u}_r = \bar{E}_r / E_r$, and $\bar{E}_i$ and $\bar{E}_r$ are the incident and reflected electric fields at the surface of the reflector, respectively. The radiation pattern integral from the entire parabolic reflector antenna system can now be calculated from the aperture field as

$$\bar{\mathfrak{R}} = [(\hat{u}_r \cdot \hat{x})\hat{x} + (\hat{u}_r \cdot \hat{y})\hat{y}]E_0 \int_0^{2\pi}\int_0^a \frac{F_f(\theta',\phi')}{r'}\exp[j\beta\rho\sin\theta\cos(\phi-\phi')]\rho\,d\rho\,d\phi'. \qquad (11.50)$$

The complete radiation field is then obtained from:

$$E_\theta = j\beta\frac{e^{-j\beta r}}{2\pi r}(\mathfrak{R}_x\cos\phi + \mathfrak{R}_y\sin\phi) \qquad (11.51)$$

and

$$E_\phi = j\beta\frac{e^{-j\beta r}}{2\pi r}\cos\theta(\mathfrak{R}_y\cos\phi - \mathfrak{R}_x\sin\phi) \qquad (11.52)$$

where $\mathfrak{R}_x$ and $\mathfrak{R}_y$ are the x and y components of the radiation integral $\vec{\mathfrak{R}}$.

Example 11.3

Consider a 28 GHz parabolic reflector antenna fed by a circular corrugated horn, positioned at the focal point. It is characterized by diameter $D = 1.219$ m, focal length/diameter, $L_f / D = 0.50$. Find

    a. the distance along the axis of the reflector from the focal point to the center of the paraboloid,
    b. the angle subtended by the edge of the rim with the axis of the reflector,
    c. the distance from the focal point to the edge of the rim,
    d. the aperture efficiency, and
    e. the overall directivity of the antenna.

Solution:

The focal length is obtained from $L_f / D = 0.5$ giving $L_f = (0.5)(1.219) = 0.609$ m. The wavelength is $\lambda = 3\times10^8 / (28\times10^9) = 0.0107$ m.

    a. The distance along the axis of the reflector from the focal point to the center of the paraboloid is obtained using the relation given by

$$z_0 = L_f - (D^2 / 16L_f) = 0.609 - [1.219^2 / (16\times0.609)] = 0.4565 \text{ m}.$$

b. The angle subtended at the rim is obtained using (11.47)

$$\theta_0 = \tan^{-1}\left(\frac{(1/2)(f/d)}{(f/d)^2 - (1/16)}\right) = \tan^{-1}\left(\frac{(0.5)(0.5)}{(0.5)^2 - (0.063)}\right) = 53.2°$$

c. The distance from the focal point to the edge of the rim is obtained from (11.43) as

$$r' = \frac{2L_f}{(1+\cos\theta')}$$

where $\theta' = \theta_0$ leads to $r' = r_0$ which, in turn, results in an expression given by

$$r_0 = \frac{2L_f}{(1+\cos\theta_0)}$$

It follows then

$$r_0 = \frac{(2)(0.609)}{[1+\cos(53.2°)]} = 0.76 \text{ m}.$$

d. The aperture efficiency is obtained using (11.48) as

$$\varepsilon_{ap} = 24\left\{\sin^2\left(\frac{\theta_0}{2}\right) + \ln\left[\cos\left(\frac{\theta_0}{2}\right)\right]\right\}^2 \cot^2\left(\frac{\theta_0}{2}\right)$$

$$= 24\left\{\sin^2(26.6°) + \ln[\cos(26.6°)]\right\}^2 \cot^2(26.6°) = 0.75 = 75\%.$$

e. The overall directivity of the antenna is obtained using (11.49) as

$$D = \varepsilon_{ap}\left(\frac{\pi D}{\lambda}\right)^2 = (0.75)\left(\frac{\pi(1.219)}{0.0107}\right)^2 = 96{,}072.9 = 49.82 \text{ dB}.$$

Cassegrain Reflector Systems

Dual-reflector systems may be utilized to improve the performance of large ground-based microwave reflector antennas for satellite tracking and communication. The arrangement is the *Cassegrain dual-reflector* system suggested by Hannan[3] in Figure 11.3, which was often utilized in the design of

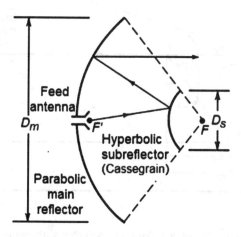

FIGURE 11.3 Schematic diagram of a Cassegrain reflector antenna.

optical telescopes and named after its inventor. To achieve the desired collimation characteristics, the main reflector must be a paraboloid, and the subreflector a hyperboloid. The use of the subreflector gives an additional degree of freedom for achieving good performance in different applications.

The basic operation of Cassegrain reflector antenna may be described by means of a schematic diagram in Figure 11.3. It consists of a feed horn, a subreflector, and a main reflector. The subreflector is situated between the focal point and the vertex of the main reflector with one of its foci coincident with the focus of the main reflector. The feeding horn is situated at the other focus of the subreflector, which also determines the overall focal length of the system. The subreflector is assumed to be in the far field of the feed system. Spherical waves emanating from the feeding system are transformed after a reflection by the subreflector into a second set of spherical waves. These appear to emanate from a virtual focus on the paraboloid reflector, which transforms them into plane wave propagating in the axial direction.

The concept of virtual feed is not useful in quantitative analysis, which can be circumvented by the concept of *equivalent parabola*. In this technique, the main reflector and the subreflector are replaced by an equivalent focusing surface at a certain distance from the real focal point. This surface is shown dashed in Figure 11.4, and it is defined (Stutzman and Thiele, 1981) as "the locus of intersection of incoming rays parallel to the antenna axis with the extension of the corresponding rays converging toward the real focal point." This equivalent system also reduces to a single-reflector arrangement.

The different parameters defined in the equivalent parabola in Figure 11.4 are related by the following expressions:

$$\tan\frac{\theta_r}{2} = \frac{1}{4}\frac{D_m}{L_m}, \qquad (11.53)$$

$$\frac{1}{\tan\theta_v} + \frac{1}{\tan\theta_r} = 2\frac{L_s}{D_s}. \qquad (11.54)$$

and the eccentricity is

$$e = \frac{\sin(\theta_v + \theta_r)/2}{\sin(\theta_0 - \theta_r)/2}. \qquad (11.55)$$

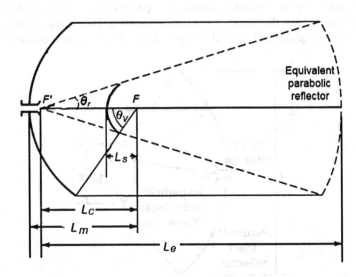

FIGURE 11.4 Equivalent parabola of a Cassegrain reflector with parameters (From Stutzman and Thiele, 1981).

Aperture and Phased Array Antennas

It is also true that

$$\frac{L_m}{D_m} = \frac{L_s}{D_s} \tag{11.56}$$

where L_m, L_s, D_m, and D_s are the focal lengths and diameters of the main and subreflector, respectively.

11.4 PHASED ARRAY ANTENNAS

The array antenna has unique characteristics that make it a candidate for consideration in radar applications because its properties are different from those of other microwave antennas. It has an aperture that is assembled from a great many similar radiating elements, such as slots or dipoles, each element being individually controlled in phase and amplitude. Accurately predictable radiation patterns and beam-pointing directions can be achieved. The array antenna has been used in many applications, especially in aircraft and satellite surveillance, ballistic missile and air defense, aircraft landing systems, tracking of ballistic missiles, and others.

The array antenna made up of individual radiating elements has a radiation pattern that is determined by the type of the individual elements used, their configurations, relative amplitude, and phases of the currents feeding them. For simplicity of discussion we begin by considering elements forming the array to be identical isotropic point sources. The resulting radiation patterns are called the *array factor*. The principle of *pattern multiplication* will then be applied for the purpose of including the effects of actual elements used in the array. In an array of identical elements, the following are the controls that can be used to shape the overall radiation pattern of the antenna:

a. the geometrical configurations,
b. the relative spacing between consecutive elements,
c. the amplitude excitation of the individual elements,
d. the phase excitation of the individual elements, and
e. the relative radiation pattern of the individual elements.

11.4.1 The Array Factor of Linear Array

Figure 11.5 represents a typical antenna array as a transmitting one for convenience but because of the reciprocity principle, the results obtained apply equally well to a receiving antenna array. The elements are assumed to be isotropic point sources radiating uniformly in all directions with equal

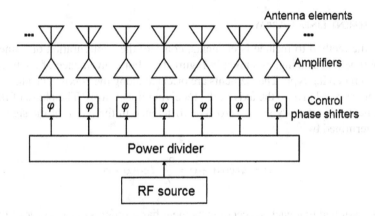

FIGURE 11.5 A typical linear array.

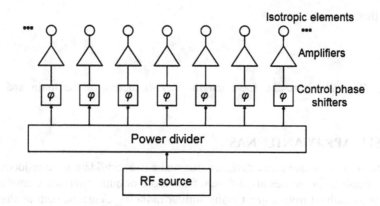

FIGURE 11.6 Equivalent configuration of the array in Figure 11.5.

amplitude and phase. Although isotropic elements are not realizable in practice, they are a useful concept in array theory. If each element of the array is replaced with an isotropic point source without disturbing the relative positions, the resulting pattern will be defined as the array factor. The effects of actual elements with the nonisotropic pattern can be determined by multiplying the array factor of the isotropic point sources by the radiation pattern of a single element. This is called *pattern multiplication*, and it applies only for arrays of identical elements.

The array factor for the array in Figure 11.5 is found from the array in Figure 11.6, which has point sources replacing the actual elements. The array factor for this transmitting array is the vector sum of all the transmitting point sources $\left\{ e^{j\xi_0}, e^{j\xi_1}, e^{j\xi_1}, \cdots \right\}$ in the array, weighted by the amplitude and phase shift $\left\{ I_0, I_1, I_2, \cdots \right\}$, introduced in the transmission line connected to each element. The array factor of the array shown in Figure 11.9 is then

$$AF = I_0 e^{j\xi_0} + I_1 e^{j\xi_1} + I_2 e^{j\xi_2} + \cdots I_2 e^{j\xi_{(N-1)}} \tag{11.57}$$

where ξ_m are the phases of an incoming plane wave at the element locations $m = 0, 1, 2, \ldots$, referenced to some point such as the origin. Hence, the phase of the wave arriving at element m leads the phase of the wave arriving at the origin by ξ_m. The simplest and one of the most practical arrays is formed by placing the elements along a line to form an array called *linear array*. The analysis of an N-element array is in order now. The general array factor for the case of equally spaced arrays will be considered.

11.4.2 N-Element Linear Array

To generalize the method to include N elements, consider the configuration of a linear array with uniform amplitude and spacing, as shown in Figure 11.7. For convenience of analysis, the array is assumed to be a receiving type. The elements are oriented along the z-axis. The angle θ is that of an incoming plane wave relative to the axis of the receiving linear array. The phase of the wave arriving at the element 1 is arbitrary set to zero. The difference in the phase of the signals in adjacent elements is determined by

$$\psi = \beta d \cos\theta + \varphi = \frac{2\pi}{\lambda} d \cos\theta + \varphi. \tag{11.58}$$

In (11.58), it is assumed that each succeeding element has a progressive phase lead of φ relative to the preceding one implying that each element is excited with progressive phase shift to steer the

Aperture and Phased Array Antennas

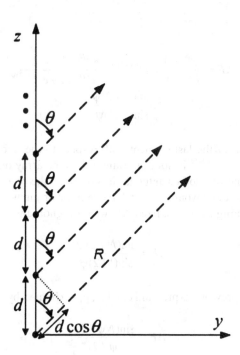

FIGURE 11.7 Far-field geometry of N-element array isotropic point sources.

beam in any direction. It is further assumed that the amplitudes and the phases of the signals at each element are weighted uniformly. An array of identical elements with identical magnitudes and with a progressive phase is called a *uniform array*. The AF of the uniform array can be obtained by considering the individual elements as point (isotropic) sources. Then, the total field pattern can be obtained by simply multiplying the AF by the normalized field pattern of the individual element (provided the elements are not coupled). The amplitudes of the signals at each element are the same and, for simplicity and convenience, will be taken to be unity. The array factor representing the sum of all the signals from the individual elements—when the total phase difference ψ consisting of the phase introduced due to the spacing between elements and the progressive phase shift introduced intentionally—can be written as

$$AF = 1 + e^{j\psi} + e^{j2\psi} + e^{j3\psi} + \cdots + e^{j(N-1)\psi}. \tag{11.59}$$

Or, in a compact form, it can be written as

$$AF = \sum_{n=1}^{N} e^{j(n-1)\psi}. \tag{11.60}$$

Multiplying both side of (11.59) by $e^{j\psi}$ yields

$$(AF)e^{j\psi} = e^{j\psi} + e^{j2\psi} + e^{j3\psi} + \cdots + e^{j(N-1)\psi} + e^{jN\psi}. \tag{11.61}$$

Using (11.59) and (11.61), we get

$$AF(e^{j\psi} - 1) = (e^{jN\psi} - 1) \tag{11.62}$$

214 Fundamental Principles of Radar

which can be written as

$$AF = \frac{e^{jN\psi} - 1}{e^{j\psi} - 1} = e^{j(N-1)\psi/2} \frac{e^{j(N/2)\psi} - e^{-j(N/2)\psi}}{e^{(j1/2)\psi} - e^{-j(1/2)\psi}}$$

$$= e^{j(N-1)\psi/2} \cdot \frac{\sin(N/2)\psi}{\sin(1/2)\psi}.$$

(11.63)

Here, N shows the location of the last element with respect to the reference point in steps of the length d. The phase factor $e^{j(N-1)\psi/2}$ is not important unless the array output signal is further combined with the output signal of another antenna. It represents the phase shift of the array's phase center relative to the origin, and it would be equal to one if the origin were to coincide with the array center. As we aim at obtaining the normalized AF, we will ignore the phase factor, which gives

$$AF = \frac{\sin(N/2)\psi}{\sin(1/2)\psi}.$$

(11.64)

For small values of ψ, the previous expression can be approximated by

$$AF = \frac{\sin(N\psi/2)}{\psi/2}.$$

(11.65)

The maximum value of AF occurs when $\psi = 0$, resulting in $AF = N$. Therefore, disregarding the phase factor and normalizing we obtain

$$(AF)_n = \frac{1}{N} \left[\frac{\sin(N\psi/2)}{\psi/2} \right] = \frac{\sin(N\psi/2)}{(N\psi)/2}.$$

(11.66)

Let $(AF)_n$ be denoted by $f(\psi)$. It follows then

$$f(\psi) = \frac{\sin(N\psi/2)}{(N\psi)/2}.$$

(11.67)

This is the normalized array factor for an N element, uniformly excited, equally spaced array, which is centered about the origin. The array factor is periodic in 2π, which is true in general. Evaluating $|f(\psi)|$ for various values of N yields the curves shown in Figure 11.8. A number of features are observed:

a. As N increase the width of the main lobe decreases.
b. As N increases there are more sidelobes in one period of $f(\psi)$. There are one main lobe and $(N - 2)$ sidelobes in one period.
c. The sidelobe peaks decrease with increasing N. The first sidelobe level of the array factor approaches −13.3 as N increased.
d. $|f(\psi)|$ is symmetric about π.

We can also illustrate the procedure (Stutzman and Thiele, 1981) for obtaining the polar plot of the generalized array factor $|f(\psi)|$ as a function of polar angle θ as shown in Figure 11.9. Below it a circle is constructed with a radius equal to βd and its center located at $\psi = \varphi$. For a given value of θ, locate the intersection of a radial line from the origin of the circle and the perimeter point at point a. The corresponding value of ψ, at point b, is on a vertical line from a. The array factor corresponding

Aperture and Phased Array Antennas

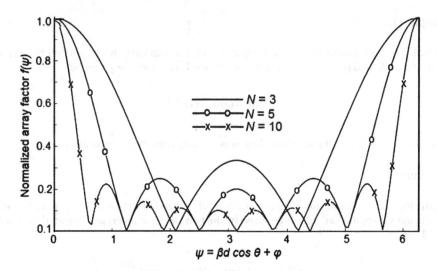

FIGURE 11.8 Array factors for different values of an array element.

to these values of ψ and θ is then c, also on the vertical line from a. Note that the distance from the $\psi = 0$ axis to point a can be written as $\psi = \beta d \cos\theta + \varphi$. The array factor has the rotational symmetry about the line of the array. Therefore, its complete structure is determined by its values for $0 < \theta < \pi$, which is designated as a *visible region*. Exactly one period of the array factor appears in the visible region for $d = \lambda/2$.

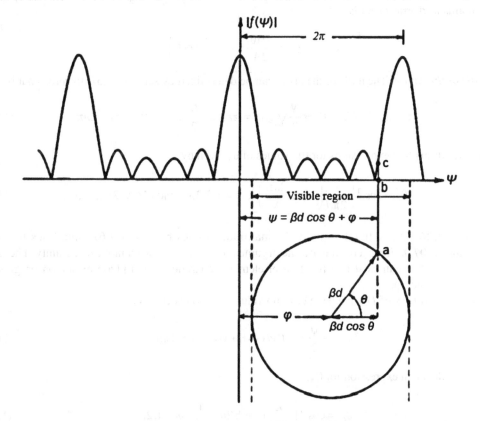

FIGURE 11.9 Construction techniques for plotting the array factor (From Stutzman and Thiele, 1981).

Example 11.4

Two isotropic antennas in an array are separated by half wavelength and excited by currents of equal amplitude and phase. Show that the normalized array factor is given by

$$f(\theta) = \cos\left(\frac{\pi}{2}\cos\theta\right)$$

where θ is the angle that the incident plane wave makes with the antenna axis.

Solution:

Assume that the center of the array is symmetrically placed at the origin. The array factor AF is equal to the sum of the two signals received by the two isotropic antennas separated by a distance $\lambda / 2$. Thus

$$AF = Ae^{-j\beta(d/2)\cos\theta} + Ae^{j\beta(d/2)\cos\theta} = 2A\cos\left(\frac{\beta d}{2}\cos\theta\right)$$

where $\beta d = \left(\frac{2\pi}{\lambda}\right)\left(\frac{\lambda}{2}\right) = \pi$, which gives

$$AF(\theta) = 2A\cos\left(\frac{\pi}{2}\cos\theta\right).$$

The maximum value of the array factor occurs at $\theta = \pi / 2$, and is given by $2A$. Therefore, the normalized array factor is

$$f(\theta) = \frac{AF(\theta)}{2A} = \cos\left(\frac{\pi}{2}\cos\theta\right).$$

Nulls of the Array: The nulls of the array can be determined by setting (11.67) to zero. That is,

$$\sin(N\psi / 2) = 0 \Rightarrow \frac{N}{2}\psi\Big|_{\theta=\theta_n} = \pm n\pi \Rightarrow \frac{N}{2}(\beta d \cos\theta_n + \varphi) = \pm n\pi. \tag{11.68}$$

This yields explicit expressions for the values of θ_n given by

$$\theta_n = \cos^{-1}\left[\frac{\lambda}{2\pi d}\left(-\varphi \pm \frac{2n\pi}{N}\right)\right], \quad n = 1,2,3,\cdots \text{ and } n \neq N, 2N, 3N, \cdots. \tag{11.69}$$

For $n \neq N, 2N, 3N, \cdots$, the AF attains its maximum values because (11.67) simplifies to a *sinc function* $\sin 0 / 0$. For a zero to exist, the argument of the arccosine cannot exceed unity. The number of nulls that can exist is a function of element separation d and phase excitation progressive phase difference φ.

Maximum Values of $f(\psi)$: The maximum values of $f(\psi)$ occur when

$$\frac{\psi}{2} = \frac{1}{2}(\beta d \cos\theta + \varphi)\Big|_{\theta=\theta_m} = \pm m\pi. \tag{11.70}$$

This results in an expression for θ_m:

$$\theta_m = \cos^{-1}\left[\frac{\lambda}{2\pi d}(-\varphi \pm 2m\pi)\right], \quad m = 1,2,3,\cdots. \tag{11.71}$$

Aperture and Phased Array Antennas

Thus the array factor $f(\psi)$ has only one maximum, and occurs when $m = 0$. It follows

$$\theta_m = \cos^{-1}\left(-\frac{\lambda\varphi}{2\pi d}\right). \tag{11.72}$$

3-dB Beamwidth of the Main Lobe: The half-power or 3-dB beamwidth of the array factor $f(\psi)$ occurs when

$$\frac{N}{2}\psi = \frac{N}{2}(\beta d\cos\theta + \varphi)\big|_{\theta=\theta_h} = \pm 1.391. \tag{11.73}$$

This results in an expression for θ_h

$$\theta_h = \cos^{-1}\left[\frac{\lambda}{2\pi d}\left(-\varphi \pm \frac{2.782}{N}\right)\right]. \tag{11.74}$$

For a symmetrical pattern around θ_m (the angle at which maximum radiation occurs), the half-power beamwidth *HPBW* is calculated as

$$HPBW = 2(\theta_m - \theta_h). \tag{11.75}$$

Maxima of Minor Lobes (Secondary Maxima): They are the maxima of $f(\psi)$, where $f(\psi) < 1$. These are illustrated in the plot of Figure 11.8, which shows the array factors as a function of $\psi = \beta d\cos\theta + \varphi$ for a uniform equally spaced linear array with $N = 3, 5, 10$. The secondary maxima occur where the numerator attains a maximum and the $f(\psi)$ is beyond its first null:

$$\sin\left(\frac{N}{2}\psi\right) = \sin\left[\frac{N}{2}(\beta d\cos\theta + \varphi)\right]\Big|_{\theta=\theta_s} = \pm 1. \tag{11.76}$$

In other words,

$$\frac{N}{2}(\beta d\cos\theta)\big|_{\theta=\theta_s} = \pm\left(\frac{2s+1}{2}\right)\pi. \tag{11.77}$$

Therefore, it follows

$$\theta_s = \cos^{-1}\left\{\frac{\lambda}{2\pi d}\left[-\varphi \pm \left(\frac{2s+1}{N}\right)\pi\right]\right\}, \quad s = 1, 2, 3, \cdots. \tag{11.78}$$

The maximum of the first minor lobe of $f(\psi)$ occurs approximately when

$$\frac{N}{2}(\beta d\cos\theta)\big|_{\theta=\theta_s} \approx \left(\frac{3\pi}{2}\right) \tag{11.79}$$

or when

$$\theta_s = \cos^{-1}\left\{\frac{\lambda}{2\pi d}\left[-\varphi \pm \frac{3\pi}{N}\right]\right\}. \tag{11.80}$$

218   Fundamental Principles of Radar

At that angle, the magnitude $\left| f(\psi) \right|$ reduces to

$$\left| f(\psi) \right| = \left[\frac{\sin\left(\dfrac{N}{2}\right)\psi}{\dfrac{N}{2}\psi} \right]_{\substack{\theta=\theta_s \\ s=1}} = \frac{2}{3\pi} = 0.212 \tag{11.81}$$

which in dB is equal to

$$f(\psi) = 20 \log_{10}\left(\frac{2}{3\pi}\right) = -13.46 \text{ dB.} \tag{11.82}$$

Broadside Array

A broadside array is an array that has maximum radiation at $\theta = 90°$ (normal to the axis of the array) of Figure 11.7. For optimal solution, both the element factor and the $f(\psi)$, should have their maxima at $\theta = 90°$. From (11.67), it follows that the maximum of the $f(\psi)$ occurs when

$$\psi = \beta d \cos\theta + \varphi = 0. \tag{11.83}$$

For broadside radiation, we have then

$$\psi = (\beta d \cos\theta + \varphi)|_{\theta=90°} = \varphi = 0. \tag{11.84}$$

The uniform linear array has its maximum radiation at $\theta = 90°$, if all array elements have their excitation with the same phase. To ensure that there are no maxima in other directions (grating lobes), the separation between the elements d must be smaller than the wavelength: $d < \lambda$. The expressions for the nulls, maxima, and half-power points for the broadside array can be derived from (11.66)–(11.74), and are listed in Table 11.1.

Ordinary End-Fire Array

An end-fire array is an array that has its maximum radiation along the axis of the array ($\theta = 0°$ or $180°$). It may be required that the array radiates only in one direction (either $\theta = 0°$ or $\theta = 180°$). For a maximum of $f(\psi)$ at $\theta = 0°$,

$$\psi = (\beta d \cos\theta + \varphi)|_{\theta=0°} = \beta d + \varphi = 0 \Rightarrow \varphi = -\beta d. \tag{11.85}$$

TABLE 11.1
Null, Main Lobe Maxima, Half-Power Points, and Minor Lobe Maxima for a Broadside Array

Parameters	Expressions
Nulls	$\theta_n = \cos^{-1}\left(\pm\dfrac{n\lambda}{Nd}\right), \quad n = 1, 2, 3,\cdots \text{ and } \quad n \neq N,\ 2N,\ 3N,\cdots$
Maxima	$\theta_m = \cos^{-1}\left(\pm\dfrac{m\lambda}{d}\right), \quad m = 0, 1, 2, 3,\ \cdots$
Half-power points	$\theta_h = \cos^{-1}\left(\pm\dfrac{1.39\lambda}{\pi Nd}\right), \quad \pi d / \lambda \ll 1$
Minor lobe maxima	$\theta_s = \cos^{-1}\left[\pm\dfrac{\lambda}{2d}\left(\dfrac{2s+1}{N}\right)\right], \quad s = 1, 2, 3,\cdots, \quad \pi d / \lambda \ll 1$

Aperture and Phased Array Antennas

TABLE 11.2

Null, Main Lobe Maxima, Half-Power Points, and Minor Lobe Maxima for an Ordinary End-Fire Array

Parameters	Expressions
Nulls	$\theta_n = \cos^{-1}\left(1 - \dfrac{n\lambda}{Nd}\right)$, $\quad n = 1,\ 2,\ 3, \cdots$ and $n \neq N,\ 2N,\ 3N, \cdots$
Maxima	$\theta_m = \cos^{-1}\left(1 - \dfrac{m\lambda}{d}\right)$, $\quad m = 0,\ 1,\ 2,\ 3,\ \cdots$
Half-power points	$\theta_h = \cos^{-1}\left(1 - \dfrac{1.39\lambda}{\pi Nd}\right)$, $\quad \pi d / \lambda \ll 1$
Minor lobe maxima	$\theta_h = \cos^{-1}\left(1 - \dfrac{(2s+1)\lambda}{2Nd}\right)$, $\quad s = 1,\ 2,\ 3,\ \cdots,\quad \pi d / \lambda \ll 1$

For the maximum radiation toward $\theta = 180°$

$$\psi = (\beta d \cos\theta + \varphi)\big|_{\theta=180°} = -\beta d + \varphi = 0 \Rightarrow \varphi = \beta d. \tag{11.86}$$

If the element separation is multiple of a wavelength, $(d = n\lambda,\ n = 1,2,3,\cdots)$, then in addition to having the end-fire maxima, there also exist maxima in the broadside directions. In addition, the end-fire radiation is directed toward both directions of the axis of the array. Thus for $d = n\lambda,\ n = 1,2,3,\cdots$, there exist four maxima; two in the broadside directions and two along the axis of the array. As with the broadside array, to avoid grating lobes, the maximum spacing between the element should be less than λ, that is, $d < \lambda$. The expressions for the nulls, maxima, and half-power points for the ordinary end-fire array can be derived from (11.66)–(11.74), and are listed in Table 11.2.

Phased Arrays

An array antenna, whose direction of the main beam is controlled primarily by the relative phase of the element excitation current, is referred to as a phased array.[4] Phased arrays are finding increasing applications in radar where extremely fast tracking is required and in communications where the radiation pattern must be adjusted to accommodate varying traffic conditions. We have seen how to direct the major radiation from an array, by controlling the phase excitation between elements in the broadside and end-fire directions of the array. Now this concept can be extended logically to design a scanning array to steer the main beam in any desired direction just by electronically controlling the progressive phase shift φ. More precisely,

$$\psi = (\beta d \cos\theta + \varphi)\big|_{\theta=\theta_0} = \beta d \cos\theta_0 + \varphi = 0 \Rightarrow \varphi = -\beta d \cos\theta_0. \tag{11.87}$$

Thus by controlling the phase shift φ with time according to (11.87), the maximum radiation can be squinted with time in any desired direction. This is the basic principle of electronic scanning phased array operation. When the scanning is required to be continuous, the feeding system must be capable of continuously varying the progressive phase φ between the elements. This is accomplished by ferrite phase shifters. The phase shift is controlled by the magnetic field within the ferrite, which in turn is controlled by the amount of current flowing through the wires wrapped around the phase shifter.

Example 11.5

An array consists of 12 isotropic elements placed along the z-axis with a spacing of a quarter wavelength between the elements. If it is desired to steer the main beam at 60° with the z-axis, find the corresponding progressive phase shift to be introduced, first null beamwidth, and half-power beamwidth, position of the maximum of the first minor lobe, and verify the location of the maximum value of the radiation pattern.

Solution:

The progressive phase shift φ can be obtained using (11.87):

$$\psi = (\beta d \cos\theta + \varphi)|_{\theta=\theta_0} = \beta d \cos\theta_0 + \varphi = 0 \Rightarrow \varphi = -\beta d \cos\theta_0$$

It follows

$$\varphi = -\left(\frac{2\pi}{\lambda}\right)\left(\frac{\lambda}{4}\right)\cos 60° = -\frac{\pi}{2}(0.5) = -0.785 \text{ radian} = -45°$$

The first null beamwidth is obtained using (11.69)

$$\theta_n = \cos^{-1}\left[\frac{\lambda}{2\pi d}\left(-\varphi \pm \frac{2\pi}{N}\right)\right] = \cos^{-1}\left[\frac{4\lambda}{2\pi\lambda}\left(0.785 \pm \frac{2\pi}{12}\right)\right]$$

$$= \cos^{-1}[0.637(0.785 \pm 0.524)]$$

$$\theta_n = \begin{cases} 33.50° \\ 80.43° \end{cases} \Rightarrow FNBW = (80.43 - 33.50) = 46.93°$$

The half-power beam width is obtained using (11.74)

$$\theta_h = \cos^{-1}\left[\frac{\lambda}{2\pi d}\left(-\varphi \pm \frac{2.782}{N}\right)\right] = \cos^{-1}\left[\frac{4\lambda}{2\pi\lambda}\left(0.785 \pm \frac{2.782}{12}\right)\right]$$

$$= \cos^{-1}\left[\frac{4\lambda}{2\pi\lambda}(0.785 \pm 0.232)\right] = \cos^{-1}[0.637(0.785 \pm 0.232)]$$

$$\theta_h = \begin{cases} 49.62° \\ 69.37° \end{cases} \Rightarrow HPBW = (69.37 - 49.62) = 19.75°$$

The position of the maximum of the first minor lobe is obtained using (11.80):

$$\theta_s = \cos^{-1}\left[\frac{\lambda}{2\pi d}\left(-\varphi \pm \frac{3\pi}{N}\right)\right] = \cos^{-1}\left[\frac{4\lambda}{2\pi\lambda}\left(0.785 \pm \frac{3\pi}{12}\right)\right]$$

$$= \cos^{-1}\left[0.637(0.785 \pm 0.785)\right]$$

$$\theta_s = \begin{cases} 0° \\ 90° \end{cases} \Rightarrow \theta_s = 90°$$

Since the + sign results in an argument of arccosine greater than unity indicates that the solution $\theta_s = 0°$ is not a feasible solution. Thus the first minor lobe maximum occurs at 90° as expected. The position of the maximum value is obtained using (7.72)

$$\theta_{max} = \cos^{-1}\left(\frac{-\lambda\varphi}{2\pi d}\right) = \cos^{-1}(-(0.637)(-0.785)) = \cos^{-1}(0.5) = 60° : \text{ verified}$$

11.4.3 PLANAR ARRAY

A *planar array* is an antenna in which all the elements, both active and parasitic, are in one plane. It provides a large aperture and may be used for directional beam control by varying the relative phase of each element. Planar arrays are more versatile; they provide more symmetrical patterns with lower sidelobes and much higher directivity (narrow main beam). They can be used to scan the main beam toward any point in space. It finds wide applications in tracking radars, remote sensing, communication systems, etc.

The array factor can be derived by referring to Figure 11.10 where $M \times N$ elements are considered to be placed along the x- and y-axis, respectively. There are M elements along any column in the y-direction and N elements along any row along the x-direction. The total number of elements is $M \times N$. The elements are placed on a rectangular lattice, d_x and d_y being the separations between any adjacent pair of elements in the x- and y-directions, respectively. It is convenient to define a direction cosine as the cosine of the angle from a coordinate axis to the line defining any arbitrary direction (θ, ϕ) in space. The two direction cosines, $\cos(\gamma_x)$ and $\cos(\gamma_y)$, are given by

$$\cos(\gamma_x) = \hat{x} \cdot \hat{R} = \sin\theta \cos\phi, \tag{11.88}$$

$$\cos(\gamma_y) = \hat{y} \cdot \hat{R} = \sin\theta \sin\phi. \tag{11.89}$$

The array factor AF of a linear array of M elements along the x-axis is:

$$(AF)_{xm} = \sum_{m=1}^{M} I_{m1} e^{j(m-1)(\beta d_x \sin\theta\cos\phi + \varphi_x)} \tag{11.90}$$

where I_{m1} denotes the excitation amplitude of the element at the point with coordinates: $x = (m-1)d_x$, $y = 0$. In (11.90) this is the element of the mth row and the first column of the array matrix. φ_x is the progressive phase shift along each column. If N such arrays are placed next to each other in the y-direction, with a progressive phase φ_y, a rectangular array will be formed as shown in Figure 11.10. It will also be assumed that the normalized current distribution along each of the

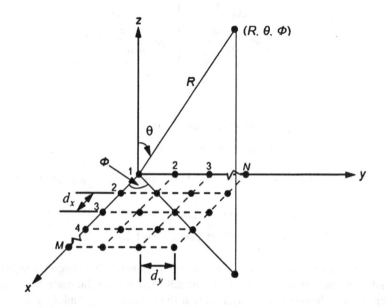

FIGURE 11.10 A planar array geometry.

y-directed arrays is the same, but the absolute values correspond to a factor of I_{1n} ($n = 1, 2, 3, \cdots, N$). Then the array factor of the entire array will be:

$$AF(\theta, \phi) = \sum_{n=1}^{N} I_{1n} \left[\sum_{m=1}^{M} I_{m1} e^{j(m-1)(\beta d_x \sin\theta\cos\phi + \phi_x)} \right] e^{j(n-1)(\beta d_y \sin\theta\sin\phi + \phi_y)} \tag{11.91}$$

Applying (11.86) and (11.87) into (11.89), we get

$$AF(\theta, \phi) = \sum_{n=1}^{N} I_{1n} \left[\sum_{m=1}^{M} I_{m1} e^{j(m-1)(\beta d_x \cos\gamma_x + \phi_x)} \right] e^{j(n-1)(\beta d_y \cos\gamma_y + \phi_y)}. \tag{11.92}$$

or

$$AF(\theta, \phi) = S_{xm}(\gamma_x) \cdot S_{yn}(\gamma_y) \tag{11.93}$$

where

$$S_{xm}(\gamma_x) = \sum_{m=1}^{M} I_{m1} e^{j(m-1)(\beta d_x \cos\gamma_x + \phi_x)}. \tag{11.94}$$

$$S_{yn}(\gamma_y) = \sum_{n=1}^{N} I_{1n} e^{j(n-1)(\beta d_y \cos\gamma_y + \phi_y)}. \tag{11.95}$$

The pattern of a rectangular array is the product of the array factors of the linear arrays in the x- and y-directions. The amplitude of the (m, n)th element can be written as $I_{mn} = I_{m1} I_{1n}$. For a uniform planar array $I_{mn} = I_0$ for all m and n, that is, all elements have the same excitation amplitudes. The normalized form of (11.92) can be written as

$$F(\theta, \phi) = \left\{ \frac{1}{M} \frac{\sin\left(\frac{M}{2}\psi_x\right)}{\sin\left(\frac{\psi_x}{2}\right)} \right\} \left\{ \frac{1}{N} \frac{\sin\left(\frac{N}{2}\psi_y\right)}{\sin\left(\frac{\psi_y}{2}\right)} \right\} \tag{11.96}$$

where

$$\psi_x = \beta d_x \cos\gamma_x + \phi_x \tag{11.97}$$

$$\psi_y = \beta d_y \cos\gamma_y + \phi_y. \tag{11.98}$$

The major lobe and grating lobes of S_{xm} and S_{yn} in (11.94) and (11.95) are located at

$$\beta d_x \cos\gamma_x + \phi_x = \pm 2m\pi, \quad m = 1, 2, \cdots \tag{11.99}$$

$$\beta d_y \cos\gamma_y + \phi_y = \pm 2n\pi, \quad n = 1, 2, \cdots. \tag{11.100}$$

The principal maxima correspond to $m = 0$ and $n = 0$. In general, the progressive phases ϕ_x and ϕ_y are independent of each other, and they can be adjusted so that the main beam of S_{xm} is not the same as that of S_{yn}. However, it is required that the beams of S_{xm} and S_{yn} intersect and their maxima be directed toward the same direction. If it is desired to have one main beam that is

Aperture and Phased Array Antennas

directed along $\theta = \theta_0$ and $\Phi = \Phi_0$, the progressive phase shift between the elements in the x- and y-directions must be

$$\varphi_x = -\beta d_x \cos\gamma_{xo} = -\beta d_x \sin\theta_0 \cos\phi_0 \tag{11.101}$$

$$\varphi_y = -\beta d_y \cos\gamma_{yo} = -\beta d_y \sin\theta_0 \sin\phi_0. \tag{11.102}$$

The explicit expressions for θ_0 and ϕ_0 can be obtained by simultaneously solving (11.101) and (11.102), and are given by

$$\phi_0 = \arctan\left(\frac{\varphi_y d_x}{\varphi_x d_y}\right) \tag{11.103}$$

$$\theta_0 = \arcsin\sqrt{\left(\frac{\varphi_x}{\beta d_x}\right)^2 + \left(\frac{\varphi_y}{\beta d_y}\right)^2}. \tag{11.104}$$

The principal maxima $(m = n = 0)$ and the grating lobes are located by

$$\begin{aligned} \beta d_x(\sin\theta\cos\phi - \sin\theta_0\cos\phi_0) &= \pm 2m\pi, \quad m = 0,\, 1,\, 2,\cdots \\ \beta d_y(\sin\theta\sin\phi - \sin\theta_0\sin\phi_0) &= \pm 2n\pi, \quad n = 0,\, 1,\, 2,\cdots. \end{aligned} \tag{11.105}$$

or

$$\begin{aligned} \sin\theta\cos\phi - \sin\theta_0\cos\phi_0 &= \pm\frac{m\lambda}{d_x}, \quad m = 0,\, 1,\, 2,\cdots \\ \sin\theta\sin\phi - \sin\theta_0\sin\phi_0 &= \pm\frac{n\lambda}{d_y}, \quad n = 0,\, 1,\, 2,\cdots. \end{aligned} \tag{11.106}$$

The values of θ and ϕ can be extracted from (11.106) as

$$\theta = \sin^{-1}\left[\frac{\sin\theta_0\cos\phi_0 \pm m\lambda/d_x}{\cos\phi}\right] = \sin^{-1}\left[\frac{\sin\theta_0\sin\phi_0 \pm n\lambda/d_y}{\sin\phi}\right] \tag{11.107}$$

$$\phi = \tan^{-1}\left[\frac{\sin\theta_0\sin\phi_0 \pm n\lambda/d_y}{\sin\theta_0\cos\phi_0 \pm m\lambda/d_x}\right]. \tag{11.108}$$

The principle of planar array theory can be demonstrated by considering the three-dimensional pattern (Balanis 1982) of a 15×15 elements array of uniform amplitude, $\varphi_x = \varphi_y = 0$ and $d_x = d_y = \lambda/2$ is displayed in Figure 11.11.

11.4.4 Circular Array

In a circular array, the elements are placed in a circular ring. It has wide applications that span radio direction finding, air and space navigation, radar, sonar, and many other systems.

Referring to Figure 11.12, let us assume that N isotropic elements are equally spaced in the x-y plane along a circular ring of radius a. The field of the circular array can be written as

$$E(R,\theta,\phi) = \sum_{n=1}^{N} a_n \frac{e^{-j\beta R_n}}{R_n} \tag{11.109}$$

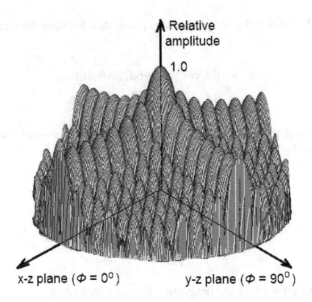

FIGURE 11.11 Three-dimensional antenna pattern of a planar array of isotropic elements.

where R_n is the distance from the nth element to the field point and a_n is the excitation coefficient of the nth element. Assuming that $R \gg a$, $R_n \approx R$ for the far-field pattern, and that the angular position of the nth element $\phi_n = 2\pi n / N$, (11.109) becomes

$$E(R,\theta,\phi) = \frac{e^{-j\beta R}}{R} \sum_{n=1}^{N} a_n e^{j\beta a \sin\theta \cos(\phi-\phi_n)}. \qquad (11.110)$$

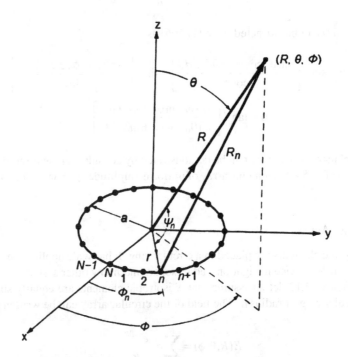

FIGURE 11.12 Geometry of an N-element circular array.

Aperture and Phased Array Antennas

In general, the excitation of the nth element can be written as

$$a_n = I_n e^{j\varphi_n} \tag{11.111}$$

where I_n and φ_n are the amplitude and phase excitations of the nth element.

Combining (11.110) and (11.111) gives

$$E(R,\theta,\phi) = \frac{e^{-j\beta R}}{R}\left[AF(\theta,\phi)\right] \tag{11.112}$$

where

$$AF(\theta,\phi) = \sum_{n=1}^{N} I_n e^{\left[j\beta a \sin\theta \cos(\phi-\phi_n)+\varphi_n\right]}. \tag{11.113}$$

Equation (11.113) represents the array factor of a circular array of N equally spaced elements. For the main beam to be directed in the (θ_0,ϕ_0) direction, the phase excitation of the nth element can be selected as

$$\varphi_n = -\beta a \sin\theta_0 \cos(\phi_0 - \phi_n). \tag{11.114}$$

Then the array factor of (11.113) can be written as

$$AF(\theta,\phi) = \sum_{n=1}^{N} I_n e^{j\beta a\left[\sin\theta \cos(\phi-\phi_n)-\sin\theta_0 \cos(\phi_0-\phi_n)\right]}. \tag{11.115}$$

Now, let the direction cosines be defined as

$$\begin{aligned}
\cos\gamma &= \sin\theta \cos(\phi - \phi_n) \\
\cos\gamma_0 &= \sin\theta_0 \cos(\phi_0 - \phi_n).
\end{aligned} \tag{11.116}$$

Applying (11.116) into (11.115) gives

$$AF(\theta,\phi) = \sum_{n=1}^{N} I_n e^{j\beta a(\cos\gamma-\cos\gamma_0)}. \tag{11.117}$$

The array factor $AF(\theta,\phi)$ expressed by (11.117) can be used to compute the array factor, once N, I_n, a, θ_0, and ϕ_0 are specified. This is usually very time-consuming even for a moderately large number of elements comprising the circular array.

11.4.5 CONFORMAL ARRAY

A conformal array antenna is the one that conforms to a prescribed geometry of a nonplanar surface to achieve a directive beam, with reasonably low sidelobes and good efficiency. The shape can be some part of an airplane, high-speed train, or other vehicle. Usually, a conformal array is cylindrical, spherical, or some other shape, with the radiating elements mounted on or integrated into the smoothly curved surface. Many variations exist like approximating the smooth surface by several planar facets. The purpose is to build the antenna so that it becomes integrated with the structure, and makes the antenna less problematic and less visible.

Conformal antennas are a form of phased array antennas. It is composed of an array of many identical small flat antenna elements, such as dipole and horn, to essentially cover the nonplanar surface. In principle the elements of an array are made to radiate a beam in some desired direction by applying the proper phase, amplitude, and polarization at each element. The mechanism for feeding the elements and steering the beam of a conformal array are generally more complicated than those of a planar array. The phase shifters also compensate for the different phase shifts caused by the varying path lengths of the radio waves due to the location of the individual antennas on the curved surface. Because the individual antenna elements must be small, conformal arrays are typically limited to high frequencies in the UHF or microwave range.

Conformal antenna arrays were developed in the 1980s by integrating them onto the curving skin of military aircraft to reduce aerodynamic drag, replacing conventional antenna designs that project from the aircraft surface.[5] Military aircrafts and missiles are the largest application of conformal antennas, but they are also used in some civilian aircrafts, military ships, and land vehicles. As the cost of the required processing technology comes down, they are being considered for civilian applications such as train antennas, car radio antennas, and cellular base station antennas, to save space and also to make the antenna less visually intrusive by integrating it into existing objects. As such, conformal arrays are used for flush-mounted airborne radar antennas, and air traffic control interrogation antennas.

The complications that arise when dealing with a general nonplanar surface have restricted its consideration to relatively simple shapes, such as the cylinder, cone, ogive, and sphere. The cylinder has geometry suitable for arrays that can scan 360° in azimuth. The circular symmetry of the cylinder ensures that the mutual coupling between elements is always the same as the beam scan in azimuth. The ogive where the radiating elements are arranged along the nose of an aircraft bears a resemblance to the cone. This would allow coverage, provide a good aerodynamic shape, and permit larger antenna apertures than the conventional nose-antenna configuration.

PROBLEMS

11.1 Find the total radiated power if the radiated power density of an antenna is given by $\hat{P} = \hat{R}(10\sin\theta / R^2)$, W/m^2, where $\hat{R}$ is the unit vector in the direction of R.

11.2 Repeat Problem 11.1 to find the radiation intensity and hence the total radiated power of the antenna.

11.3 Repeat Problem 11.1 to find the directivity of the antenna.

11.4 The radiation intensity of a resonant half-wavelength dipole antenna is $P(\theta,\phi) = 15\sin^3\theta$ W. Find the directivity of the antenna.

11.5 Consider a 2.4 m diameter reflector antenna with a focal length to diameter ratio of $L_f / D = 0.6$ operating at a frequency of 6 GHz. Find the aperture efficiency.

11.6 For the reflector antenna of Problem 11.6, find the distance from the focal point to the edge of the rim.

11.7 For the reflector antenna of Problem 11.6, find the overall directivity of the antenna.

11.8 A 10 m diameter reflector antenna operating at 600 MHz has a directivity of 35 dB. Find the aperture efficiency and the effective aperture.

11.9 Consider Example 11.4 except for the sources with equal amplitudes and opposite phases, and show that the normalized array factor is given by

$$f(\theta) = \sin\left(\frac{\pi}{2}\right)\cos\theta.$$

11.10 Consider Example 11.4 except for the sources with equal amplitudes and 90° out of phase, and show that the normalized array factor is given by

$$f(\theta) = \cos\left[\frac{\pi}{4}(\cos\theta - 1)\right].$$

Aperture and Phased Array Antennas

11.11 Consider Example 11.4 except for the spacing between the elements is one wavelength, and show that the normalized array factor is given by

$$f(\theta) = \cos(\pi \cos \theta).$$

11.12 Work Example 11.5 for a broadside array.

11.13 Work Example 11.5 for an ordinary end-fire array.

11.14 Find the HPBW, FNBW, and the position of the maximum of the first minor lobe of a 10-element uniform phased array consisting of isotropic sources placed on the z-axis. The spacing between the elements is $\lambda / 3$ and the main beam is steered at 75° from the z-axis.

REFERENCES

1. IEEE, "IEEE Standard Definitions of Terms for Antennas," *IEEE Transaction on Antenna and Propagation,* AP-22, no. 1 (January 1974).
2. C. A. Balanis, *Antenna Theory Analysis and Design* (New York: Harper & Row Publishers, 1982).
3. P. W. Hannan, "Microwave Antennas Derived from the Cassegrain Telescope," *IRE Transactions on Antennas and Propagation,* AP-9 (1961): 140–153.
4. L. Stark, "Microwave Theory of Phased Array Antennas—A Review," *Proceedings of the IEEE,* 62 (1974): 1661–1701.
5. L. Josefsson and P. Persson, *Conformal Array Antenna Theory and Design* (Hoboken, NJ: Wiley Inter-Science Publication, IEEE Press, 2006).

12 Radar Height Finder and Altimeter

"I think what you do is, you keep your sensors open. And it's - the more that you do the job, the more you come to understand in a kind of intuitive way that you're always – you know, your radar is on. And the thing is going around and around and around. And it's not picking up any blips."

—**Stephen Hawkins**

12.1 INTRODUCTION

Height finding in radar from ground-based positions is always a derived rather than a direct measured quantity, since the radar can only measure range and angle of the target's arrival. Ground-based radars derive the height by utilizing the range of the target and its elevation angle above the local horizon. The accurate calculation must account for such effects as the orientation and location of the antenna, curvature of the earth, multipath reflection from the surface, the local terrain below the target, and refractions from the atmosphere. Effects of the equipment errors can also be offset by incorporating internal calibration measurements into the range and angle estimations. Radar height finders that determine the target height with respect to the earth's surface and radar altimeters in aircraft and airborne radars that measure the relative height of the aircraft above the surface or in relation to other objects will be discussed.

In addition to flat-earth approximation, better approximations to the target height above the sea level—for targets at somewhat greater range and altitudes—require considerations of the earth's curvature, atmospheric refractions, and ground multipath reflections.

12.2 DERIVATION OF RADAR HEIGHTS

Flat-earth approximation: For very close targets, a sufficiently reasonable estimate of the target height is obtained by considering the earth to be approximately flat. According to Figure 12.1, the estimated target height h_t for the flat earth approximation is given by

$$h_t = R_s \sin\theta \tag{12.1}$$

where R_s is the slant range and θ is the measured or estimated target elevation angle.

Spherical-earth approximation: A somewhat better approximation to estimate the target height that accounts for the earth's curvature can be derived by considering Figure 12.2. For the radar located at a height of h_r above the surface of the earth, it can be shown from the law of cosines that

$$(r_0 + h_t)^2 = R_s^2 + (r_0 + h_r)^2 - 2(r_0 + h_r)R_s \cos\left(\theta + \frac{\pi}{2}\right) \tag{12.2}$$

229

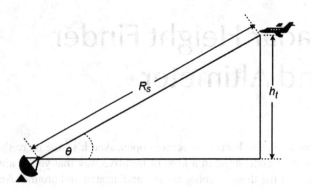

FIGURE 12.1 Geometry of the flat-earth approximation.

where r_0 is the radius of the earth as defined in Figure 12.2. By simple mathematical manipulation, it follows that, to a first approximation for targets at low altitude assuming $2r_0 \gg h_t$, (12.2) reduces to

$$h_t \approx \frac{R_s^2 + 2r_0 h_r + 2r_0 R_s \sin\theta}{2r_0 + h_t} \qquad (12.3)$$

By applying the same approximation a second time to (12.3), we get

$$h_t \approx h_r + R_s \sin\theta + \frac{R_s^2}{2r_0}. \qquad (12.4)$$

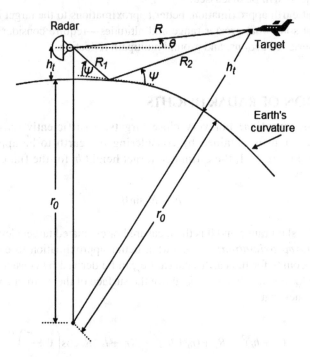

FIGURE 12.2 Geometry of the spherical earth.

Radar Height Finder and Altimeter

This demonstrates that the height calculated with the spherical-earth approximation exceeds that calculated by using the flat-earth approximation, increasing by a factor $R_s^2 / 2r_0$. The exact value of the target height can be obtained from (12.2) as follows:

$$h_t = \left[R_s^2 + (r_0 + h_r)^2 + 2(r_0 + h_r)R_s \sin\theta \right]^{1/2} - r_0. \tag{12.5}$$

Effects of atmospheric refractions:[1] Atmospheric refraction of the radar beam produces errors in radar measurements of range and elevation angle. In free space, radar signals travel in straight lines; however, EM waves are generally bent or refracted downward in the earth's atmosphere especially in the troposphere. Refraction is a term used to describe the deviation of radar wave propagation from a straight line. The bending or refracting of the radar waves in the atmosphere is caused by the small variations of the propagation velocity with the altitude due to the variation of the index of refraction, which is defined as the ratio of the velocity of propagation in free space to the velocity in the medium. The index of refraction normally decreases with increasing altitude causing downward bending and extending the propagation path, as shown in Figure 12.3(a). In this case refraction imposes limitations on the radar capability to measure the position of the target thereby introducing an error in measuring the elevation angle. In the tropospheric portion of the atmosphere the index of refraction n depends on variations in pressure, temperature, and water vapor content. This can be represented by[2]

$$(n-1) \times 10^6 = N = \frac{77.6p}{T} + \frac{3.73 \times 10^5 e}{T^2} \tag{12.6}$$

where N is called the atmospheric refractivity, p is the barometric pressure in millibars, T is the temperature of air in Kelvin, and e is the partial pressure of water vapor in millibars.

Normally, the refractivity gradient close to the earth's surface is almost constant. However, temperature changes and humidity lapses may cause the refractive index to become very

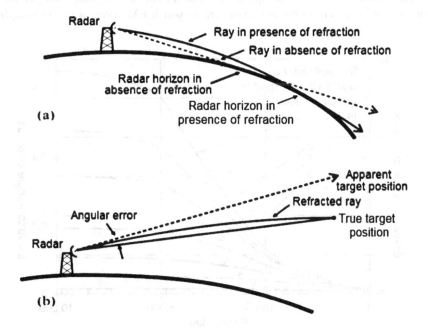

FIGURE 12.3 Geometric considerations: (a) effect of refraction for targets at low altitude and (b) effect refraction for the target at high altitude.

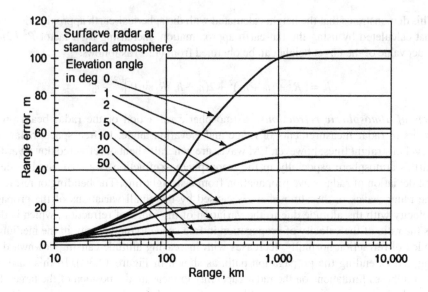

FIGURE 12.4 Range errors as a function of range and elevation angle (From Blake, 1972).

large, which results in the bending of the EM wave around the curvature of the earth. This phenomenon is known as ducting, is illustrated in Figure 12.3(b). Consequently, the distance of the target to the horizon is extended.

As noted, atmospheric refraction produces errors in radar measurements of range and elevation angle, which are used to derive the height of the target. These errors decrease with the increasing elevation angle, and are independent of frequency as shown by Curry (2012) in Figures 12.4 and 12.5. The common way of dealing with refraction is to replace the actual earth radius r_0 by an equivalent earth of radius $r_e = \kappa r_o$ and to replace the actual atmosphere by a homogeneous atmosphere in which EM waves travel in straight lines

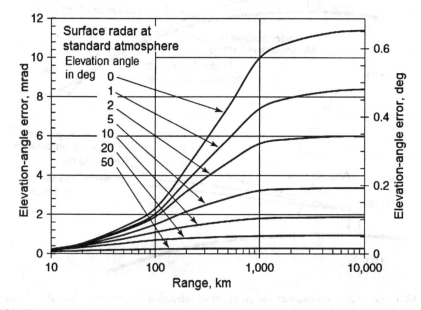

FIGURE 12.5 Angular errors as a function of range and elevation angle (From Blake, 1972).

Radar Height Finder and Altimeter

rather than curved lines. It may be shown by Snell's law in a spherical geometry that the value of κ is given by (5.41), which is reproduced here for convenience:

$$\kappa = \frac{1}{1 + r_o(dn / dh)} \tag{12.7}$$

Assuming $r_o = 6,370$ km, and a typical value of the refractivity gradient $(dn / dh) = -3.9 \times 10^{-8}$/m at microwave frequencies, a typical value κ becomes 4/3. Using an effective earth radius $r_e = (4 / 3)r_0$ produces what is known as the *four-thirds earth model*. In general, the effective earth radius is expressed as

$$r_e = r_o(1 + 6.37 \times 10^{-3}(dn / dh)). \tag{12.8}$$

Blake (1972) derives the height finding equation for the four-thirds earth model given by

$$h_t = h_r + 6076 R_s \sin\theta + 0.6625 R_s^2 (\cos\theta)^2 \tag{12.9}$$

where h_t and h_r are the expressed in feet and R_s is in nautical miles. All variable are defined in Figure 12.2 except that r_0 is now replaced by r_e.

For accurate height calculations it is possible to correct for variations in surface refractivity in otherwise normal atmospheric refraction conditions. In normal radar operation, the surface refractivity is measured periodically at the radar site, where it is used in conjunction with measured target elevation θ and range R_s. The final height is computed as a function of elevation angle, range, and surface refractivity along with the partial derivatives of the height function with respect to the three variables as follows:

$$h_t = h_t(R_k, \theta_k, N_k) + \frac{\partial h_t}{\partial r_k}(R - R_k) + \frac{\partial h_t}{\partial \theta_k}(\theta - \theta_k) + \frac{\partial h_t}{\partial N_k}(N - N_k) \tag{12.10}$$

where R_k, θ_k and N_k are the closest values of range, elevation angle, and surface refractivity to the measured values of R, θ, and N.

Terrain effects: Masking the radar LOS by terrain or the sea surface can limit the observation of low-altitude targets by radars at low altitude. For radar altitudes less than 10 km, the horizon range r_h can be found using the four-thirds earth model as

$$r_h = \sqrt{(r_e + h_r)^2 - r_e^2} = \sqrt{h_r^2 + 2r_e h_r} \tag{12.11}$$

where the variables are defined in Figure 5.11 in Chapter 5. The radar horizon range as a function of radar altitude above a smooth surface of the earth can be obtained using (12.11) as shown in Figure 12.6 (Curry 2012).

Multipath effects: One of the fundamental effects limiting the height accuracy in all height finding is due to the multipath from surface reflections. When the radar beam illuminates a flat reflecting surface, a reflected signal path is created in addition to the direct path. When the grazing angle is small, the range difference ΔR between the reflected path and the direct path signal is approximately given by (5.5), and is reproduced here for convenience:

$$\Delta R \approx \frac{2h_t h_r}{R}, \qquad h_t \gg h_r \tag{12.12}$$

The direct and reflected signals add coherently at the radar receiver, either increasing or decreasing the received signal, depending on their relative phase. This effect is called

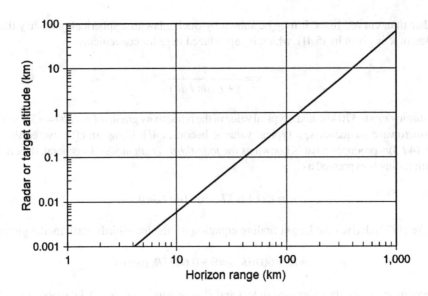

FIGURE 12.6 Radar or target altitudes as a function of horizon range.

multipath propagation as discussed in Chapter 5. Multipath propagations can produce significant angle measurement errors at a low elevation angle. With sufficient range resolution, the direct signal can be separated from the reflected signal. The time separation Δt between these two signals depends on the target height, and can be expressed as

$$\Delta t = \frac{2 h_r h_t}{cR}. \tag{12.13}$$

It follows then

$$h_t = \frac{cR\Delta t}{2 h_r}. \tag{12.14}$$

Normally, the errors for a smooth reflecting surface are about half of the elevation beamwidth, while the errors for a rough reflecting surface are typically 0.1 of the elevation beamwidth. Measurement errors from multipath fluctuate slowly as the observation geometry changes, and should be treated as bias errors due to equipment calibrations.

12.3 HEIGHT-FINDING RADARS

Because radar height is a derived quantity from the basic radar measurements of range and elevation angle, various radar techniques of height finding are employed to determine the relative height of targets with respect to radar on the earth's surface. The practice of using the earth's surface for height finding is quite common because the antenna can be conveniently used at lower radio frequencies, and for a broad elevation beam. The first height-finding radar was developed in 1939 by the US Naval Research Laboratory (NRL) and used the range to estimate its height, based on a knowledge of the shape of the radiation pattern near the horizon. Another early height-finding radar such as the British Chain Home (CH) series, employed in World War II for the defense of Britain, made height measurements simply by comparing signals received from a pair of vertically mounted antennas. One of the earliest forms of radar height finding, called *lobe switching*, was developed in 1937 by the US Army Signal Corps for directing antiaircraft gunfire. This system uses two separate

Radar Height Finder and Altimeter

beams, one above the other. By switching between the two beams, the operator can keep the antenna directed at the target. Conical scanning and lobe switching are special cases of a general technique of *sequential lobing,* which is typically limited to making a measurement on a single target at a time. A widely used early radar dedicated to finding the height of a target was the *nodding antenna,* which appeared in the mid-1940s and employs a narrow beam antenna structure in the vertical plane containing a target. Another technique that employs a stack of horizontal fan beams is called the *stacked-beam radar.* The *V-beam* radar used in military surveillance operations utilized two fan-shaped antenna beams: a vertical fan beam and a slant fan beam inclined at 45° with respect to the vertical beam. A radar height finder developed by the Japanese is based on phase interferometry to find height in air traffic control applications.

In this section some of the radar techniques for height finding that employ the relative height of targets with respect to radars on the earth's surface are discussed.

12.3.1 Nodding Height Finder

The nodding height finder is one of the earliest and most common forms of height-finding radars used in ground-based aircraft surveillance systems. In this type of radar an entire narrow-beam structure is mechanically scanned (nodded) in the vertical plane containing the target on a continuous basis to measure the elevation angle. As the radar beam intercepts the target continuously, the return echoes received in the main lobe are displayed manually by an operator by means of a range-height-indicator (RHI) type of display, as shown in Figure 12.7. This allowed the operator to precisely estimate the target height by placing the height cursor on the estimated center of the blip by a process called *beam splitting,* thereby obtaining a direct reading of the target height.

It must be noted that the nodding height finders have little or no surveillance capability; they must be used in conjunction with surveillance radar. The surveillance radar and the nodding height finder are normally collocated and operated so that the operator can observe detection by the surveillance radar and then command a height determination by the height finder, which, in turn, would slew to the commanded azimuth and obtain a height and range measurement. This method of operation is

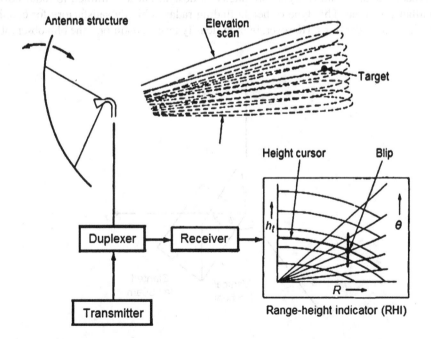

FIGURE 12.7 Nodding height-finder representations (From Skolnik, 1970).

relatively slow, warranting limitations in continuous use in military applications. This type of height finder is not suitable for rapid height determinations on multiple targets located at various azimuth positions. However, the height data rate demands are usually met with a nodding height finder in case of a low to medium number of targets. The beam-splitting proficiency of the RHI operator in this type of radar is an important factor that essentially determines the height accuracies achievable. A good RHI operator can be expected to provide reasonably acceptable data.

12.3.2 V-Beam Radars

One of the early radars, referred to as a three-dimensional radar described by Brown (1970), is technically not a three-dimensional radar because it lacks resolution in the elevation dimension. This shortcoming limits its use in high-density target situations. Brown utilized the V-beam principle, which employs two fan beams generated by a common rotating antenna structure, one a vertical fan beam and another a slant fan beam inclined 45° with respect to the vertical beam. The two beams are energized simultaneously by the same or separate transmitters, but each beam has its own receiver. The vertical fan beam, operating as conventional two-dimensional radars, provides the azimuth-range coordinates in the surveillance volume, while the slanted fan beam provides a second set of detections as the antenna is rotated, as shown in Figure 12.8. The azimuth separation of the center of the two sets of detections corresponding to a single target is measured either manually by an observer or automatically by pulse counting circuitry, which determines the blip centers by pulse-count average. This azimuth separation ϕ is found to be proportional to the height of the target, and the height h is determined by $h = D \sin\phi$, where D is the ground range to the target given by $D = \sqrt{R_s^2 - h^2}$. Thus the height can be expressed in terms of the slant range R_s and the azimuth separation ϕ by eliminating D from the previous two relations:

$$h = \frac{R_s \sin\phi}{\sqrt{1 + \sin^2\phi}} \tag{12.15}$$

The V-beam radar is known by its simplicity in design, but it is limited to radar coverage of the surveillance volume. This type of height-finding radar is highly satisfactory for conditions of relatively light target densities. However, in high-density target conditions, the blip observation may

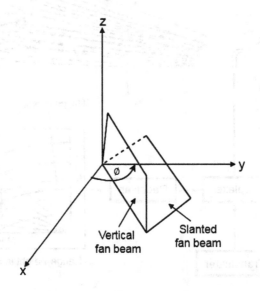

FIGURE 12.8 V-beam configurations.

Radar Height Finder and Altimeter

become difficult when it is likely to encounter many targets at the same range and azimuth, but at different altitudes. Because of this, other forms of height-finding radars for improving the height accuracy, even though they are more expensive and complex in design, are sometimes desired.

12.3.3 Stacked-Beam Radar

The stacked-beam height-finding radar is also a three-dimensional radar that provides information by simultaneously measuring the three basic position coordinates of a target such as range, azimuth, and elevation. This type of radar employs three-dimensional radar to obtain volumetric coverage by employing vertical stack fixed-elevation pencil beams, which continuously rotate in azimuth to perform search and target position estimation. It provides a higher gain antenna and a greater resistance to jamming and other forms of electronic countermeasures. An example of a widely used stacked-beam radar is the AN/TPS-34. It is an S-band transportable radar, which was deployed in the 1970s, and has been extensively used in air surveillance in the US Air Force Tactical Air Command System (TACS). Another example of a stacked-beam radar is the S713, an L-band transportable radar with an eight-beam stack.

Usually, a stacked-beam antenna structure consists of a section of paraboloid reflector fed by a vertical stack of fixed feed horns to produce a series of overlapping beams in the vertical plane, as shown in Figure 12.9. On transmission all feed horns are excited in phase from a single transmitter to produce a composite envelope that approximates a cosecant-squared transmit beam.

The main lobe of the transmit beam must be broad enough in elevation to cover all receive beams. On reception each feed horn is processed in a separate receiver to preserve the directional and gain

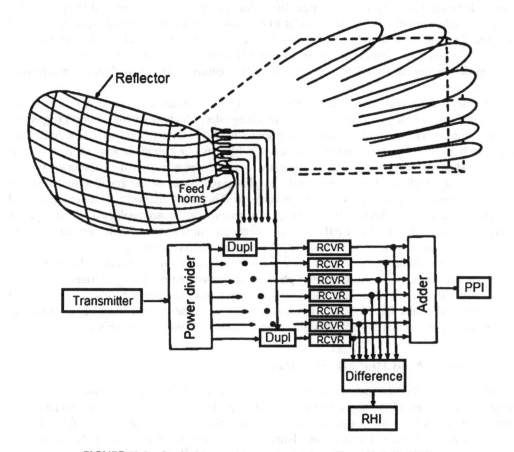

FIGURE 12.9 Stacked-beam radar representations (From Skolnik, 1970).

properties of the individual receiving beams. Elevation angle can be obtained in such a radar by simultaneous amplitude comparison of the target signals on adjacent beams.[3] Stacked-beam radars are normally equipped with circuitry that selects only the two adjacent elevation beams in which the target signal is the strongest during each successive range element. The independent received signals are normally passed through logarithmic amplifiers, which produce a net signal proportional to the difference between logarithmic outputs of two adjacent beams. Azimuth and range are determined from the range-azimuth search operation as in conventional two-dimensional radars. Height finding is accomplished by interpolating received signal strengths in adjacent elevation beams of the stack to determine target elevation angle. The accuracy of this height-finding approach depends on relative spacing of the feed horns and the elevation angle of the target's arrival.

By including anticlutter MTI and/or Doppler processing in the beam stack, the performance of the stacked-beam radar can be improved to some extent because the lowest beam in the stack, which is especially susceptible to surface clutter returns, causes degradation in the received signals. In some benign surface clutter applications, it is economical to implement the stacked-beam radar without MTI or Doppler processing in the beam stack.

12.3.4 FREQUENCY SCANNED RADARS

Three-dimensional scanner radars represent another basic class of combination search and height-finding techniques suitable for obtaining continuous volumetric surveillance in high air-traffic situations by employing an antenna that rapidly scans, either mechanically or electronically, and the elevation coverage as it rotates in azimuth. The azimuth and elevation angular accuracy achievable in mechanically scanned three-dimensional radars is usually limited to applications that permit relatively modest values of radar range, and elevation and azimuth accuracy. One of the three-dimensional techniques that are used in height-finding systems for volumetric air surveillance is *frequency scanning*. From an operational point of view, frequency scanners, applied as elevation scanners in volumetric surveillance, offer features difficult to obtain in mechanical elevation scanners.

Frequency scanners essentially lack inertia, and employ the frequency-dependent phase characteristics inherent in a long transmission line or in waveguides, which, when folded into a serpentine configuration, provide a precise method for positioning the resulting beam as a function of the excitation frequency. By controlling the frequency, a different phase gradient across the aperture is produced when the beam is steered electronically to the desired elevation angle. Frequency scanning may be accomplished from pulse to pulse by changing the transmitter and receiver frequency sequentially from one pulse to the next to produce "within-pulse" frequency scanning combined with pulse compression.[4] Another type of frequency scanner can be accomplished by employing a continuously variable chirp LFM pulse transmission, wherein the target height is determined by frequency discriminators in the receivers at each elevation-beam position.

The operating frequency of the frequency scanner is severely limited because the beam position is usually derived from the frequency-phase shift characteristics of a fixed transmission line or waveguide. By using a wider bandwidth phased array, a more frequency agility can be obtained along with the desirable flexible scan. The elevation angle achievable in this class of height-finding radar is not as good as that of stacked-beam radars employing simultaneous lobing.

12.3.5 PHASED ARRAY HEIGHT-FINDING RADARS

In a phased array, the scanning of the pencil beam is accomplished by electronically changing the progressive phase of the phase shifters placed at the feed outputs of the array antenna. This approach provides the most flexible of the various three-dimensional radar height-finding techniques, allowing full use of the frequency band independent of the waveform and beam position. Currently, this is becoming more commonplace in military applications due to the varying nature of target environments. The AN/TPS-5 L-band radar is an example of a long-range transportable three-dimensional tactical radar with phase scanning to steer the beam in the elevation plane. The HADR is a ground-based three-dimensional S-band transportable phased array radar that employs phase scanning in elevation and mechanical rotation in azimuth to obtain the height of the desired target.

12.3.6 DIGITAL BEAMFORMING HEIGHT-FINDING RADAR

Digital beamforming in radar height finding is becoming more attractive because of its ability to have full adaptive control of the beam patterns for ECCM purposes. Digital beamforming involves placing a receiver on each element of a vertical array of elements. By digitally weighting and linearly combining the analog-to-digital (A/D) converted receiver outputs, a stack of receive beams in the elevation plane can be easily generated. However, the major challenge in digital beam forming height-finding radar is to develop techniques to preserve monopulse ratios in the presence of an adaptive array cancellation of jamming where the accuracy of height finding depends on a precise and unambiguous knowledge of the relative patterns of the adapted beams.

12.3.7 INTERFEROMETRY HEIGHT-FINDING RADAR

Another type of height-finding radar, called the *interferometry height-finding radar*, developed by the Japanese is based on phase interferometry to find the height in air traffic control applications.[5] It does not provide significant resolution in the elevation dimension compared with its elevation coverage. It consists of two individual antennas spaced to obtain a narrow beamwidth to accommodate the accurate angle measurement. The phase difference between the signals of the two antenna elements of the interferometer is proportional to the sine of the angle of arrival θ of the received target echo, as shown in Figure 12.10. This type of angle measurement is similar to the phase-comparison monopulse radar, except that the size of the individual antennas is small compared to the separation between them. Grating lobes resulting from the wide spacing between the antennas can cause ambiguities in the measurement.

The ambiguities due to grating lobes can be resolved by employing more than two antennas with unequal spacing. In this implementation, the interferometer employs a set of four horizontal line arrays vertically displaced in a staggered fashion about a conventional two-dimensional reflector-type antenna. Also, there is some measurement error caused by multipath due to ground reflections. However, by shaping the underside of the two-dimensional transmitting antenna to minimize the energy incident on the ground surface can reduce such errors.

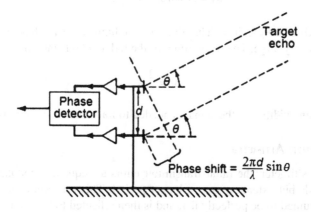

FIGURE 12.10 Phase relationships in phase interferometry (From Skolnik, 1990).

12.4 RADAR ALTIMETERS

A *radar altimeter* is, according to the International Telecommunications Union, defined as "the radio navigation equipment, on board an aircraft or spacecraft, used to determine the height of the aircraft or the spacecraft above the earth's surface or another surface." Thus, radar altimeters are used on aircraft or spacecraft to measure altitude of the aircraft or the spacecraft above the earth's surface or another surface. The altimeter transmits electromagnetic energy down to the ground and measures the time taken by the return echo after being reflected from the ground. The altitude is then calculated from the knowledge of the travel time and the velocity of propagation of the wave. Radar altimeters also provide a reliable and accurate method of measuring height above water, when flying long sea tracks.

In 1924, American engineer Lloyd Espenschied invented the radio altimeter. In 1938, Bell Labs put Espenschied's device in a form that was adaptable for aircraft use. In 1938 in cooperation with Bell Labs, United Airlines fitted a radar-type device to some of its airliners as a terrain avoidance device. A significant development in radar altimetry makes it useful in commercial and military applications. Currently, radar altimeters are frequently used by commercial aircraft for approach and landing, especially in low-visibility conditions and automatic landings, allowing the autopilot to know when to begin the flare maneuver. Radar altimeters are also used in *ground proximity warning systems* (GPWS), warning the pilot if the aircraft is flying too low or descending too quickly. However, radar altimeters cannot see terrain directly ahead of the aircraft. Radar altimeters are also used in military aircraft to fly quite low over the land and the sea to avoid radar detection and targeting by anti-aircraft guns or surface-to-air missiles. A related use of radar altimeter technology is terrain-following radar, which allows fighter bombers to fly at very low altitudes.

To-date treatments of many different types of radar altimeters are available in the literature. In this chapter, we will discuss some of the radar altimeters[6], currently in use for most practical purposes.

12.4.1 BEAM-LIMITED ALTIMETER

The radar altimeter operates in the microwave part of the spectrum so the method is based on the assumptions that the atmosphere is very transparent at 13 GHz, since there is little stray radiation coming from the earth. The illumination pattern on the surface of the ocean is very broad for antennas of reasonable size. The angular resolution θ_r of a circular aperture having a diameter d is given by

$$\sin \theta_r = 1.22\lambda / d \tag{12.16}$$

where λ is the wavelength of the radar, as shown in Figure 12.11. The diameter d_s of the illuminated pattern on the ocean surface is given by

$$d_s = 2h \sin \theta_r \cong 2.44h \frac{\lambda}{d} \tag{12.17}$$

where it is assumed that $\tan \theta_r \cong \sin \theta_r$. The illumination diameter or the beamwidth of the radar is quite large when h is very large. The travel time for the radar echo is measured to an accuracy of

$$\Delta t = \frac{2h}{c} \tag{12.18}$$

which implies that bandwidth B of the signal is needed to form such a sharp pulse $B = 1 / \Delta t$.

12.4.2 PULSE-LIMITED ALTIMETER

In the pulse-limited altimeter, the radar transmitter emits a frequency-modulated chirp having a much lower amplitude but extending over a longer period. This transmitted signal hits the ocean surface, which is assumed to be perfectly flat, and is then reflected to the radar receiver where it is convolved with a matched filter to regenerate the desired pulse.

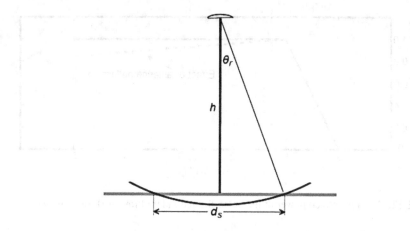

FIGURE 12.11 Footprint of a band-limited radar altimeter (From Sandwell, 2011).

Figure 12.12 shows an illustration of the vertical cross section of the radar pulse that interacts with a flat surface. The radar altimeter measures the return power of the reflected radar pulse, which is interpreted to estimate the distance between the radar altimeter and the reflecting surface. From the vertical cross section of the radar pulse, the radius of the outer edge of the illumination pattern is obtained using the Pythagorean theorem:

$$h^2 + r_p^2 = (h + \ell_p)^2 = h^2 + \ell_p^2 + 2h\ell_p \tag{12.19}$$

where r_p is the leading edge of the pulse, h is the vertical distance of the radar altimeter from the ocean surface, and ℓ_p is the increment of h due the pulse as illustrated in the figure. Equation (12.19) can be approximate to obtain an explicit expression of r_p as

$$r_p \approx \sqrt{2h\ell_p} = \sqrt{hct_p} \tag{12.20}$$

where it is assumed that the pulse is of square shape of length $\ell_p = ct_p/2$ with c as the velocity of propagation in free space.

The power measured from the footprint of the radar signal reflected from the ocean surface during the pulse of length t_p, as derived mathematically by Sandwell, is based on the time evolution of

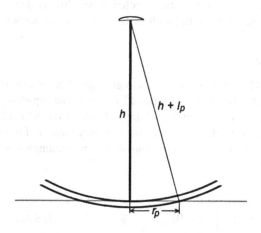

FIGURE 12.12 Footprint of a pulse-limited radar altimeter (From Sandwell, 2011).

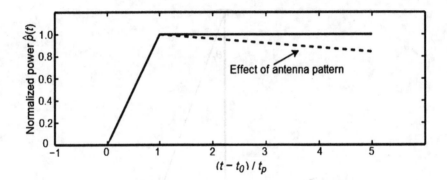

FIGURE 12.13 Normalized power of the pulse-limited radar altimeter (From Sandwell, 2011).

the footprint and on the assumption that the power is linearly proportional to the area of the ocean illuminated. Thus, the power is expressed as

$$P(t) = \begin{cases} 0 & t < t_0 \\ \pi r^2(t) & t_0 < t < t_0 + t_p \\ \pi[r^2(t) - r^2(t-t_p)] & t > t_0 + t_p \end{cases} \quad (12.21)$$

where three components are accounted for: (1) the time before the leading edge of the pulse arrives, (2) the time between the leading and trailing edge arrival times, and (c) the time after the trailing edge arrives. Using (2.20) for r_p and normalizing by the peak power yields:

$$\hat{P}(t) = \begin{cases} 0 & t < t_0 \\ \dfrac{(t-t_0)}{t_p} & t_0 < t < t_0 + t_p \\ 1 & t > t_0 + t_p \end{cases} \quad (12.22)$$

where $\hat{p}(t)$ represents the normalized power measured from the footprint of the radar signal reflected from the ocean surface. The plot shown in Figure 12.13 demonstrates that the pulse begins to increase linearly at time zero and continues until the time the full pulse reaches the ocean surface at $(t-t_0)/t_p = 1$. However, the power of the reflected pulse actually decreases with time as a function of the illumination pattern of the radar altimeter on the ocean surface.

12.4.3 SAR Altimeter

In SAR altimeter, it emits far more pulse signals to give the effect of covering the same footprint as pulse-limited but with better resolution.[7] Figure 12.14 shows the top-down footprints of the pulse-limited and SAR illustrating the decrease of footprint size of the SAR altimeter compared to the conventional pulse-limited altimeter. To determine the power signal of the SAR altimeter as a function of time, it is assumed that the footprint consists of two rectangles of width w. It follows also from Sandwell (2011) that

$$P(t) = \begin{cases} 0 & t < t_0 \\ 2wr(t) & t_0 < t < t_0 + t_p \\ 2w[r(t) - r(t-t_p)] & t > t_0 + t_p \end{cases}. \quad (12.23)$$

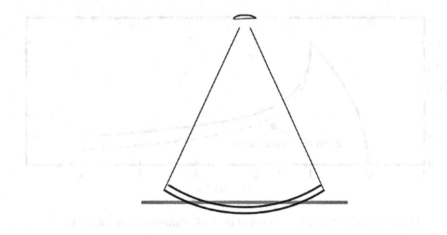

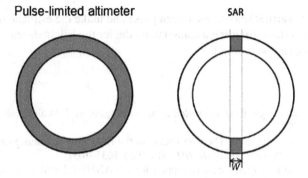

FIGURE 12.14 Pulse-limited and SAR footprints (Sandwell, 2011).

Following the same procedures as in the pulse-limited altimeter, the normalized power as a function of time relative to the arrival time of the leading edge $t' = t - t_0$ can be expressed as

$$\hat{p}(t) = \begin{cases} 0 & t' < 0 \\ \sqrt{t'/t_p} & 0 < t' < t_p \\ \sqrt{t'/t_p} - \sqrt{t'/(t' - t_p)} & t' > t_p \end{cases} \quad (12.24)$$

The plot shown in Figure 12.15 demonstrates the power function $\hat{P}(t)$ of the SAR altimeter for the radar pulse. On the leading edge, the power increases as the square root of time. The main difference is the trailing edge where the SAR pulse decreases as the square root of time while the pulse-limited altimeter has a uniform power with time until the pulse radius approaches the beam-limited footprint of the radar.

There are two major advantages of the SAR altimeter over the traditional pulse-limited altimeter: first, the radar footprint is covering less area so the radar pulse emitted requires much less power, covers the radar footprint using more frequent, but less power-consuming radar pulses; second, the return waveform has a more complex signature that includes both the leading and trailing edges, which provide a more accurate constraint[8] on the arrival time.

One insidious problem lies in the fact that the actual ocean surface, of course, has roughness due to waves and swells. The return power will be convolved with the height distribution of the waves

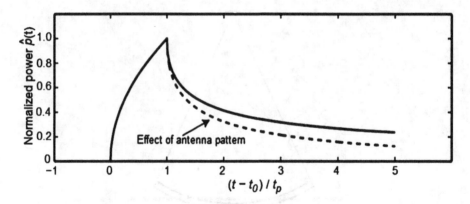

FIGURE 12.15 Normalized power of the SAR altimeter (Sandwell, 2011).

within the footprint to further smooth the return pulse and make the estimate of the arrival time of the leading edge less certain. By using a Gaussian model for the height distribution, the error in the estimated arrival time can be reduced.

REFERENCES

1. B. P. Brown, "Radar Height Finding, in *Radar Handbook,* ed. M. I. Skolnik (New York: McGraw-Hill Book Company, 1970).
2. E. K. Smith, and S. Weintraub, "The Constraints in the Equation for Atmospheric Refractive Index at Radio Frequencies," *Proceedings of the IRE,* 41 (1953): 1035–1037.
3. E. E. Herman, "The Elevation Angle Computer for the AN/SPS-2 Radar," *U.S. Naval Research Lab Report 3896,* November 19, 1951.
4. K. Milne, "The Combination of Pulse Compression with Frequency Scanning for Three-Dimensional Radars," *Radio Electronic Engineer,* 28, no. 2 (1964): 89–106.
5. M. Watanabe, T. Tamama, and N. Yamauchi, "A Japanese 3-D Radar for Air Traffic Control," *Electronics,* 44 (1971): 68–72.
6. D. T. Sandwell, "Radar Altimetry: Lecture Notes," *Scripps Institution of Oceanography,* University of California, CA, 2011.
7. R. K. Raney, "The Delay/Doppler Radar Altimeter," *IEEE Transactions on Geoscience and Remote Sensing,* 36 (1998): 1578–1588.
8. D. T. Sandwell and W. H. F. Smith, "Retracking ERS-1 Altimeter Waveforms for Optimal Gravity Field Recovery," *Geophysical Journal International,* 163 (2005): 79–89.

13 Radar Electronic Warfare

"Mankind must put an end to war before war puts an end to mankind."

—**John F. Kennedy**

13.1 INTRODUCTION

The world has witnessed numerous wars, such as the conflicts in Lebanon, Golan Heights, the Falkland Islands, and the bombing of the Iraqi nuclear facility by Israeli fighter planes, to name a few. All these conflicts caused severe suffering and pain on mankind. But the rapid progress in science and technology has changed the face of modern-day warfare completely by encountering electronic warfare with far-reaching consequences.

Today, everybody is familiar with fighter aircraft, warships, missiles, and submarines. The majority of people have seen them in action either directly or via television or movies. But there is another kind of invisible fight involving the use of radio and radar emissions, which is always going on in the atmosphere. This silent battle of electromagnetic energy is commonly called *electronic warfare* (EW). EW is not rigorously "electronic"; it is not conducted utilizing electrons, but it is essentially electromagnetic using the entire range of the electromagnetic spectrum. Because of this some people call it electromagnetic warfare. Winston Churchill, during World War II, coined the terms *Wizard War* and *Battles of the Beams*. However, to support nomenclature of EW, we can argue that electronic circuits are integral components in EW equipment.

The rudimentary concept of EW is to exploit the enemy's electromagnetic emissions to invade the secrecy of the enemy's strategy in battle, intentions and capabilities, and to use *countermeasures* accordingly. Countermeasures include all those means of exploiting an adversary's activities to determine intentions or reduce the enemies' effectiveness. Countermeasures may also be applied against weapons operating over the entire spectrum. The EW battlefield is ecumenical, and its intensity depends on national interests and perceptions of potential threats. In fact, EW is a catalyst toward the maintenance of regional and global balances that deter the outbreak of armed conflict. In EW what works today may not work tomorrow due to the rapid developments in EW systems to cater threats. Because of this and the difficulty in studying countermeasures revolving around classified information, constant updating and refinement of EW equipment are essential.

13.2 HISTORICAL BACKGROUNDS[1]

Radar was developed into a very useful military weapon during World War II. The widespread development and employment of radar rapidly resulted in the introduction of EW primarily by the United States and the United Kingdom. EW eventually gained significant sophistication and maturity, and has subsequently played a prominent role in the wars in Vietnam and the Middle East. Efforts to enhance radar performance in the face of enemy interference involved both improvements in radar technology and in the creation of methods to reduce the effects of enemy interference.

Germans employed the electromagnetic wave for communication jamming, which is considered the first real action of EW in 1914. During World War I, both sides used EW with electronic deception in its simplest forms. In fact, the sophisticated forms of EW began to develop only during World War II. The sophisticated German Würzburg radars created a sensation; and the British began to equip their aircrafts with both noise jammers and passive ECM equipment as a countermeasure.

245

During the Vietnam War in 1965, the Soviet SA-2 was employed in the stormy battlefield and came out successful in downing the US fighter aircraft. To counter these serious threats, some projects were started by the United States to develop an adequate EW capability to meet the new challenge of aircraft losses. The 1973 Middle East War encountered most of the latest Soviet surface-to-air missiles (SAM) and antiaircraft artillery (AAA) system in action in addition to the employment of a different region of the electromagnetic spectrum for target tracking and guidance. After sustaining heavy air losses in the first few days of the Arab-Israeli War, Arabs managed to adapt countermeasures to suppress the radar-controlled SAMs and AAAs. The 1973 war thus placed EW into the forefront of modern military actions warranting of possessing a complete range of EW equipment even in peace time.

Another EW conflict in Falklands War in 1982 initiated the development of airborne early warning radars in the wake of the British Sheffield that was destroyed by a sea-skimming French-built missile. Israelis made use of a special type of deception technique, called *decoys* in the Lebanon-Israel War in 1982 to destroy numerous enemy aircrafts. This led to an incredible victory of the Israeli Air Force. Activities in EW are continuing so that all military radars must include the most updated, modern means to combat enemy interference. The importance of EW is immensely needed in terms of its operational efficiency and survivability under an adverse electromagnetic environment.

13.3 ELECTRONIC WARFARE DEFINITIONS

It is important to define certain terms relating to radar EW. The US Department of Defense (DOD) accepted the definitions for EW and associated components. According to DOD, countermeasures in the RF region of the electromagnetic spectrum are collectively grouped under an umbrella termed *electronic warfare*, and the field of EW is most commonly subdivided into three broad categories: *electronic support measures* (ESM), *electronic countermeasures* (ECM), and *electronic counter-countermeasures* (ECCM).

The study of EW is facilitated by organizing it into divisions, subdivisions, and sub-subdivisions as warranted. Each of the divisions or categories of EW embrace many techniques, and can be organized in a number of ways. Various authors use different methods of organizing each EW division. The official DOD EW organization includes only three divisions, but various authors[2] have a different organization system. Figure 13.1 depicts one of the organizations to reflect various components of EW and their relationships with one another. DOD definitions[3] of EW and related components are presented.

Electronic warfare (EW): EW is a military action involving the use of electromagnetic energy to determine, exploit, reduce, or prevent the hostile use of the electromagnetic spectrum as well as action that retains the friendly use of the electromagnetic spectrum.

Electronic support measure (ESM): ESM is that division of EW involving actions taken to search for, intercept, locate, record, and analyze radiated electromagnetic energy, for the purpose of exploiting such radiations to support military operations. Thus, ESM is an important source of EW information to carry out electronic countermeasures and electronic counter-countermeasures. ESM involves, in general, the gathering of EW information through *electronic intelligence* (ELINT), *communications intelligence* (COMINT), and ESM receivers. ESM is the tactical reception of signals to locate, identify, and analyze signals so their effects can be avoided or countered.

Electronic countermeasures (ECM): ECM is that division of electronic warfare involving actions taken to prevent or reduce an enemy's effective use of the electromagnetic spectrum. ECM itself consists of two types. The first is *jamming* (sometimes called *denial* ECM)—where an electromagnetic signal provides so much jamming signal that the radar's signal processor and displays are occupied almost totally with the jamming to the

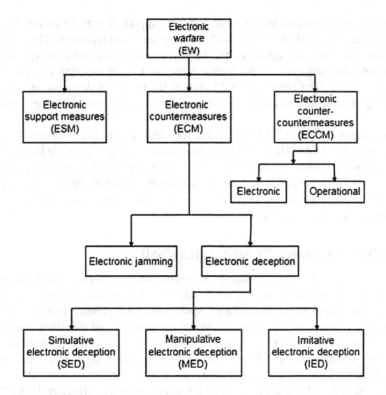

FIGURE 13.1 Electronic warfare (EW) and associated components.

exclusion of the targets. The second, called *deception*, is where emitted ECM signals are meant to *simulate* or *imitate* desired signals, but with changes, or are meant to *manipulate* the system into reporting. Essentially, deceptive jamming attempts to fool the radar into reporting targets at the wrong range, or wrong Doppler, or wrong angles. Manipulative ECM attempts to induce actions on the part of its victim that are detrimental to the victim's well-being. Simulative ECM makes an excellent radar design tool, and provides radar performance assessment. Digital techniques are widely used for ECM simulation.

Electronic counter-countermeasures (ECCM): ECCM is that division of electronic warfare involving actions taken to insure friendly effective use of the electromagnetic spectrum despite the enemy's use of electronic warfare.

Electromagnetic compatibility (EMC): EMC is not a division of EW, although it is closely related to ECCM in the techniques used. Johnston[4] defines it as "the ability of communications electronics equipment, subsystems, and systems to operate in their intended operational environments without suffering or causing unacceptable degradation because of unintentional electromagnetic radiation or response." EMC covers those problems resulting from the unintentional presence of electromagnetically coupled signals in electronic equipment. The interference signals may be generated intentionally or unintentionally. In many respects, problems in EMC are analogous to those in ECM. In EMC, the unwanted signals are unintentionally coupled, while in ECM, they are deliberate.

13.4 EFFECTS OF EW ELEMENTS

As mentioned, the field of EW is most commonly subdivided into three broad categories: ESM, ECM and ECCM, and definitions are straightforward. However, it is helpful to describe the effects of each element to form the basis for a possible means of element organization.

ESM systems as defined by the DOD to provide support of both ECM and ECCM systems. ESM functions in support of ECM are listed.[5] Most of these are represented by radar homing and warning equipment, and intercept receivers. Most jammers include an intercept receiver to place the jammer onto the frequency of the victim receiver, and the function of signal identification and threat assessment. The effects of ECM include the denial of detection and measurement of target, tracking of an invalid target, overloading of the computer, and a delay in detection/tracking initiation.

The basic purpose of the ECCM is to enable the radar to perform its intended purpose despite the presence of hostile ECM. The ECCM effects include warning of possible hostile activity, angular locations of hostile jammers, information on jammer targeting and avoidance, and information for ECCM selection. The application of an ESM receiver as part of the frequency agile radar can randomly change frequency before transmission. The ESM receiver functions are usually accomplished by the regular radar receiver.

13.5 ELECTRONIC SUPPORT MEASURES (ESM)

As already defined, ESM is that division of electronic warfare involving actions taken to search for, intercept, locate, record, and analyze radiated electromagnetic energy, for the purpose of exploiting such radiations to support military operations. ESM is based on the use of intercepting (which primarily involves interception, frequency estimation, and direction finding), and relies heavily on a previously compiled directory of both tactical and strategic electronic intelligence (ELINT). ESM is entirely passive, being confined to identifying, analyzing, and locating sources of hostile radiations. ESM is for "tactical" purposes that require immediate actions as contrasted with similar functions, which are performed for intelligence gathering, such as SIGINT, which has ELINT, COMINT, and RINT as its constituent parts. As indicated before, ESM can be used in support of both ECM and ECCM with common functions in both applications. The ESM receiver is used to control the deployment and operation of ECM; the link between ESM and ECM is often automatic.

ESM is thus employed to accomplish three primary functions: interception of radar emissions, analysis of radar signals, and direction finding to the origin of radar signals. The primary use of ESM in tactical situations is in the simplest *radar warning receiver* (RWR), which intercepts radar signals, warns that a radar is emitting radar signals, and attempts to ascertain the functions of the radar. It advises the presence of threats such as missile radar supplying the relative bearing on cockpit-based display. It is unsophisticated low-frequency equipment that is present to cover the bandwidth of expected threats, and it exploits the range information to indicate the threat before it comes into firing range. Receivers then increase the complexity through tactical ESM to the full ELINT capability. In some radar design, radar's process gain exceeds the process gain of the interceptor to the same illumination signal so that the echo detection takes place at a longer range than interception. Radars of this type that emit waveforms coded in such a way as to be difficult to detect without knowing the code, are known as *low probability of intercept* (LPI)[6] radars. Diversity of signals, such as pulsed, CW, and interpulse-modulated, must all be accommodated with a high *probability of intercept* (POI) and a low *false-alarm rate* (FAR). The probability of intercept in RWRs deteriorates, particularly when many emitters are present in the dense environment. It needs filtering or sorting of emissions to classify each signal to know the important parameters like the amplitude, pulse width, frequency, angle of arrival, coherency, polarization, and pulse train characteristics of the radar.

Many advanced ESM receivers have been developed based on various design approaches. These ESM receivers have excellent multiple signal handling capability in a dense emitter environment. Each receiving system has its own relative advantages and disadvantages as discussed next.

Radar Electronic Warfare 249

Crystal video receivers: Receivers are low-cost, small in size, and are excellent in limited applications. However, the receivers suffer from many limitations: the analyzer has to handle a wide open system, and cannot readily handle a complex and dense signal, they have poor sensitivity and are susceptible to ECCM, and are basically incapable of handling frequency agile systems.

Superheterodyne receivers: These receivers have the advantage of high sensitivity and good frequency resolution. Unfortunately, this type of receiver has a poor probability of intercept. This condition is much worse if the emitter is also frequency-agile or frequency-hoping. One method of mitigating this problem is to employ *acousto-optic Bragg cell receivers,* which utilize the Bragg refraction of optical-guided beams by *surface acoustic-wave* (SAW) filters to perform spectral analysis. Pulse width, frequency agility, and speed in searching are unreliable parameters that are inherent.

Microscreen receivers: Receivers have the advantage of a good probability of detection and the ability to handle wideband signals and frequency agile signals, but suffer from their limitations of requiring a channelizer, minimum pulse width, and a very IF bandwidth.

Channelized receivers: Receivers have the advantages of high selectivity, high probability of detection, not being susceptible to jamming. However, they suffer from their limitations of limited frequency accuracy, poor resolution, and the requirement of a channelizer.

Surface acoustic wave receivers: *Receivers* have the advantages of a high probability of detection, reasonable sensitivity, minimum pulse width, proper handling of frequency-agile signals, and not being susceptible to jamming. The limitations of the receivers include the moderate time needed to resolve pulses that are close. It follows, therefore, that there is hardly any single receiver that can be employed for all purposes. In practice, either a hybrid approach or a combination of two or more receivers is used to exploit the relative advantages to handle effectively the dense signal environment.

Assuming a monostatic radar with the same antenna for transmitting and receiving, the radar equation for a target in the presence of noise is given in Chapter 3 as

$$R_{\max} = \left[\frac{W_t G^2 \lambda^2 \sigma}{(4\pi)^3 FKT_0 L (S_o / N_o)_{\min}} \right]^{1/4}$$

(13.1)

where the symbols have their usual meanings except for W_t representing the total energy for the dwell time. The intercept SNR at the ESM receiver can be expressed as

$$\left(\frac{S}{N} \right)_i = \frac{P_t G_t G_i \lambda^2 G_{pi}}{(4\pi)^2 R^2 F_i KT_0 B_i L_i}$$

(13.2)

where the subscripts t and i refer to the transmitter and the interceptor, respectively. The interceptor processing gain is denoted by G_{pi}. The maximum range at which the target is intercepted is then

$$(R)_{i,\max} = \left[\frac{P_t G_t G_i \lambda^2}{(4\pi)^2 F_i KT_0 B_i L_i (S / N)_{i,\min}} \right]^{1/2} .$$

(13.3)

Given the intercept SNR in (13.2), the POI can be found assuming that integration is not possible in ESM receivers.

Example 13.1

Consider an X-band radar operating at 10 GHz has the following specifications:

Transmit peak power	25 kW
Pulse repetition frequency	21 kHz
Pulse width	1 µs
Transmitter antenna gain	40 dB
Transmitter antenna sidelobes	−5 dB
Bandwidth	5 MHz
Losses	0 dB
Noise figure	3 dB
Pulses per dwell	60
Process gain to echoes	16 dB
Detection S/N	15 dB

The intercept receiver's specifications are:

Antenna gain	10 dB
Interceptor receiver bandwidth	200 MHz
Noise figure	10 dB
Process gain to radar signal	0.4 dB
Interception S/N	20 dB

a. Find the range at which the radar detects a target of RCS 2 m².
b. Find the range at which the interceptor intercepts the radar's illumination.
c. If the interception can take place in the radar antenna main lobe, find the intercept range.

Solution:

a. The dwell energy in 60 pulses is $W_t = 25,000 \times 60 \times 10^{-6} = 1.5$ J. Using (13.1) gives

$$R_{max} = \left[\frac{W_t G^2 \lambda^2 \sigma}{(4\pi)^3 FKT_0 L(S/N)_{min}} \right]^{1/4} = \left[\frac{(1.5)(10^4)^2(0.03)^2(2)}{(4\pi)^3(2)(1.38 \times 10^{-23})(290)(1)(31.62)} \right]^{1/4}$$

$$\Rightarrow R_{max} = 152.27 \text{ km}$$

where $G = 40$ dB $= 10^4$, $F = 3$ dB ≈ 2, $S/N = 15$ dB $= 31.62$, $\lambda = 3 \times 10^8 / 10^{10} = 0.03$ m.

b. Assume that the intercept and radar specifications given previously are simultaneously met so that we can apply (13.3) as

$$(R)_{i,max} = \left[\frac{P_t G_t G_i \lambda^2 G_{pi}}{(4\pi)^2 F_i K T_0 B_i L_i (S/N)_{i,min}} \right]^{1/2} = \left[\frac{(20 \times 10^3)(.316)(10)(0.03)^2(0.4)}{(4\pi)^2(10)(1.38 \times 10^{-23})(290)(200 \times 10^6)(1)(10^2)} \right]^{1/2}$$

$$\Rightarrow R_{i,max} = 13.42 \text{ km}$$

where $G_t = -5$ dB $= 0.316$, $G_i = 10$ dB $= 10$, $F = 10$ dB $= 10$, $S/N = 20$ dB $= 10^2$.

c. In this case the interception takes place in the radar's main lobe where $G_t = 40$ dB $= 10^4$. Then it follows that the new intercept range is

$$R'_{i,max} = R_{i max} \left[\frac{10^4}{.316} \right]^{1/2} = 2465.58 \text{ km.}$$

13.6 ELECTRONIC COUNTERMEASURES (ECM)

Electronic countermeasure (ECM), as already defined, is that division of EW involving actions that are taken to prevent or reduce the enemy's effective use of the electromagnetic spectrum. ECM is thus the means of interfering with the enemy's electromagnetic activity. These means may be used to either deny the information, or to provide false information or to overload the enemy's computing capacity with so much false data, which degrades the performance of this system. This ECM mission may be achieved either by jamming or deception.

The basic operational objectives of an ECM are to: deny information (detection, position, track initiation, track update, and classification of one or more targets) that the radar seeks; surround desired radar echoes with so many false targets that the true information cannot be extracted; destroy hostile electronic systems to deny the key elements of the hostile radar sets and command, control, and communication (C3) structure; and saturate the threat system's data processing and operator capability.

ECM tactics and techniques may be classified in many ways to prevent or reduce the enemy's effective use of the electromagnetic spectrum. An encyclopedia of ECM tactics and techniques can be found in the literature.[7] Here, it is intended to limit the description to the most common types of ECM, as illustrated in Figure 13.2.

13.6.1 ACTIVE ECM

Active ECM involves degradation of the effectiveness of the enemy system by generating and transmitting electromagnetic energy. Denial ECM techniques are generally brute force ECM schemes, intended to prevent from detecting/tracking targets. Active ECM may be achieved either by denial jamming or deceptive jamming. Denial jamming attempts to deny the use of the electromagnetic spectrum to the radar, usually by providing so much noise signals to basically jam the radar's signal processor and displays from any effective operation. Deceptive jamming attempts to fool the radar into reporting the incorrect range, velocity, and azimuth information.

Denial Jamming

Receiver noise generally limits the sensitivity of most microwave radars. Raising the noise level by external means further degrades the radar's sensitivity. Noise is the fundamental limitation to radar performance and, therefore, can be an effective countermeasure. The objective of noise jamming is to inject an interference signal into the enemy's electronic system so that the actual signal

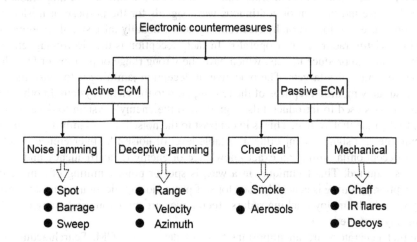

FIGURE 13.2 Illustration of the most common types of ECM.

is completely submerged by interference. This type of jamming is also called *denial jamming* or *obscuration jamming*, which generally employs some form of noise waveform. Some form of noise is generally preferable for denial jammers, since noise is more universal in its effectiveness against most types of radar emission. Denial noise jammers are conceptually very simple. The primary advantage of noise jamming is that only the minimal details about the enemy equipment need be known. Within the general class of noise jamming, there are three different techniques for generating noise-like interference:

Spot jamming: Any technique where the jammer's energy is entirely concentrated within the radar bandwidth is called *spot jamming*. It is also called *point jamming* where all the power output of the jammer is concentrated in a very narrow bandwidth, which can be effective against non-agile radars. The spot jammer can be a potent threat to the radar if it is allowed to concentrate large power within the radar bandwidth. This can be accomplished by changing the radar frequency pulse-to-pulse in an unpredictable fashion over the entire tuning band available to the radar. Spot jamming is usually directed against specific radars, and requires a panoramic receiver to match the jamming signal to the radar signal.

Barrage jamming: In *barrage jamming*, all the power output of the jammer is spread over a bandwidth much wider than that of the radar signal to accommodate radar frequency agility. In other words, it involves the massive and simultaneous jamming of the whole frequency band. It is less effective than spot noise because most of the jamming energy is rejected by receiver filters.

Sweep jamming: This is also similar to barrage jamming. In this case, the power output of the jammer is swept back and forth over a very wide bandwidth. It is generally true that the bandwidth of sweep jamming is wider than that of barrage jamming, but the relative bandwidth is often determined by the hardware used. The actual difference between barrage and sweep jamming lies in the modulation techniques and size of the frequency band covered. Barrage jamming often uses an amplitude-modulated signal covering a 10-percent frequency band. Sweep jamming often uses a frequency-modulated signal, and the frequency is swept back and forth over a wide frequency bandwidth. Both barrage and sweep jamming are used when the exact frequency of the enemy system is not known. One major disadvantage of this form of jamming is that it requires much more output power than spot jamming.

Deception Jamming

Deception is the intentional and deliberate transmission or retransmission of amplitude, frequency, phase, or otherwise intermittent or continuous wave signals for the purpose of misleading in the interpretation or use of information by electronic systems. Usually false signals (*spoofers*) are used to mislead the victim radar receiver operator. In fact, deception is the electromagnetic emissions or other techniques to produce targets, which have the wrong range or position or Doppler shift or which produce many false targets. The objective of deception jamming is to mask the real signal by injecting suitably modified replicas of the real signal into the victim system. In other words, this type of jamming is used to introduce false signals into the enemy's system to deceive or confuse, and hence to degrade that system. This is in contrast to the noise type of jamming, whose objective is to obscure the real signal by injecting a suitable level of noise-like interference into the victim system. For deception jamming, an exact knowledge of enemy radar frequency and transmission parameters is required. This technique, in a way, is spot or point jamming of a more intelligent nature. Deception jamming is generally used for self-protection applications against terminal-threat weapon types, which employ tracking radars. Techniques are numerous and many of the effective ones are mostly classified.

False target generators are an important form of deception ECM. Search/acquisition radars are susceptible to electronic deception by various schemes, which create the apparent presence of

Radar Electronic Warfare

253

multiple targets when there may be only one real target. The intent is merely to cause the defensive radar network to commit defensive resources improperly.

Deception jamming can either be *manipulative* or *imitative*. Manipulative implies the alteration of friendly electromagnetic signals to accomplish deception, while imitative consists of radiation into radar channels, which imitates a hostile emission. Manipulative ECM attempts to induce actions on the part of victims that are detrimental to the victim's well-being.

Within the general class of deception jamming, three main electronic techniques to return false signals have been developed. These signals have characteristics similar to those of the radar, thereby deceiving the radar into erroneous conclusions about range, velocity, or azimuth. The techniques of *range deception, velocity deception,* and *angle deception* can be applied against surveillance radars.

Range deception: Range deception jamming is used to foil missile-guiding radar systems where the tracking radar guides the missile (or other defensive measures) to the target in range by locking a range gate onto the target. It may also be used against search/acquisition and tracking radars. For the search/acquisition radar, range deception is really a false target generator. The most popular form of range deception against tracking radars is the range gate. This range gate delays the target echo, and its position is relayed to the missile to be used for intercept information. A range deception jammer, called a *range gate stealer,* attempts to break the tracking lock on itself by capturing the radar's range gate with a false echo and then moving it off to a false range location. Currently, false target generators use the poor antenna sidelobe levels of some surveillance/acquisition radars.

Velocity deception: Special deception jammers are used against Doppler tracking radars. Doppler tracking radars use the Doppler-shifted target return for tracking in a manner analogous to the range tracking gates in a pulse tracking radar. In the deceptive velocity jammer operation, the CW illuminator signal is detected by the jammer and an exact false, strong Doppler-shifted signal is sent back to the radar. The radar locks on to the incorrect Doppler signal and the jammer slowly sweeps the false signal's frequency farther from the actual Doppler frequency of the target. When the radar has been led far enough away in frequency, the jammer is turned off and the radar is once more left without a target. One can also create false Doppler signals using various means like pseudo noise.[8]

Azimuth deception: This is another deceptive ECM technique that degrades a tracking radar's ability to develop the correct azimuth and/or elevation data of a target, and can be used against angle tracking radars. Two methods of angle tracking in target tracking radars are conical scan and monopulse. An azimuth deception jammer, called *inverse gain radar repeater,* is normally used to deceive conical tracking radars. Several variants of inverse gain jamming have been conceived, the principal variant being sometimes called *swept audio frequency modulated jamming.* This vulnerability of a simple conical scan greatly contributed to the widespread switch to monopulse angle tracking, which is more complex and expensive and heavier. The conical scan is still more attractive for missile-borne tracking systems.

13.6.2 Passive ECM

Passive ECM is any interference technique that does not involve emitting an electromagnetic jamming signal. Examples include chaffs, decoys, chemicals, and any technique involving tactics only, such as hiding targets. This involves deception of an enemy's system by employing confusion reflectors. This may be achieved either by chemical or mechanical means. This type of jamming is also sometimes called *expendable countermeasures.* In its broadest sense, it not only uses the expendable passive ECM devices but also expendable active devices. These latter devices may be either jammers or deceivers, depending on the particular effect desired.

Chemical Jamming

Chemical jamming includes smoke and chemical agents dispersed in the atmosphere to confuse the enemy.

Smoke: It is the oldest countermeasure known to confuse enemy gunners by intentionally generating large clouds of billowing smoke, behind which friendly forces could then deploy thereby avoiding enemy fire. In the battle of Jutland in World War I, the German naval forces retreated under the cover of smoke to protect their decimated flotilla from further losses. The use of smoke, particularly against laser threats, has caused a resurgence of interest in recent times. Aerosols are the best chemical agents that are used as smoke, dust, mist, or fog.

Aerosols: Aerosols are fine solid or liquid particles dispersed in the atmosphere for countermeasure applications. The aerosol particle size and type are chosen in such a way that it allows both the scattering and/or absorption of radiations from electro-optical system targets and sometimes absorbing microwave signals as well. The scattering effect usually is the dominant source of attenuation for a countermeasure application. The most common types of smoke in use are those using either white phosphorus or fog oil.

Mechanical Jamming

This involves deception of an enemy's electronic system by use of mechanical jamming, which is synonymous with *chaff, decoys, flares,* and other reflectors that require no prime power.

Chaff: The chaff is made of elemental passive reflectors or absorbers that can be floated or otherwise suspended in the atmosphere for the purpose of confusing, screening, or otherwise adversely affecting the victim electronic system. Examples are metal foil strips packaged as a bundle, metal coated dielectrics (aluminum, silver, or zinc over fiberglass or nylon being the most common), string balls, ropes and semiconductors. Chaff consists of dipoles cut to approximately a half wavelength of radar frequency. It is usually packaged in cartridges that contain a broad range of dipole lengths designed to be effective over a wide frequency band, and is dispensed from aircraft, ships, or vehicles. The width/diameter ratio of the chaff affects its bandwidth as well. The chaff has properties very similar to those of weather clutter, thus appearing on enemy radar screen either as a blot masking the real target or as hundreds of false targets around the real one. This effectively breaks the track of a radar-guided missile. In 1973, Israeli boats bused rapid blooming chaff to screen themselves from the radars of Syrian gunboats equipped with Styx missiles. A list of most of the standard US chaff packages with their manufactures has been published.[9]

An analysis of the action of chaff shows that for maximum signal return one should make its length a multiple of one-half wavelength of the radar signal. This length maximizes the electrical resonance effect. It is also observed that the thinner the chaff, the more pronounced and frequency-specific the resonance effect is. Also, the radiated energy is strongest broadside to the individual chaff element. When this unit is dispensed in the atmosphere, it creates a radar echo similar to that of a small aircraft. If a stronger echo is needed, then two or three units are dispensed simultaneously. Fall rate of the chaff is both material and shape dependent. Fall rate is an approximately linear function of altitude.

The radar cross section of the chaff, which it presents to the radar is very important when the chaff is used to conceal a target. Current trends to a smaller radar cross section of an aircraft reduce the amount of chaff required to screen a target. Since the target is moving, a single chaff may not suffice for screening, but may be adequate for deception. Radar cross section alteration may be accomplished by shaping of the target. The use of additional reflectors may cause change of the apparent target RCS at the victim radar.

Radar Electronic Warfare   255

Since the chaff particles have considerable aerodynamic drag, their forward velocity quickly drops to near zero. Because of its low velocity, chaff can be regarded as an airborne clutter. Radars such as CW, pulse Doppler, and MTI that can reject clutter are not seriously affected by chaff. Thus, they can continue to track a target within a chaff cloud as long as the target has a radial component of velocity.

In the situation of defending relatively slow systems such as surface ships, there is so little difference in velocity between the potential target and the chaff that CW, pulse Doppler, and MTI radars have difficulty in separating the target from the chaff clutter. In shipboard defense, chaff rockets can be fired to burst at a specific location, hopefully within the field of view of the weapon RF seeker, creating an alternate target that is more lucrative than the ship itself. The disadvantage of this situation is that it requires an elaborate fire control system and movable launcher to position the chaff burst precisely. The chaff cloud combines with the ship to form one very large target with a combined centroid somewhere in the chaff cloud. An RF homing weapon that seeks the centroid of its target will thus fly harmlessly past the ship and through the chaff cloud.

Flares: A flare is a pyrotechnic target launched from an aircraft or other vehicles causing infrared homing missiles or other optical devices to be decoyed away from the true target. The flares are dispersed when the heat-seeking missile approaches its target to divert the missile from its target. Most dispensers used for chaff can also be used to drop infrared flares capable of confusing heat-seeking missiles. In addition to protecting tactical aircraft, flares also play a role in protecting strategic bombers. Early infrared weapons were very vulnerable to decoy flares, but most recent designs use flares or dual-operating frequencies.

Decoys: Decoys, which are another type of passive ECM, are a class of physically small radar targets whose RCS are generally enhanced by using reflectors or Luneburg lenses or active repeaters to simulate fighter or bomber aircraft. Decoys are made to appear as realistic targets, and could also be outfitted with a small jammer to mimic jammers on the target aircraft. Decoys might be carried on board attacking bomber aircraft and launched outside the normal radar detection ranges.

13.7   ELECTRONIC COUNTER-COUNTERMEASURES (ECCM)

ECCM is the art of reducing the effectiveness of an EW threat with the objective of making the cost of effective EW prohibitive for the enemy. The primary objective of the ECCM techniques when applied to a radar system is to allow the accomplishment of the radar-intended mission while countering the effects of the enemy's ECM. ECCM is a generic term that includes anything or any action resulting in the degradation of enemy ECM activities. It is certainly not limited to electronic techniques, but can include tactics, deployment, operational doctrines, and so on. In greater detail the benefits (Johnston 1984) of using ECCM techniques include prevention of radar saturation, enhancement of signal-to-jamming ratio, constant false alarm rate, discrimination of directional interference, rejection of invalid target, maintenance of target tracks, and radar system survivability. ECCM is mostly concerned with techniques that are embodied in the design of electronic equipment, while ECM usually requires a separate unit. There are two broad classes of ECCM: *electronic techniques* and *operational doctrines*. This leads to four areas of radar design in ECCM: *radar parameter management, signal processing techniques, design philosophy,* and *operational doctrines.*

13.7.1   RADAR PARAMETER MANAGEMENT

Components of the basic radar parameters where ECCM can be implemented include the antenna, transmitter, and receiver.

Antenna-Related ECCM

The first and probably the most important area of the radar to be considered for implementing ECCM is the antenna, since it represents the transducer between the radar and the environment, and is the first line of defense against jamming. Techniques for space discrimination include antenna coverage and scan control, reduction of main-beam width, sidelobe cancelers, sidelobe blanking, and adaptive array systems. Some of these techniques are related during the transmission, while others are useful during reception.

Either blanking or turning off the receiver can be essentially maintained in all sectors, except where the jammer is centered. Certain deception jammers depend on anticipation of the beam scan or on knowledge or measurement of the scan rate. Random electronic scanning prevents these deception jammers from synchronizing to the antenna rate, thus defeating this type of jammer. High gain antenna can be employed to spotlight a target and burn through the jammers. An antenna having multiple beams can also be used to allow deletion of the beam containing the jammer and still maintain detection capabilities with the remaining beams. Although they add complexity, cost, and possibly weight to the antenna, the reduction of beamwidth, coverage, and scan control can be accomplished.

The detection can be degraded if the jamming signal enters the sidelobes. For this reason, low sidelobes are desirable for both receiving and transmitting. Sometimes the increase in main-beam width resulting from low sidelobes worsens the problem of main-beam jamming. This warrants a careful consideration of choosing the appropriate main-beam and sidelobes.

Antenna sidelobes cause problems in a radar system. They provide a path for jamming and interference to enter the radar. In an airborne system, sidelobe clutters interfere with tail-aspect targets. Two techniques used to counteract the effects of sidelobes are *sidelobe blanking* (SLB) and *sidelobe cancelling* (SLC). Both require the use of an auxiliary antenna or antennas.

Sidelobe blanker (SLB): This is also a radar ECCM and anti-interference technique[10] that prevents some of the unwanted energy entering the sidelobes of a radar antenna from adversely affecting the radar operation. It is useful in both benign and hostile environments. In the benign environment, the SLB is useful as an EMC in reducing the effects of interference from nearby radars that operate at adjacent frequencies. Since the SLB employs a blanking gate, it is useful only against low duty cycle pulse or swept frequency jamming. High duty cycle and noise jamming effectively blanks the main channel most of the time, rendering the radar ineffective.

A simplified diagram of the basic elements of an SLB is shown in Figure 13.3. A method of achieving this is to employ two antennas: one is the radar main antenna with its main lobe and sidelobes; and the second is the sidelobe blanking antenna (also known as the guard antenna), whose gain is less than that of the main lobe of the radar antenna by perhaps 3 to 4 dB, and greater than sidelobes of the radar antenna. The receiver output from each antenna is fed to a receiver, with two receivers gain-matched to one another. Outputs

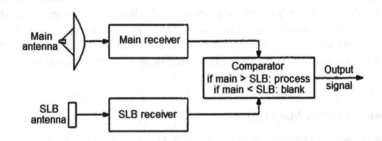

FIGURE 13.3 Principle of sidelobe blanking operations.

from the two receivers are compared on range-bin or Doppler-bin basis. If the output of the radar receiver exceeds that of the sidelobe blanking receiver, the signal must have been received from the main lobe of the radar antenna. This occurs because the main lobe is the only part of the radar antenna's pattern where the gain exceeds that of the guard antenna. This signal is then processed through the system. If the guard receiver output is greater than that of the radar receiver, the signal must have been received from a radar antenna sidelobe, and it is, therefore, blanked on a single sweep basis.

Sidelobe canceler (SLC): This is another ECCM technique for use in surveillance or tracking radars that suppress high duty cycle and unwanted noise jamming energy received through the sidelobes of the antenna. This is accomplished by employing an array of auxiliary antennas used to adaptively estimate the direction of arrival, and, subsequently, to modify the receiving pattern of the radar antenna to nulls in the jammer's direction. The conceptual scheme of an SLC system is shown in Figure 13.4 where only connection *a* is included in the closed-loop implementation techniques.

The auxiliary antennas provide replicas of the jamming signals in the radar antenna sidelobes. The auxiliaries may be individual antennas or groups of receiving antennas of a phased array antenna. The amplitude and phase of the signals by the auxiliaries are controlled by a set of suitable weights $W = \{W_1, W_2, \cdots W_N\}$. By adaptively controlling the phase and amplitude of the auxiliary channel signal and performing linear combination of the auxiliary signals and the main channel signal, a null in the composite antenna pattern response can be produced in the direction of the jammer. Through continuous adaptive control of the antenna pattern, the null can be made to track the jammer. Owing to the stochastic nature of the jamming signals in the radar and in the auxiliary channels, the radar signals are denoted by $V = \{V_1, V_2, \cdots V_N\}$. The jamming signal in the channels may be regarded to have zero mean value. The adaptive sidelobe canceler system is relatively complex.

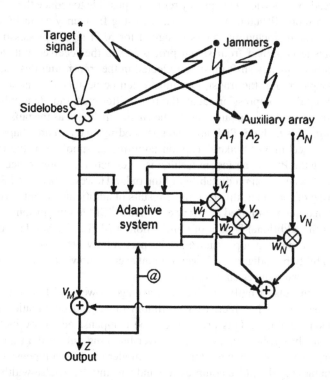

FIGURE 13.4 Principle of SLC operations (From Skolnik, 1990).

There is another very simple technique that employs the same antenna and receiver configuration as the SLB, except that a gain matching and canceling process takes place. Except the target signal, all false signals entering the sidelobe of the main antenna get canceled at the output. The technique is quite effective against a single noise jammer only, which poses a serious limitation in practical applications.

Transmitter-Related ECCM

The different types of ECCM area related to the proper use and control of the *power, frequency,* and *waveform* of the radiated signal.

Power: One brute-force approach to defeating active ECM or noise jamming is simply to increase power of the transmitter. This technique, when coupled with *spotlighting* the radar antenna on the target, results in the increase of the radar's detection range. Spotlighting or *burn-through* mode might be effective. In addition to being relatively inefficient in radar power management, such approach is not effective against chaff, decoys, repeaters, spoofers, and so on. For a ground or surface radar, power is often considered the fundamental ECCM parameter. With this view, ECM becomes a power battle, with the outcome going to the stronger, more powerful opponent. Airborne jamming equipment is limited in size and weight and therefore has a power limitation. Thus, the power advantage lies with the ground or surface radar. In the case of one surface unit versus another, both operate under the same constraint, and the outcome is not obvious.

Frequency: The use of complex, variable, and dissimilar transmitted signals can place a maximum burden on ESM and ECM. Frequency agility usually refers to radar's ability to change the frequency of the transmitter on a pulse-to-pulse basis, while frequency diversity refers to a larger frequency change on a longer time scale. Frequency agility and diversity represent a form of spread-spectrum ECCM in which the information bearing signals is spread over as wide a frequency region as possible to reduce the detectability and make jamming more difficult. A second way of using frequency as an ECCM technique is the Doppler radar, including radars designed for MTI signal processing. The actual ECCM advantage is gained from signal processing in the receiver, but the intention to use the Doppler frequency shift must be reflected in the transmitter design. For example, in a pulse-Doppler radar, the transmitter must often be designed to radiate a very stable frequency. In a pulse compression radar, the transmitter must radiate a pulse with an FM slide, sometimes called a chirp pulse due to the change in "tone" as the pulse is transmitted.

Waveform coding: This includes PRF jitter, stagger, coding, and perhaps shaping of the radar pulse. All these techniques make deception jamming or spoofing of the radar very difficult. Changing the PRF in a random fashion is an effective counter to deception because deception ECM depends on predictability of the radar. However, because PRF is related to the basic timing of the radar, this technique results in additional complexity and expense. Random PRF has been employed as a very effective ECCM feature in some radar for many years and has the additional benefit of elimination of MTI radar blind speeds. Intrapulse coding to achieve pulse compression may be particularly effective in improving target detection capability by radiation of larger amount average power by improving range resolution, which, in turn reduces chaff returns.

An increase in pulse length will increase average power and thus increase detection probability. The trade-off is increased minimum range and degradation of the radar's range resolution capability. This problem can be compensated for by including a pulse compression capability; however, due to receiver blanking during the transmit cycle, the minimum range will stay relatively long. Some modern radars compensate for these difficulties by employing the pulse compression and varying their pulse width depending on the mode of operation and expected target range.

Radar Electronic Warfare

Receiver-Related ECCM

The radar receiver plays a very important role in an ECCM environment in that it is almost a court of last resort. The receiver has two main roles in rejecting ECM: first, it should have effective band-pass filters to reject as much broadband jamming as possible; second, it amplifies both echoes and jamming linearly to prevent intermodulation products, which the signal processor may not be able to remove. The radar receiver is a descendent of communication receivers.

A radar receiver that has wide dynamic range, excellent stability, reasonable linearity, freedom from spurious responses, and is capable of avoiding saturation of the receiver signal processor, is normally used. Several other characteristics also include its capability of handling a rapid rate of change of input signals and freedom from desensitization. Galelian[11] presented an excellent review of basic considerations of radar receivers design. One method to overcome desensitization and reduce the levels of interfering signals is to incorporate appropriate filters at the RF amplifier/mixer input. A balanced mixer is useful in reducing the effects of local oscillator noise in the hostile situation.

Dicke fix: Dicke fix (Johnston 1979) was conceived as an ECCM against fast sweep jammers of the backward wave oscillator type. This is an ECCM technique to counter continuous wave jamming, and swept spot-noise jamming. The Dicke fix or wideband limiter device can provide some clear unjammed ranges where the radar can operate efficiently. Dicke fix is, thus, a technique that is specifically designed to protect the receiver from fast sweep jamming. The basic configuration consists of a broadband limiting IF amplifier, followed by an IF amplifier of an optimum bandwidth. The limit level is preset at approximately the peak amplitude of receiver noise. The bandwidth may vary from 10 to 20 MHz, depending on the jamming environment. This device provides excellent discrimination against fast sweep jamming without an appreciable loss of sensitivity.

Constant false alarm rate (CFAR): CFAR(Johnston 1979) is a technique that is necessary because of the limitation of the computer in automatic systems. It prevents the computer from being overloaded by lowering the capability of the radar to detect desired targets. Also, this is a radar receiver ECCM technique wherein the receiver adjusts its sensitivity as the intensity of the undesired signal varies. This makes the functioning of radars possible in an environment where interference due to signals from clutter, rain, jammers, and other radiating sources are present. These undesired signals can obscure real targets on the radar display or overload a computer so as to degrade decisions on absolute detection threshold criteria. The CFAR technique keeps the detection of false alarm rate constant when the radar is receiving these undesired signals. CFAR does not usually permit the detection of a target if the target is weaker than the jamming, but it does attempt to remove the confusing effects of the jamming. Thus, CFAR does not give immunity from jamming; it merely makes the operation in the presence of jamming more convenient by making the receiver less sensitive.

A CFAR receiver, no matter whether it is an automatic device or an operator controlling the receiver gain, maintains the false-alarm rate constant by reducing the probability of detection. When the threshold level is raised to maintain a constant false-alarm rate, marginal echo signals that might normally be detected do not cross the higher threshold and are lost.

Log receiver: Log (logarithmic) receiver is an ECCM technique, which is useful in preventing receiver saturation in the presence of variable intensities of jamming noise, rain, clutter, and chaff. The video output of the logarithmic receiver is proportional to the logarithm of the envelope of the RF input signal over a specified range. Log receivers have the ECCM advantage of permitting the radar receiver to target returns that are larger than jamming noise, chaff, or clutter levels. However, the disadvantage lies in the fact that low-level jamming signals will be amplified more than higher level target signals, thereby reducing the signal-to-jamming ratio.

13.7.2 Signal Processing Techniques

A large number of ECCM tactics or methods have been developed over the years as a result of continuous improvements in radar signal processing techniques. Signal processing techniques are usually functions that are incorporated into the radar receiver. Although certain signal processing techniques may place constraints on the transmitter, many of them have been added to the receiver after the radar has been built. These techniques are called ECCM or *anti-jamming* (AJ) fixes, since they were initially developed as retrofits to improve existing equipment.

Doppler radars, including radars with MTI signal processors, although not designed specifically for ECCM purposes, are quite ECM resistant. Since Doppler radars operate on the frequency shift caused by a moving target, they automatically filter out returns from nonmoving targets and consequently eliminate many unwanted signals, such as those from chaff. They will even discriminate between returns from objects of different velocities such as an aircraft in a chaff cloud. This technique can also make deception more difficult, since the deceiver must imitate the proper frequency shift.

The presence of a jamming signal in a radar with automatic threshold detection can increase the rate of false alarms to an intolerable extent. If the radar output data is processed in an automatic device such as a computer, the device might be overloaded by the added false alarms due to jamming. Thus it is important that the receiver present a constant false-alarm rate. Receivers designed to accomplish this are called CFAR receivers. Their disadvantage lies in the likelihood that some weak targets will remain below the threshold and be lost.

If an operator were monitoring the radar output, the effect of the additional false alarms could be reduced by having the operator turn down the gain of the receiver during the presence of jamming. In an automatic threshold detector, the same effect may be obtained by using the average noise level to provide an automatic gain control, much as an operator would by adjusting a manual gain control. Because the automatic CFAR circuits react faster, they are superior to an operator in keeping the false-alarm rate constant, especially when the radar is subject to noise jamming from only a few azimuth sectors.

13.7.3 Radar Design Philosophy

A general rule of thumb for ECCM radar design is to incorporate unpredictable operating parameters. The more orderly a radar is in its operation, the easier it is to predict what the radar is going to do or how it is going to operate; consequently, the job of applying an ECM technique effectively becomes simpler. ECM becomes more difficult, however, if characteristics of the victim radar are constantly changing. The parameter that may most easily be varied to confuse the ECM operator is the frequency. The capability for operator variation of pulse length, PRF, modulation, and antenna characteristics is commonly built into radars to make ECM more difficult.

The most common way to introduce unpredictability into radar design is through frequency diversity. Early radars were all designed to operate in a few specific frequency bands, where narrow-band jamming would render them all ineffective. New radar systems are designed so that each different radar type operates in a different frequency band.

13.7.4 Operational Doctrines

Another aspect of ECCM is the relationship between automatic equipment and the human operator. The trained radar operator fulfills a useful and necessary role in a countermeasure environment and cannot be completely replaced by automatic detection and data processors. An automatic processor can be designed to operate only against those interfering or jamming signals known beforehand; that is, any capability against such signals must be programmed into the equipment beforehand. New jamming situations not designed into the data processor might not be readily handled. In the

Radar Electronic Warfare

matter of operational and other ECCM, the operator has probably been the most important radar ECCM. An operator has the ability to adapt to new and varied situations and is more likely to be able to cope with and properly interpret a strange new form of interference than can a machine. Therefore, a skilled operator is the most important counter-countermeasure for maintaining radar operation in the presence of deliberate and clever countermeasures.

A good operator can do a great job in detecting targets amid jamming. This requires extensive training in real jamming. There are a number of operation methods that should be considered. Some of these are the use of dummy transmitters, intermittent radar operation, variation of radar operating procedures, and minimizing of radar operation. Radar operation in chaff also requires a careful attention. Johnston[12] in his paper presented several ECCM operating techniques in a chaff environment. These include observation of the chaff cloud on the PPI, then looking outside the cloud for stray aircraft, variation of radar operating frequency, and reduction of radar pulse width. In the case of radar deployment, the radar operators should be remotely located from the operating position to protect them from bomb damage. ESM can be of assistance to the radar operators by monitoring of aircraft communications as a means of alerting the radar operator. In other operational ECCMs, there should be a close liaison between development engineers and the military field forces. In some cases new environments arise requiring changes in the radar design.

REFERENCES

1. A. Price, "History of US Electronic Warfare," *Association of Old Crows*, 1 (1984).
2. US Department of Defense (DOD), "Electronic Warfare Definitions," *Electronic Warfare Magazine*, vol 2, pages 4 and 29 (1970).
3. US Department of Defense, Joint Chief of Staff, "Dictionary of Military and Associated Terms," *JCS Pub-1*, September 1974.
4. S. L. Johnston, ed., *Radar Electronic Counter-Countermeasures* (Norwood MA: Artech House, 1979).
5. S. L. Johnston, "Radio Defense System Electronic Countermeasures," *Course Notes,* Continuing Education Institute Course ENG 216.
6. R. G. Wiley, *Electronic Intelligence: Interception of Radar Signals* (Norwood, MA: Artech House, 1985).
7. J. S. Lake, "Observable Countermeasures," *Proceedings of the Military Microwaves Conference (MM-84)* (1984): 391–396.
8. K. O. Bryant, "Programmable 20-bit Pseudorandom (PRF) Generator," *US Patent #3,662,386*, May 9, 1972.
9. E. H. Eustace, ed., *International Countermeasures Handbook*, 3rd ed. (Palo Alto, CA: EW Communications, Inc., 1977).
10. L. Maisel, "Performance of Sidelobe Blanking Systems," *Transactions on Aerospace and Electronic Systems*, 2 (1968): 174–180.
11. J. C. Galelian, "Designing Radar Receivers to Overcome Jamming," *Electronics Magazine, 36* (May 1963), 50–54.
12. S. L. Johnston, "Radar Electronic Counter-Countermeasures against Chaff," *Proceedings of the International Conference on Radar*, 517–522, Paris, France, May 1984.

14 Over-the-Horizon Radar

"Love doesn't think like that. All right, it's blind as a bat—Bats have radar. Yours doesn't seem to be working."

—**Iris Murdoch**

14.1 INTRODUCTION

For conventional radar applications, frequencies at VHF band or lower are seldom used because of their narrow bandwidths, wide beamwidths, and the potential interference from other users in the same frequency bands. However, the HF region of the spectrum has a unique property that allows propagation of the radar wave to longer distances beyond the curvature of the earth using sky wave propagation where the ionosphere can be thought of as providing a virtual mirror. In addition, modest extensions of the line-of-sight (LOS) range can be obtained in this frequency band exploiting anomalous propagations such as ducting and refraction from troposphere of the atmosphere. Although the HF band is officially defined as extending from 3 to 30 MHz, the lower frequency limit might lie just above 3 MHz, and the upper limit can be extended to 40 MHz or more.

Called *over-the-horizon radar* (OTHR), it is typically a radar system that takes advantage of the peculiar interaction between the HF-band signals and the ionosphere to operate on a very wide area of surveillance, much larger than that of conventional microwave ground-based radars. The underlying propagation processes and interactions are what make OTHR signals, different than other forms of EM propagation, more desirable for many applications. The features of large operating wavelengths and the HF band have a dominant effect on transmitter, antenna, and receiver design, and make the discipline different in kind from conventional radars operating at microwaves, where antenna aperture greater than 10 m^2 and bandwidths greater than 10% are very rare. The maximum detection range of the OTHR system is not limited by the curvature of the earth, allowing detect targets completely obscured by the horizon. It is necessary to employ a relatively reduced amount of transmitted power if compared with conventional counterparts; and this is perhaps the greatest strength of OTHR systems.

The other features—limiting OTHR applications—include high noise level and also high clutter-to-signal ratio, variable path attenuation in sky-wave propagation, and congested radio spectrum, which particularly at night can make clear channel identification a major problem.

14.2 HISTORICAL NOTES

Some of the first prototypes of radar systems, developed in the early '20s, were designed to operate in the HF band. In fact, the peculiar phenomena of interaction between the ionosphere and the radar wave with frequency included in the HF band in the spectrum has been known and investigated since the first experiments of Marconi, Tesla, Popov, and others in which they tried to establish a radio link between two locations without LOS.

During World War II, Britain created the first radar systems operating in the HF band for military purpose despite the high level of environmental noise, and succeeded in detecting the German bombers approaching the English coast. The *Chain Home* system, designed by Robert Watson-Watt, was an HF radar system designed to cover and protect the British coast demonstrating usefulness of HF radar systems.

Many of the radar development programs, conducted in numerous countries including the United States, Russia, and the United Kingdom, operated in the HF band. The first experiments were conducted in Russia at the end of the '40s and in the United States in the early '50s. In 1949 the Soviets created the first experimental OTHR called *Veyer* followed by *Duga-1, Duga-2,* and *Duga-3.*

In the United States much of the early research on OTHR was carried out at NRL extensively under the direction of William J. Thaler. Their first experimental system, *Multiple Storage, Integration, and Correlation* (MUSIC), became operational in 1955, and could detect a rocket launched 970 km away at Cape Canaveral and nuclear explosions in Nevada at 2,700 km. A greatly improved system was later developed in 1961 as *Magnetic-Drum Radar Equipment* (MADRE)[1] at Chesapeake Bay, which demonstrated its ability to detect aircraft as far as 3,000 km using as little as 50 kW of transmitter power. In the fall of 1961 MADRE began tracking aircraft on the North Atlantic air traffic route and beyond the conventional radar horizon. The transmitter and receiver of the MADRE were at the same location using the same horizontal antennas. The large fixed array was first employed to detect and track aircraft over-the-horizon (OTH) and later was used in tracking surface vessels.

Stanford University designed and built the Wide Aperture Research Facility (WARF) in the early 1960s. The WARF[2] is a high-resolution two-site OTHR system where the transmitter and the receiver are at two locations allowing the use of a frequency-modulated continuous waveform (FMCW). Both the transmit and receive antennas used by WARF were quite different from those used with MADRE. Where the MADRE used horizontally polarized antenna arrays, the WARF uses vertically polarized array antennas. The improvement in azimuthal resolution gave WARF much better performance against small, slow targets. Additional radars include the Russian Steel Yard OTHR built in 1976, which caused noticeable amounts of interference for ham radio operators in the United States, nicknamed *Russian Woodpecker.* This Woodpecker still has applications in national defense in addition to scientific research and law enforcement.

Cobra Mist developed jointly by the US Air Force and the Royal Air Force used a simple pulse waveform similar to that used by MADRE but with 10 dB higher average power with an antenna consisting of 18 *log-periodic arrays.* Cobra Mist had poor resolution in range and azimuth, and did not meet the expectations. The US Navy with the assistance of the ITT Electro-Physics Laboratory (EPL) installed an OTHR system in the mid '70s, which used a *phase-coded pulse waveform* to obtain good resolution with a high duty factor. It was capable of frequency hopping between unoccupied channels.

In 1970 the USAF-RADC (Rome Air Development Center) installed a bistatic OTHR system in Maine designed to guarantee total coverage of US east and west coastlines. In the same year the Australian Defence Science and Technology Organization (DSTO) started a series of experiments with HF radars, and conducted a project— *Jindalee Operational Radar Network* (JORN)—to study the capability of an OTHR system to detect airliners. Following the success of the first experimental phase, the JORN project led to the creation of an OTHR system to be employed as a test bed in 1982.

The US Navy with the assistance of Raytheon developed relocatable OTHR, also known as ROTHR to able to transport it to a previously prepared site, assembled, and put in operation rapidly. The construction of the apparatus began in 1988, and the radar became fully operational in 1991. The whole system once installed in Alaska was moved to Chesapeake and converted for civilian applications. France started a project called *Systeme de Traitement Universel de Diagnostic Ionospheriques* (STUDIO) and installed an OTHR to become operational in early 2000 capable of monitoring airliners above Corsica and Sardinia Islands. In 1997 Papazoglou and Krolik[3] devised a technique for using matched field processing to measure the altitude of aircraft targets being tracked by OTHR using a single coherent dwell of radar data. Later in 1999 they extended the technique to allow using multiple short dwells of radar data. In 2002 Anderson et al.[4] used a state-space fading model of the multipath propagation to make the target altitude estimate more robust to random changes in the ionosphere. The last decade of the millennium witnessed the entry of many other countries in the scenario of OTHR research and development such as China, France, Iran, India, and Ukraine.

Over-the-Horizon Radar

14.3 CLASSIFICATION OF OTHR SYSTEMS

OTHR systems can be classified on the basis of different propagation phenomena. Consequently, OTHR systems use two most commonly techniques: *skywave systems* that exploit ionospheric reflection of HF signals for long-range detection, and *surface wave systems,* which use HF radio waves due to diffraction and/or ducting, and follow the curvature of the earth to detect targets beyond the horizon.

14.3.1 SKYWAVE OTHR SYSTEM

The most common type of OTHR system, sometimes called skywave radar, uses skywave propagation in which HF radar waves are reflected from the ionosphere. Skywave propagation beyond the horizon is made possible by the existence in the atmosphere of approximately spherically stratified ionized layers concentric with the earth's surface. The radar installation in the skywave system operates like an HF communication system, radiating energy in a narrow azimuthal beam, but with a broad vertical beam extending typically from 5° to 25° elevations. The radar signal is transmitted toward the ionosphere that absorbs part of the energy; but the major part of the energy is reflected from the ionosphere, the amount being dependent on the actual electron density in the ionosphere, frequency and the angle of incidence of the HF signal. The specific frequency to be used by a skywave radar is a function of the desired range and the character of the ionosphere. Since the ionosphere varies with many factors such as time of the day, season, and solar activity, the optimum frequency must be selected from time to time to ensure that refraction in the ionospheric layers results in the maximum illumination of the target.[5] Thus, such radars can operate over a wide portion of the HF band. The *maximum usable frequency* (MUF) and the *least usable frequency* (LUF) are the two terms used to describe the frequency-dependent characteristics of the ionosphere that determine its ability to support HF skywave propagation. Both MUF and LUF are dependent on ionospheric conditions, and provide a portion of the HF band available to support skywave propagation at any given time and space.

The ionosphere may consist of more than one refracting region resulting in multipath propagations that can affect the performance of the OTHR. This effect of multipath can be reduced significantly by properly selecting the operating frequency and by using the narrow elevation beamwidths. The movement of the ionosphere can make OTHR operations more complex, which can be circumvented by using a second transmitter broadcasting directly up at the ionosphere to measure the movement and then adjusting the returns of the main radar in real time. The dynamic behavior of the ionosphere results in the uncertainty in the prediction of the propagation path and consequently the origin of the echo received by the OTHR system.

The signal reflected from the surface or sea is sometimes very large compared to the signal reflected from the target. The easiest way to distinguish targets from the background noise is to use Doppler effects. By filtering out background noise, the moving targets can be detected. More scatters that affect the performance of the skywave radar include meteor, high atmospheric noise, broadcasting transmissions, and man-made impulsive interference. Skywave radars can also suffer significant polarization losses and focusing/defocusing problems associated due to ionospheric effects, and are sometimes susceptible to deliberate jamming. A skywave radar is analogous to an airborne radar looking down at the earth's surface and trying to detect surface targets. Doppler velocity discriminations are essential because they must identify the targets against a very strong surface clutter.

Skywave radars are currently used in the roles of aircraft tracking, ship tracking and sea-surface wind and wave-height mapping, and hurricane tracking.

14.3.2 SURFACE-WAVE OTHR SYSTEM

The type of OTHR that propagates energy at HF around the curvature of the earth by diffraction and/or ducting is commonly called surface-wave or ground-wave radar. This type uses much lower frequencies in the longwave band. Radio waves at these frequencies can diffract around the curvature of the earth and any obstacles, and thus travel beyond the horizon. Echoes reflected from the target return to the radar receiver by the same path. These surface waves provide longest range over the sea. A surface-wave radar avoids the use of ionospheric reflection/refraction exploiting rather the low attenuation characteristics of vertically polarized waves over sea water at frequencies below 30 MHz. In this type of OTHR, low-frequency systems equipped with a highly sensitive receiver are generally used for tracking ships, rather than aircraft.

Detection is somewhat easier than with skywave propagation since ionospheric effects are not present and clutter returns from aurora generally can be eliminated by time gating. The surface-wave radar must be located on the coast, or on an island or ships, as even a short distance of overland path can cause severe attenuation of the transmitted and received surface wave. Overland applications are thus not desirable. The surface-wave radar has a far shorter range than with skywave radar because of the propagation loss, which increases exponentially as a function of the range, and provides a range against low-altitude aircraft targets at perhaps maximum range of 200 to 400 km. Advantages of the surface-wave radar are many: freedom from ionospheric reflection enabling the Doppler velocity discrimination and tracking more accurately; capable of avoiding long-range skywave clutter; freedom in selecting the operating frequency; and no need to follow the diurnal and seasonal ionospheric variations.

Surface-wave radars find applications for wave height and wave energy direction mapping, detection of low-flying aircrafts, monitoring of ship traffic and icebergs, and monitoring of the continental coastal zones.

14.4 IONOSPHERIC EFFECTS ON THE OTHR SYSTEM

HF skywave propagation beyond the horizon is made possible by the existence of the ionosphere consisting of approximately spherically stratified ionized layers concentric with the surface of the earth. The variability of the layers with time of day, season, and sunspot cycle heavily affects the propagation HF radar signals. The ionosphere is one of the layers of the atmosphere that contains charged particles. This region exists because ultraviolet rays from the sun ionize the atmospheric gases. The ionization increases, but not linearly, with height and tends to have maximum values at particular heights. The electron-density distribution plays a significant role in controlling the propagation of HF waves. When the radio wave is obliquely incident and traverses a path where the electron density is increasing with altitude, the ray is bent away from the ionosphere. If the gradient in electron density is sufficient, the wave will reflect back to the earth, providing long-distance illumination and thus making skywave possible. The ionospheric effects on radio waves vary significantly on the frequency. Low frequencies below HF band tend to be absorbed by the atmosphere, and the high frequencies above HF band tend to cut through the ionosphere. Thus, the OTHR using skywave propagation is feasible only in the HF band. Additional factors that can affect the radar performance include ionization irregularities and backscatters.

Figure 14.1 provides examples of the effect of an ionized layer in several ray paths demonstrating the skywave propagation to beyond the horizon. A ray penetrates farther into the layer as the angle the ray makes with the horizon is increased. Thus the rays transmitted at angles greater than the critical angle do not return to the earth, and effectively escape into the atmosphere. The maximum distance that can be covered in a single hop or reflection is roughly 2000 km. OTHR systems employ HF and one or more hops to obtain radar coverage of regions that are thousands of miles away and that are practically impossible to cover with conventional microwave radars. The radially stratified layers of the ionosphere that are considered necessary to model for transmission paths are designated as D, E, and F layers as discussed next (Kelso 1964). The different layers in the ionosphere and the corresponding typical electron densities of nominal values are illustrated in Figures 14.2 and 14.3, respectively.

D Layer: This layer occupies the lowest altitude ranging from 60 to 90 km, where the electron density increases rapidly with altitude in the daytime at which time the layer bends and absorbs low frequency less than 3 to 70 MHz. The maximum ionization in the D region occurs near the sunspot solar activity.

Over-the-Horizon Radar

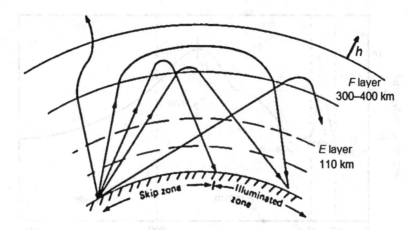

FIGURE 14.1 Typical behaviors of ray paths passing through the ionized layers.

E Layer: This layer has characteristics very similar to those of the *D* layer, but exists at a higher altitude extending between about 90 and 120 km. In addition there may be an anomalous ionization referred to as sporadic *E* layer, which is seasonally and diurnally variable, and suffers variation with latitude.

F Layer: This is the most important layer for skywave propagation, and occupies the highest-altitude region. It is also the region of highest electron density. It may be considered as splitting into two regions, called F_1 and F_2, especially in summer day times. The F_1 layer occupies a region extending between 120 and 160 km, and is directly dependent on solar radiation. The F_2 layer is variable in both time and geographical location, and extends beyond 160 km.

OTHR system planning heavily depends on modeling of the ionosphere in terms of the refractive index n^6 and the electron density N_e^7 of the ionosphere, which are expressed, neglecting the earth's magnetic field, as

$$n = \left[1 - \left(\frac{f_c}{f}\right)^2\right]^{1/2} \tag{14.1}$$

where f_c is the critical or plasma frequency, and

$$f_c = \left(\frac{N_e e^2}{4\pi^2 m \varepsilon_0}\right)^{1/2} \approx 8.98\sqrt{N_e} \tag{14.2}$$

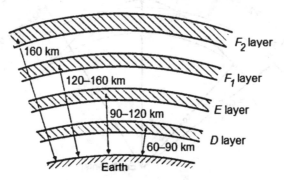

FIGURE 14.2 The layers of the ionosphere.

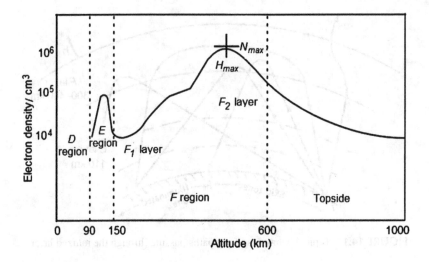

FIGURE 14.3 Typical electron densities of nominal values (cm^{-3}).

where f is the transmitter frequency in Hz, ε_0 is the free-space permittivity, e and m are the charge and mass of an electron, respectively, and N_e is the electron density per cubic meter at a desired height, given by

$$N_e = N_m \left\{ 1 - \left(\frac{r - r_m}{y_m} \right)^2 \left(\frac{r_b}{r} \right)^2 \right\} \quad \text{for } r_b \leq r \leq \frac{r_m r_b}{r_b - y_m} \qquad (14.3)$$

where r is the radial distance from the center of the earth, r_m is the radial distance from the center of the earth where the maximum electron density N_m occurs, r_b is the radial distance from the center of the earth to the base of the designated layer, y_m is the semi-thickness of the designated layer. The parameters are illustrated in Figure 14.4.

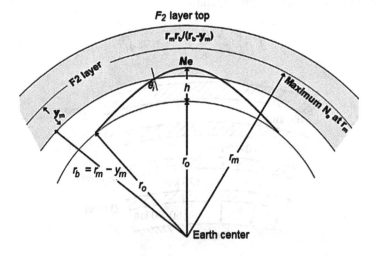

FIGURE 14.4 Illustrations of variables used in the expression of N_e.

14.5 RAY PATH TRAJECTORIES

It is essential to model for the trajectory of ray paths due to the presence of ionized ionosphere and the troposphere with variable refractive index that decreases with increasing altitude. The electron-density distribution is the major control over the propagation path of HF radio waves. Surface illumination over the horizon is enabled by refraction from the ionosphere. If the gradient of the electron density is sufficient, the radio wave is reflected back to the earth. In ground-wave propagation, paths considered are due to direct LOS and by sea-surface reflection when the radar and target are above the horizon and the illumination in the shadow region by anomalous propagation of waves. The useful models for such rays are available for three cases pertinent to thin, thick, and multiple layers. In these models the effect of the earth's magnetic field is assumed negligible.

14.5.1 The Thin Layer Model

In the *thin layer model* it is assumed that the electromagnetic energy emitted from an antenna makes an angle of elevation θ_e with LOS, and propagates in free space up to the layer where it is reflected specularly, which occurs if the transmitted frequency f satisfies the following relation:

$$f < f_c \sec \theta_i \qquad (14.4)$$

where f_c is the critical frequency expressed by (14.2), and θ_i is the angle of incidence. The critical angle θ_c, which is defined as the minimum angle of elevation at which the ray escapes through the layer, is given by

$$\theta_c = \cos^{-1}\left(\frac{f_c}{f}\right). \qquad (14.5)$$

Applying the sine law to the triangle TOB of the geometry shown in Figure 14.5, we can write an expression relating the parameters as

$$\frac{p/2}{\sin \theta_o} = \frac{r_b}{\sin(\theta_e + \pi/2)} = \frac{r_o}{\sin \theta_i} \qquad (14.6)$$

where p is the total slanted path length, D is the ground range, r_o is the radius of the earth, and r_b is the distance of the layer from the center of the earth.

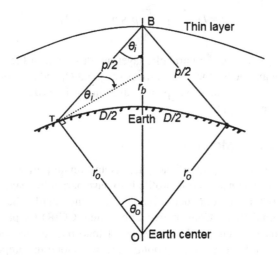

FIGURE 14.5 Geometry of thin layer model.

Solving (14.6) and observing that $\theta_o = (\pi/2 - \theta_i - \theta_e)$, we can write the explicit expressions for the parameters as follows:

$$\theta_i = \sin^{-1}\left[\left(\frac{r_o}{r_b}\right)\cos\theta_e\right] \tag{14.7}$$

$$D = 2r_o \tan\theta_o \approx 2r_o\theta_o \tag{14.8}$$

$$p = \frac{2r_o}{\sin\left(\dfrac{\pi}{2} + \theta_e\right)}\sin\theta_o. \tag{14.9}$$

These relationships are directly applicable to propagation by the E layer of thin sheets of ionization. The thin layer model can also be applied to propagation by the F_2 layer at frequencies well below the MUF.

14.5.2 The Thick Layer Model

The analysis of propagation by way of the thick ionospheric layer follows when the frequency of transmission or the angle of incidence is sufficiently large for the rays that penetrate deep into the layer before being refracted back to the earth. A single *thick layer model* applies if the ionization beneath the main refracting layer is lower compared with that in the layer itself. Bouguer's rule, described as the Snell's law for spherically stratified layers, is usually applied to the vertical variation of electron concentration $N(h)$:

$$nr\sin\theta = r_b\sin\theta_i = \text{constant} \tag{14.10}$$

where r is the radius from the center of the earth to a point on the ray, r_b is the radius to the bottom of the layer where the refractive index n departs from unity and follows the parabolic law. If the curvature of the earth and the spherically stratified ionosphere are taken into account, the approximate solutions[8] for the group path p', and the ground range D, respectively, are

$$p' = p_1' + p_2' = 2xy_m \arctan\left[\frac{x\cos\theta_i}{1-(y_m/r_b)x^2\sin^2\theta_i}\right] + \frac{2r_o\sin\theta_o}{\sin\theta_i} \tag{14.11}$$

$$D = D_1 + D_2 \approx \frac{r_o}{r_b}p_1'\sin\theta_i + 2r_o\theta_o \tag{14.12}$$

where $x = f/f_c$, semi-thickness of the parabolic layer as illustrated in Figure 14.4, p_1' and p_2' are parts of the group path in and below the layer respectively, and D_1 and D_2 represent the corresponding portions of the ground range. The group path p' used in (14.11) is defined as the product of free space velocity and the time of flight.

14.5.3 The Multiple Layers Model

The concept of discrete stratified layers in ionosphere is an oversimplification. Bradley[9] has shown that a good model of electron concentration profile is to have parabolic maxima and linear regimes in between. The errors obtained from such a model are found small. The model is now recommended by the International Radio Consultative Committee (CCIR) for prediction purposes, and ray parameter expressions are mentioned in the Atlas of Ionospheric Characteristics.[10] It has been very effectively used in a modified *quasi-parabolic form* by Pielou[11] in computing spatial coverage and multipath occurrence.

Over-the-Horizon Radar

14.6 PRINCIPLES OF OTHR SYSTEMS

The discussions underlying the principles of OTHR systems are addressed in this section especially focusing on OTHR equations, waveforms, and target detections.

14.6.1 OTHR Range Equation

The factor affecting the design of an OTHR are slightly different than those affecting radars that use microwave frequencies. Overall factors that influence the detection of target include frequency and waveform selection, radar cross section, path losses, multipath effects, noise, interference, gains of transmit and receive antennas, spatial resolution, sky clutter, and operating environments. It is assumed that a thin layer model of ray trajectory can be used in radar range equations. The propagation techniques in skywave OTHR systems are illustrated as shown in Figure 14.6.[12] Having found the ionospheric attenuation, path length, and other factors, we are in a position to estimate the simple radar equation—mostly based on (3.15) and applied to commonly used OTHR system analysis—given by

$$(SNR)_o = \frac{S_o}{N_o} = \frac{P_{av} G_t G_r \lambda^2 \sigma T_c F_p}{(4\pi)^3 R_t^2 R_r^2 F k T_0 B_n L_s} \tag{14.13}$$

where $(SNR)_o$ = output SNR
 P_{av} = average transmitted power, W
 G_t = transmitter antenna gain
 G_r = receiver antenna gain
 λ = wavelength of operation at the carrier frequency, m
 σ = target cross section, m^2
 T_c = coherent processing time, s
 F_p = factor to account for the two-way propagation effects
 F = noise factor defined by input to output SNR
 k = Boltzmann's constant given by $K = 1.38 \times 10^{-23}$, J/K
 T_0 = absolute noise temperature, K
 B_n = noise bandwidth of the radar receiver, Hz
 R_t = distance of the ray traveling from the radar transmitter to the target, m
 R_r = distance of the ray traveling from the target to the radar receiver, m
 L_s = system losses including the transmitter and receiver subsystem losses

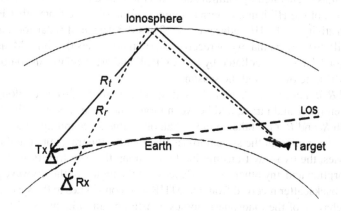

FIGURE 14.6 Propagation techniques in the skywave OTHR system.

In (14.13) the power received by the radar, $S_i = kT_0B_nF(S_o / N_o)$, has been applied. The parameters used in (14.5) are explained briefly as follows:

Antenna gains, G_t and G_r: The transmitting and receiving antenna gains are shown separately in (14.13), since it is sometimes convenient in an OTHR system to have separate antennas for two functions. To achieve narrow beamwidths, the radar antenna must be a physically large phased array. It is also possible to utilize a common aperture for transmit and receive, with equal transmit and receive beamwidths. A common convention for an OTHR system is to include earth effects in the antenna performance characterization. A narrow, steerable beam in the elevation plane is desired to avoid multipath propagation. The conductivity and the dielectric constant are factors that determine the performance of the antenna, and must be taken into account in the antenna design. The electrical properties of the earth can influence the antenna performance for vertical polarization rather than for horizontal.

Wavelength (λ): The wavelength of operating frequency in the HF spectrum must be constrained and selected not to interfere with other users. Since the ionosphere at HF band is very time-dependent and the target cross section is frequency dependent, an adaptive frequency management is necessary.

Radar cross section (σ): The radar cross section of targets at HF is often different in the resonance region where the radar cross sections are generally larger than at microwaves. At the lower end of the HF band where the wavelength is large compared to the target dimensions, the cross section will be in the Raleigh region where σ decreases rapidly with decreasing frequency. If the frequency is sufficiently low, the cross section of small aircraft or missile might be smaller than microwave values. Clutter levels will be large relative to most targets and, therefore, are important in radar design. In such case the surface scattering coefficient $\sigma°$ is multiplied by the resolution cell A and the cosine of the grazing angle.

Coherent integration time (T_c): The coherent integration time is included to emphasize that an OTHR system that is usually a Doppler-processing radar requires a dwell time T_c seconds if a frequency resolution of $1 / T_c$ Hz to separate targets from large earth clutter is needed.

Propagation factor (F_p): The propagation factor includes the energy loss along the ionospheric path, the mismatch loss due to a change in polarization resulting from Faraday rotation in the ionosphere, ionospheric focusing gain or loss, and losses due to the dynamic nature of the path. Since many targets have RCSs that vary with polarization, a proper management of polarization will improve performance of OTHR.

Noise factor (F): The receiver noise for the OTHR system operating at HF band includes the ambient noise generated by natural sources as well as the combined interference from the many users of the HF band. External sources of interference from other HF users can limit the sensitivity of the HF radar receiver. Thus the design of radar for maximum performance will differ from that when receiver thermal noise sets the limit. Measured values of noise factor F have been collated by the (CCIR)[13] and are available in contour maps as a function of the time of day and the season.

Range (R_t and R_r): The ranges R_t and R_r used in (14.13) are the distances along the virtual paths between radar and target, and between target and radar, respectively. The apparent range in both R_t and R_r may take on more than one value since multiple paths may exist.

Losses (L_s): Losses represent the system losses that include the transmitter and receiver subsystems' losses, the two-way transmission losses along the path traversed including ionospheric absorption and any other losses. Because of the nature of the skywave propagation, much of the backscatter received from an OTHR also contributes to the system losses. The statistical behavior of the ionospheric losses constitutes an unknown real-time operation of the OTHR system.

Over-the-Horizon Radar

Equation (14.13) can be used to write the explicit expression of the maximum detection range of the OTHR system as

$$R_{max} = \left[\frac{P_{av} G_t G_r \lambda^2 \sigma T_c F_p}{(4\pi)^3 F k T_0 B_n L_s (SNR)_{o,min}} \right]^{\frac{1}{4}} \tag{14.14}$$

where R_{max} is obtained by assuming that the OTHR transmitter and receiver are collocated implying $R_t = R_r$. Define the receiver sensitivity δ_r as the product of minimum output SNR and the output noise power in the bandwidth of the receiver, which is

$$\delta_r = F k T_0 B_n (SNR)_{o,min}. \tag{14.15}$$

Equation (14.14) is then written as

$$R_{max} = \left[\frac{P_{av} G_t G_r \lambda^2 \sigma T_c F_p^2}{(4\pi)^3 L_s \delta_r} \right]^{\frac{1}{4}}. \tag{14.16}$$

The detection range $R_{footprint}$ along the curvature of the earth becomes

$$R_{footprint} = 2 \left[\left(\frac{R_{max}}{2} \right)^2 - (h_{F_2})^2 \right]^{\frac{1}{2}}. \tag{14.17}$$

The parameters are illustrated in the geometry[14] shown in Figure 14.1 and also in Figure 14.7.

The simple OTHR equation expressed by (14.5) can be easily modified to apply for the surveillance radar, which covers a specified angular sector in a specified time. In the HF radar, the elevation angle θ_e is often not available as a design parameter because of the large-cost high antenna structure. Substituting $G_t = G_r = G = \pi^2 / \theta_a \theta_e$ into (14.5), where θ_a is beamwidth of the azimuth angle, $T_c = t_s \theta_a / \theta_T$, t_s = scan time, and θ_T is the total azimuth angle coverage, we get

$$R^4 = \frac{\pi P_{av} \lambda^2 \sigma F_p t_s}{64 F k T_0 B_n L_s (SNR)_o \theta_a \theta_e^2 \theta_T}. \tag{14.18}$$

Equation (14.18) can be written as

$$R^4 \propto \frac{P_{av}}{\theta_a} = H \frac{P_{av}}{\theta_a}. \tag{14.19}$$

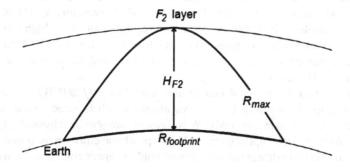

FIGURE 14.7 A geometry illustrating the footprint range calculation.

274   Fundamental Principles of Radar

where H is a constant of proportionality given by

$$H = \frac{\pi\lambda^2 \sigma F_p t_s}{64 F k T_0 B_n L_s (SNR)_o \theta_e^2 \theta_T}.$$ (14.20)

Equation (14.19) implies that the ratio of average power to azimuth beamwidth is a measure of performance of the HF surveillance radar.

14.6.2 OTHR WAVEFORMS

Simple pulsed waveform: The Cobra Mist[15] is a single-site, ground-based air-search radar located at Orford Ness in England, which employed a simple pulsed waveform similar to that used in MADRE. The MADRE located at Chesapeake utilizes 100 µs pulsed waveform with 25 kW average power, while the Cobra Mist used 10 dB higher average power. Due to poor resolution in range and azimuth, the Cobra Mist was removed from service after two years of operation. The Valensole[16] HF skywave radar is a monostatic system with 100 kW transmitted average power using a pulsed waveform.

Pulse Doppler radar: In the pulse Doppler technique used by the Chinese pulse Doppler over-the-horizon backscatter (PD-OTH-B) radar, the system can successfully solve the critical synchronization problems of time, phase, and spatial processing, and handle sophisticated two-dimensional signal processing techniques. The Doppler velocity discrimination is used to determine the relative velocity of targets, and is capable of identifying echoes against a very much stronger surface return.

Phase-coded pulse: A phased-coded pulse waveform can be used to provide a good resolution and long range with a high duty factor. The bandwidth of the phased-coded waveform can be used to determine the range resolution and for the Doppler resolution determination. There are several classes of code sequence commonly used in phase-coded pulse waveforms that include the Barker sequence and pseudorandom sequence. The New Transhorizon Decametric System Applying Studio Methods (Nostradamus)[17] developed by the French operates in HF band utilizing the phased-coded pulse waveform, and is capable of isolating and tracking multiple targets simultaneously.

Frequency-modulated continuous-wave (FMCW): To avoid the disadvantages—low-duty ratio, high peak power, and high output data rate—of pulse radar systems for use in the HF band, the scheme using FMCW has been adopted. FMCW can be used to determine both range and relative velocity of targets with reasonable accuracy. The WARF designed and built by Stanford University has a high-resolution performance utilizing FMCW, and is more compatible with other users in the HF band than the pulsed waveforms used by MADRE. The radar in Maine developed by the Continental United States OTH Backscatter (CONUS-OTH-B)[18] also utilized an FMCW similar to that used in the WARF but with much higher power. By utilizing FMCW waveforms, OTHR systems can be made compatible with simple solid-state transmitters, and have high time-bandwidth products, which makes the systems very attractive. In addition, FMCW radars provide good range resolution, can reject external interference, and can recover the range information from the *IF* signal.

Frequency-modulated interrupted continuous waveforms (FMICW): FMICW is basically an interrupted format of FMCW waveform, which is gated on and off typically with a pseudorandom sequence. FMICW waveforms are obtained by multiplying FMCW waveforms with the interrupt sequence. This type of waveform is used essentially in an experimental OTHR to detect surface targets[19] and to permit common-site working. This scheme[20] is considered to have much potential for use in the HF band.

14.6.3 Detection of Targets

As mentioned, the longer-range beyond the horizon by utilizing skywave propagation, and the relatively short but still OTH distances by utilizing surface wave propagation can be achieved in the HF band. Targets of interest are the same as for conventional radars operating at microwaves, and include high-speed flying targets such as aircrafts and low-speed targets such as ships or low-altitude flying targets.[21] In addition, OTHR systems are used to detect various forms of high-altitude atmospheric ionization such as that due to aurora, meteors, and missiles. The purpose for detecting targets in skywave propagation and surface wave propagation makes use of different parameters and factors to achieve the great effectiveness of OTHR systems.

An improvement in the SNR is achieved by appropriately selecting the frequency. This generates a strong return signal reflected back from the target through the ionosphere. In addition, a transmitter operating in the narrow-band, high PRF is chosen, which improves the detection accuracy and avoids Doppler aliasing. On the contrary, low-speed targets such as ships or low-altitude flying targets are detected quite differently. The parameters such as wideband and low PRF signal, in ship detection, are chosen so that the Doppler resolution and the SCR are significantly improved. Despite unique advantages as mentioned previously, OTHR systems can also operate by combining the two detection processes to detect aircraft and ships simultaneously at the cost of some degradation in performance.

In such a case, wideband waveforms and high PRF might be used. Spectral analysis techniques are also used for target detection processes. However, because of the effects due to the dispersive transmission path and polarization rotation with frequency, the bandwidths are limited to the order of only 100 kHz without correction, and must be accounted for in the spectral analysis used for target detection. The spectral processing before directional determination is clearly useful for discrete targets, but it has also been used for surface and ocean mapping and ship location.[22]

14.7 CURRENT OTHR SYSTEMS IN USE

OTHR systems have been developed since World War II to detect cruise missiles, flying targets, and ocean ships at very long ranges beyond the radar horizon by several countries such as the United States, the United Kingdom, Canada, Australia, China, France, Italy, and the Soviet Union. Some descriptions are already highlighted in Section 14.2. However, we are in a position to provide more details in this section.

A. *United States:* The United States has been developing OTHR systems since World War II and is still working in upgrading the systems. NRL investigated intensively from 1961 to detect ships and aircrafts by HF radar with the MADRE, which is a pulse-Doppler radar, the transmitter and receiver being located at the Chesapeake Bay. It relied on the comparison of returned signals stored on magnetic drums, which added an acoustic delay line that stored the received signal for exactly the amount of time needed for the next signal pulse to arrive. This radar filtered out clutter returns to display the aircraft returns. The magnetic drum, recently introduced, provides a convenient and easily controlled variable delay system. The radar operated typically with average powers from 5 and 50 kW and was even capable of detecting the rocket launch from Cape Canaveral and atomic tests in Nevada.

The WARF designed and constructed by the Ionospheric Dynamics Laboratory, and now operated by SRI International is a bistatic radar using FMICW, which offers better azimuth resolution performance against small and slow targets. The radar employs an endfire 16-elements log-periodic transmitting antenna operating in the 6–28 MHz frequency range, and a broadside receiving antenna array consisting of 492 elements with an aperture of 2.8 km. In addition to operation at the full capability of the transmitting and receiving system, the installation can operate in sweep frequency mode as well.

The first Anglo-American OTHR system known as Cobra Mist is a ground-based air search radar employing an enormous 10 MW transmitter. It was capable of detecting aircraft over the western Soviet Union from its location in Suffolk. Due to an unexpected source of noise causing instability in the system operation, it was removed from service in 1973 after two years of operation.

The US Navy with the assistance of EPL developed a monostatic radar that uses a phased-coded pulse waveform to provide good range resolution with a high duty factor. Another OTHR system developed by the US Air Force with the assistance of GE in 1971 is a bistatic experimental radar located in Maine to demonstrate the feasibility. It is basically a prototype of the CONUS-OTH-B system using FMCW waveform similar to that used in the WARF. The CONUS-OTH-B ground-based early warning system provides surveillance of aircraft at extended ranges. It is designed for the tracking of aircraft and cruise missiles flying at any altitude. A prototype relocatable over-the-horizon radar (ROTHR) bistatic system was installed in 1991 on the isolated Aleutian Island of Amchitka, Alaska, to monitor the eastern coast of Russia, and later moved to Virginia to counter the illegal drug trade, covering Central America and the Caribbean.

B. *The United Kingdom:* Cobra Mist, technically known as AN/FPS-95, was the code name for an Anglo-American experimental OTHR station at Orford Ness, Suffolk, England. Cobra Mist was part of a small number of "Cobra" long-range surveillance radars operated by the United States and was originally intended to be mounted in Turkey to offer coverage of most of the European Soviet airspace. When Turkey objected to the site, it was moved to the United Kingdom to offer a view of most of Eastern Europe. In particular, it would be able to track missile launches from the Northern Fleet Missile Test Center at Plesetsk. Although not as useful as the original site in Turkey, the UK site was nevertheless acceptable and the USAF accepted the new location. The AN/FPS-95 antenna consisted of 18 individual strings radiating outward from a single point near the eastern shore of Orford Ness. Each string was 620 m long, supported on masts from 13 m to 59 m high, with multiple active elements hung from the strings. The strings were arranged 8 degrees 40 minutes apart, covering an arc from 19.5 to 110.5 degrees clockwise from true north. Beneath the antenna was a large wire mesh screen acting as a reflector. The mesh extended past the hub to the east. Through the early part of 1972, the testing found a considerable amount of unexpected noise, which appeared as frequency shifting of the signal. This made targets appear in all of the filters even when it seemed there was nothing of the sort in that area. A lengthy series of investigations into the source of the noise found no convincing explanation leading to the shutting down of the project in 1973.

The Alenia Marconi medium-range HF OTHR system developed by the United Kingdom was deployed for early warning surveillance and for detecting surface and flying air targets. This system is shoreline-located, and uses surface wave transmission for detection beyond the radar horizon. Another OTHR system, known as Overseer, is a surface wave radar system used mainly for ship tracking. It provides continuous surveillance with track outputs of about 500 surface vessels and 100 low-flying aircrafts.

C. *Canada:* A team from Defence Research and Development Canada (DRDC) has been working on a project at Shirleys Bay to improve the effectiveness in monitoring maritime traffic along the longest coastline making use of HF surface wave to follow the curvature of the earth. The radar designed and developed by DRDC, known as *surface-wave radar* (SWR), to detect the surface vessels and low-altitude aircraft below the radar horizon by using the method of shore-based radar. HF SWRs developed by Raytheon Company are shore-based systems to detect and track ships and aircrafts.

Over-the-Horizon Radar   277

D. *Australia:* The *Jindalee Operational Radar Network* (JORN) is an OTHR network that can monitor air and sea movements across 7.7 million square kilometers with a coastline of approximately 37,000 kilometers. It has a normal operating range of 1,000 km to 3,000 km, and can also monitor maritime operations, wave heights, and wind directions. It is also capable of detecting stealth aircraft, as typically these are designed only to avoid detection by microwave radar. Project DUNDEE was a cooperative research project, with US missile defense research, into using JORN to detect missiles. The JORN was anticipated to play a role in future Missile Defense Agency initiatives, detecting and tracking missile launches in Asia.

JORN is a bistatic OTHR system, and is controlled from the Jindalee Facility Alice Springs (JFAS), where two separate locations, the main OTHR transmitter being located at Harts Range and the receiver system at Mount Everard. The Alice Springs radar was the original Jindalee test bed on which the design of the other two stations was based. It continues to act as a research and development test bed in addition to its operational role. JORN uses a radio frequency band between 5 and 30 MHz, which is far lower than used by most other civilian and military radars that operate in the microwave frequency band. Unlike most microwave radars, JORN does not use pulsed transmission, nor does it use movable antennas. Transmission is FMCW, and the transmitted beam is aimed by the interaction between its "beam-steering" electronics and antenna characteristics in the transmit systems. Radar returns are distinguished in range by the offset between the instantaneous radiated signal frequency and the returning signal frequency. Returns are distinguished in azimuth by measuring phase offsets of individual return incidents across the kilometers-plus length of the multielement receiving antenna array. Intensive computational work is necessary to JORN's operation, and refinement of the software suite offers the most cost-effective path for improvements.

As JORN is reliant on the interaction of signals with the ionosphere, ("bouncing"), disturbances in the ionosphere adversely affect performance. The most significant factor influencing this is solar changes, which include sunrise, sunset, and solar disturbances. The effectiveness of JORN is also reduced by extreme weather, including lightning and rough seas.

TIGER is part of an international network of similar HF radars called SuperDARN (Super Dual Auroral Radar Network) operated by ten nations to provide simultaneous coverage of both southern and northern polar regions. TIGER explores the impact of solar disturbances on earth by monitoring the location of aurora and related phenomena occurring in the ionosphere ranging 100 to 300 km above the earth.

E. *China:* China deployed as many as three OTH skywave radar systems and at least one surface-wave OTHR to use in targeting aircraft carriers. These systems are used in an early warning capacity.

While protesting the deployment of the Terminal High Altitude Area Defense (THAAD) missile defense system in South Korea, China had installed OTH Tianbo radar in Inner Mongolia. Its main objective was to detect an opponent's missile launch and the localization of an intercontinental ballistic missile. Within a minute, Tianbo can confirm the target to strike, as it detects the launched missile. The OTHR, which was installed in January 2017, has a range of about 3,000 km, enabling it to detect not only South Korea and Japan but even the Western Pacific. According to the Chinese media, the Tianbo radar can also detect US F-35B stealth fighters deployed at the Iwakuni base in Japan. It can also monitor aircraft carriers and warships within its radius.

China has deployed an *over-the-horizon backscatter radar* (OTH-B) to provide surveillance of the South China Sea. The precise location of this facility remains unclear. China began development of HF ground wave OTH sensors in November 1967 to provide targeting data for their embryonic antiship cruise missile program. In the 1970s the experimental ground wave OTH radar, with an antenna length of 2,300 meters, was deployed. Foreign export restrictions, however, prevented China from obtaining technology that was needed for further improvements.

A large OTH-B radarfaces south near the southern coast of China. In the 1970s the experimental OTH radar could pick up surface ships at 250 km. The fact that Guangzhou is the headquarters of the South China Fleet indicates a major complex of tactical and strategic space and land-based communication and long-range radars in that area.

China's skywave system radar system can detect US aircraft and ships at a long distance from the coastline of the country. Skywave and ground-wave radars are collectively referred to as OTHR. OTHR has two basic types: the use of ionospheric shortwave reflection effect so that radio waves spread to the distant radar, known as skywave OTHR; the use of longwave, medium, and shortwave diffraction effect in the earth's surface so that radio waves spread along the curve, is known as the ground-wave OTHR. OTHR is one of the technologies that Western countries have imposed on China's major blockades and embargoes. The former Soviet Union had only given some guidance to China theoretically, and China's own world of radar has truly reached its combat readiness level based entirely on its ability. The effect of ground-wave OTHR is short, but it can monitor the area that cannot be covered by skywave OTHR.

F. *France:* The French developed an OTHR system called Nostradamus during the 1990s and brought this method in service in 2005. The French military applied the Nostradamus OTHR in the detection of stealth aircraft and other low observable targets such as cruise missiles with low radar and infrared signature. The Nostradamus radar system is monostatic and employs a set of 288 bi-cone antenna elements distributed over the arms of a three-branch star, with a buried infrastructure to shelter the transmission and reception electronics. Nostradamus detects any aircraft flying 700 to 2,000 km away. Indeed this new radar concept is based on very low-frequency waves (6 to 30 MHz) that bounce off the ionosphere, which allows it to detect targets beyond the horizon. Whereas transhorizon radars usually require huge linear antenna networks to beam the signals, the special surface distribution of Nostradamus makes it possible to control the electronic beams both in azimuth (360°) and elevation.

G. *Italy:* The CONDO-R is an OTH surface surveillance radar system designed to exploit anomalous electromagnetic wave propagation conditions utilizing atmospheric ducting that enables detection beyond the horizon. It detects not only the surface ships but also low-altitude flying air targets. The modern digital signal processing and automatic control of the beam position inside super-propagation ducts make this system very attractive for many applications. A mobile coastal radar system, designated as TPS-828, capable of meeting the coastal surveillance needs for long-range surface and OTH surveillance was designed and developed.

H. *The Soviet Union:* Their first experimental model appears to be the Veyer (Hand Fan), which was built in 1949. The next serious Soviet project was Duga-2, built outside Nikolayev on the Black Sea coast near Odessa. Duga (Russian: Ayra) was a Soviet OTHR system used as part of the Soviet antiballistic missile early-warning network. The system operated from July 1976 to December 1989. Two operational Duga radars were deployed, one near Chernobyl and Chernihiv in the Ukrainian SSR (present-day Ukraine), the other in eastern Siberia. Aimed eastward, Duga-2 first ran on November 1971, and was successfully used to track missile launches from the Far East and Pacific Ocean to the testing ground on Novaya Zemlya.

This was followed by their first operational system Duga-3, known in the west as *Steel Yard*, which first broadcast in 1976. Built outside Gomel near Chernobyl, it was aimed northward and covered the continental United States. Its loud and repetitive pulses in the middle of the shortwave radio bands led to it being known as the *Russian Woodpecker* by amateur radio (ham) operators. The random frequency hops disrupted legitimate broadcasts, amateur radio operations, oceanic commercial aviation communications, and utility transmissions, and resulted in thousands of complaints by many countries worldwide. The signal became such a nuisance that some receivers such as amateur radios and televisions actually began including *Woodpecker Blankers* in their circuit designs in an effort to filter out the interference. The Soviets eventually shifted the frequencies they used, without admitting they were even the source, largely due to its interference with certain long-range air-to-ground communications used by commercial airliners. A second system was set up in Siberia, also covering the continental United States, as well as Alaska.

In early 2014 the Russians announced a new system, called *Container*, that was to see over 3,000 km. *Podsolnuh*, also known as *Sunflower*, allows the detection, tracking, and classification of up to 300 offshore and 100 air objects, determining their coordinates and providing them with targeting complexes and systems of armament for ships and air defense systems.

REFERENCES

1. J. M. Headrick and M. I. Skolnik, "Over-the-Horizon Radar in the HF Band," *Proceedings of the IEEE,* 62, no. 69 (1974).
2. J. R. Barnum, "Ship Detection with High-Resolution HF Skywave Radar," *IEEE Journal of Ocean Engineering* (Invited Paper), OE-11, no. 2 (1986): 196–209.
3. M. Papazoglou and J. Krolik, "Electromagnetic Matched-Field Processing for Target Height Finding with Over-the-Horizon Radar," *IEEE International Conference on Acoustic, Speech, and Signal Processing (ICASSP)* (1997): 559–562.
4. R. H. Anderson, S. Kraut, and J. L. Krolik, "Robust Altitude Estimation for Over-the-Horizon Radar Using a State-Space Multipath Fading Model," *IEEE Transaction on Aerospace and Electronic Systems,* 39, no. 1 (2003): 192–201.
5. K. Davie, *Ionospheric Radio Propagation* (New York: Dover Publications, 1966).
6. R. Eave, *Principles of Modern Radar* (New York: Van Nostrand Reinhold Company, Inc., 1987).
7. P. L. Dyson and J. A. Bennett, "Exact Ray Path Calculations Using Realistic Ionospheres," *IEE Proceedings—Microwaves, Antennas and Propagation,* 139, no. 5 (1992): 407–413.
8. E. D. R. Shearman, "The Technique of Ionospheric Investigation Using Ground Backscatter," *Proceedings of the IEE,* 103B, 8 (1956): 210–223.
9. P. A. Bradley, "Long-Term HF Predictions for Radio Circuit Planning," *Radio and Electronic Engineer,* 45 (1975): 31–41.
10. Joachim, M. *CCIR Atlas of Ionospheric Characteristics (Report 340),* ITU, Geneva, 1983.
11. J. M. Pielou, "Sky-Wave Radar Propagation Predictions for HF Radar System Planning," *IEE Proceedings of the International Conference on Antennas and Propagation,* CP 248 (1985): 510–514.
12. A. A. Kosolov et al., *Over-the-Horizon Radar* (Boston: Artech House, 1987).
13. CCIR, *Characteristics and Applications of Atmospheric Radio Noise, Data Report 322-2,* Geneva, ITU, 1984.
14. Y. Zahng, M. G. Amin, and G. J. Frazer, "High-Resolution Time-Frequency Distributions for Maneuvering Target Detection in Over-the-Horizon Radars," *Proceeding of the IEE Radar Sonar Navigation,* 150, no. 4 (2003): 299–304.
15. J. F. Thomason, "Development of Over-the-Horizon Radar in the Unites States," *Proceedings of the IEEE International Conference on Radar* (2003): 599–601.
16. M. Six, J. Parent, A. Bourdillon, and J. Dellue, "A New Multibeam Receiving Equipment for the Valensole Skywave HF Radar: Description and Applications," *IEEE Transactions on Geoscience and Remote Sensing,* 34, no. 3 (1996): 708–719.
17. S. Saillant, "NOSTRADAMUS Over the Horizon Radar–First Subway Experimentation," *IEEE International Conference on Radar* (1992): 78–81.

18. "AN/FPS-118 Over-the-Horizon Radar," *Jane's Radar and Electronic Warfare Systems, Land-Based Air Defense Radars* (2001).
19. R. H. Khan, and D. K. Mitchell, "Waveform Analysis for High-Frequency FMICW Radar," *IEE Proceedings on Radar and Signal Processing,* 138, no. 5 (1991): 411–419.
20. L. R. Wyatt, G. D. Burrows, and M. D. Moorhead, "An Assessment of an FMICW Ground-Wave Radar System for Ocean Wave Studies," *International Journal of Remote Sensing,* 6 (1985): 275–282.
21. G. S. Sales, *OTH-B Radar System: System Summary,* Philips Laboratory, Air Force Systems Command, Hanscom Air Force Base, MA, May 1992.
22. "Special Issue on High Frequency Radar for Ocean and Ice Mapping and Ship Location," *IEEE Journal of Oceanic Engineering,* 11 (1986): 145–332.

15 Secondary Surveillance Radar

"The most essential gift for a good writer is a built-in, shock-proof, shit detector. This is the writer's radar and all great writers have had it."

—**Ernest Hemingway**

15.1 INTRODUCTION

The rapid wartime development of radar had obvious applications for *air traffic control* (ATC) as a means of providing continuous surveillance of air traffic disposition. This type of radar, now called a *primary radar*, can detect and report the position of anything that reflects its transmitted radio signals. For air traffic control purposes this is both an advantage and a disadvantage. Its targets do not have to cooperate; they only have to be within its coverage and be able to reflect radio waves. It only indicates the position of the targets without proper identification. Primary radar is still used by ATC today as a backup/complementary system to secondary radar, although it has limited coverage and information.

The need to identify aircraft more easily and reliably led to another wartime radar development, the *Identification Friend or Foe* (IFF) system, which had been created as a means of positively identifying friendly aircraft from unknowns. This system, which became known in civil use as *secondary surveillance radar* (SSR), or in the United States as the *air traffic control radar beacon system (ATCRBS)*, heavily relies on a piece of equipment aboard the aircraft known as a *transponder*. There are two similar and overlapping forms of SSR: IFF used to locate an aircraft and to segregate friendly aircraft from unknowns, and ATCRBS used to identify and report the altitude of civil and military aircraft in the ATC system. IFF and ATCRBS use the same frequencies and, in many cases, the same transponders.

SSR[1] is a radar system used in ATC that makes use of a transmitter to interrogate a transponder-equipped aircraft, providing a two-way data link on separate transmitting and reply frequencies. The transponder is a radio receiver and transmitter pair, which receives on 1030 MHz and transmits on 1090 MHz. This radar not only detects and measures the position of aircraft, but also requests additional information from the aircraft such as its identity and altitude. Unlike primary radar systems that measure the bearing of targets using the detected reflections of radio signals, SSR is essentially a cooperative system that relies heavily on targets equipped with a radar transponder that replies to each interrogation signal by transmitting a response containing encoded data. Both the civilian SSR and the military IFF have become much more complex than their wartime ancestors, but remain compatible with each other, not to allow military aircraft to operate in civil airspace. Given its primary military role of reliably identifying friends, IFF has much more secure (encrypted) messages to prevent "spoofing" by the enemy, and is used on many types of military platforms including air, sea, and land vehicles. Today's SSR can provide much more detailed information, for example, the aircraft altitude, as well as enabling the direct exchange of data between aircraft for collision avoidance.

The SSR system offers many advantages over primary surveillance radar: (a) SSR does not suffer clutter and weather-return problems, (b) the power output requirements of the radar transmitter and aircraft transponder are quite modest because of one-way transmission in each case, (c) scintillation and dependence on a radar cross section are effectively eliminated, and (d) coding utilized in interrogation and reply path provides discrete target identification and automatic altitude reporting.

However, SSR has its disadvantages that include the dependence entirely on cooperation by friendly aircraft and susceptibility to various types of interference.

SSR based on IFF technology was originally developed during World War II. The combined efforts of civil and military agencies involved in the development of ATC systems have since led to the development of universally accepted standards. The International Civil Aviation Organization (ICAO), a branch of the United Nations, may be considered as a near-ideal means of generating ATC data, and standardizing performance parameters of SSR. The American Radio Technical Commission for Aeronautics (RTCA) and the European Organisation for Civil Aviation Equipment (EUROCAE) produce minimum operational performance standards for both ground and airborne equipment in accordance with the standards specified in ICAO Annex 10. Both organizations frequently work together and produce common documents. As civil aviation authorities operating SSR stations are growing, the inherent problems are becoming more widely known and solutions to all are being found.[2] However, a new approach was proposed by Ullyatt.[3] ICAO published a rationalization, known as *Mode S*, of these two schemes by retaining the existing frequencies for interrogation and reply. Aeronautical Radio, Incorporated (ARINC) is an airline-run organization concerned with the form, fit, and function of equipment carried in aircraft. Its main purpose is to ensure competition between manufacturers by specifying the size, power requirements, interfaces, and performance of equipment to be located in the equipment bay of the aircraft.

15.2 PRINCIPLES OF SSR

Interrogations: An SSR ground station interrogates transponder-equipped aircraft with coded interrogation pulses at a frequency of 1030 MHz—whose spacing denotes whether identity or altitude replies are being requested—as its antenna rotates, or is electronically scanned with a beam shape narrow in azimuth and wider in elevation. The interrogation signal transmitted from an SSR to aircraft consists of three pulses, which modulate the 1030 MHz carrier, and are internationally designated as P_1, P_2, and P_3. The pulses are illustrated in Figure 15.1. The time spacing between P_1 and P_3 is switchable, and is used to ask the transponder the reply mode expected. The interrogator can be set to cycle through various spacing patterns to initiate replies from all modes desired. The transponder is supposed to reply only to interrogations for the modes set into it. If P_1 and P_2 spacing is 8 μs, the interrogator is asking for identity, and is said to be working in mode A. If the spacing is 21 μs, the interrogator is asking for aircraft altitude, and the system is working in mode C. P_2 is not required for initiating a reply, but is used for sidelobe suppression, as described next. Various other modes are also available, but modes A and C are most common to both civil and military aircraft. Mode B gave a similar response to mode A and was at one time used in Australia. Mode D has never been used operationally. At present, five modes, as described earlier, are used in North America, and are listed in Table 15.1.

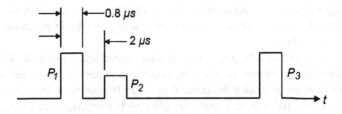

FIGURE 15.1 SSR interrogation pulses.

Secondary Surveillance Radar

TABLE 15.1
SSR Modes of Operation

Mode	P_1–P_3 Pulse spacing	Purpose
A	8 μs	Identify
B	17 μs	Identify
C	21 μs	Altitude
D	25 μs	Undefined
S	3.5 μs	Multipurpose

The new mode known as mode S, operating at the same frequency as the standard SSR interrogator and transponder, has selective interrogation characteristics. This is a development of the basic SSR to alleviate several problems in the SSR. The mode S ground interrogators and airborne transponders are fully compatible with the conventional mode A and mode C units. However, S units working together have much greater capabilities. The initial part of the interrogator signal is such that the standard SSR modes will be recognized by the normal airborne transponder, while the second part of the mode S interrogator signal consists of a message of up to 112 data bits within which 24 bits are allocated to aircraft address. This permits the controller to interrogate a specific aircraft.

Replies: An aircraft transponder within LOS range "listens" for the SSR interrogation signal and transmits a reply on 1090 MHz that provides aircraft information. The reply sent depends on the interrogation mode. The aircraft is displayed as a tagged icon on the controller's radar screen at the measured bearing and range.

An aircraft without an operating transponder still may be observed by primary radar but would be displayed to the controller without the benefit of SSR-derived data. It is typically a requirement to have a working transponder in order to fly in controlled airspace, and many aircraft have a backup transponder to ensure that condition is met. When relative pulse amplitude, width, and spacing criteria are met, replies are made in the format shown in Figure 15.2 within 3 μs of the receipt of the second P_3 pulse of the interrogation pair. The pulse train of reply is constructed according to internationally agreed standards with 0.45 μs pulse duration and a 1 μs space between pulses. The framing pulse F_1 and F_2 are always present. Data pulses are divided into four three-pulse groups, *A*, *B*, *C*, and *D*. Group *A* encodes the most significant octal digit, while group *D* encodes the least significant. The central *X* pulse is not currently in use. In octal-based code, the 12-bit code provides up to 4,096 separate aircraft identifications. Altitude is signaled by a *Gilham code pattern* in mode C giving increments of 100 feet, known as *flight levels*. However, as airspace becomes congested, it is important to monitor aircraft to move from its assigned flight level. When both aircraft identification and height are required, the interrogations may be interlaced between mode A and mode C. The conventional approach, to mitigate the problems with azimuth when the replies are spread over the horizontal beamwidth, uses the *sliding window* technique to estimate azimuth.

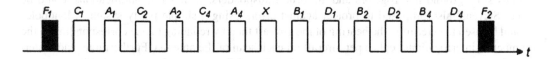

FIGURE 15.2 SSR reply pulses.

15.3 DEFICIENCIES IN SSR

The main problems in SSR are rather different, which mainly arise from the use of up-link and down-link of frequencies that are common to all SSR systems. A number of problems in SSR and their causes are described in this section to give a better understanding of modern solutions.

FRUIT: False Replies Unsynchronized in Time (FRUIT) results from false replies of multiple airborne transponders operating at the same frequency of 1090 MHz and using the same signal format. Airborne antennas are essentially omnidirectional and so replies to one interrogator can be received by others. These unwanted replies are known as FRUIT, which could indicate an aircraft even though it does not exist. As air transport expands, the amount of FRUIT generated will also increase accordingly. To prevent such problems, each interrogator in the home station must have its own distinct interrogation period allowing replies from unwanted interrogators to be decorrelated in range. Because of identical FRUIT characteristics and their random arrivals, FRUIT represents a source of corruption to desired replies when they are simultaneously present at the home station interrogator.

Garble: Garbling is a fundamental problem in the design of the classical SSR system, and denotes the overlap of two or more airborne transponders' replies. Aircrafts are often closely spaced in range and azimuth but at different heights, and can cause considerable problems in detection and decoding. Replies from multiple aircrafts will overlap if their range separation is within the equivalent of the 20.3 μs reply length. The most serious garbling situations occur when the azimuth separation is very small such that replies from both aircraft are received from all interrogations across the beam. With advanced reply processing techniques and algorithms, it may sometimes be possible to extract some or all the replies from the received signal.

In principle, one distinguishes two classes of the overlapping: *asynchronous garbling* and *synchronous garbling*. In the former, two replies overlap in time such that its time grids are not congruent. In the latter, multiple replies arrive in the system at the same time overlapping so that its time grids are congruent. Both these classes are further characterized by groups: *garble by interleaving* in which pulses of one reply lie within the spaces between the others and *garble by overlap* in which pulses of each reply overlap. Modern signal processing can resolve any problems with garbling.

Capture: Capture occurs when an airborne transponder is already replying to one ground interrogation, and is unable to respond to others at the same time. For a mode A or C interrogation the transponder reply may take up to 120 μs before it can reply to a further interrogation. Under these conditions the transponder will not respond to distant interrogators; the nearer ones will have captured it from others at longer range. Lost replies jeopardize the measurement of azimuth, and in severe cases can cause unreliable data decoding, thus reducing the detection efficiency.

15.4 SOLUTIONS TO SSR DEFICIENCIES

The deficiencies in the SSR system were recognized quite early in the use of SSR. Ullyatt proposed improvements in SSR by addressing problems that include essentially new interrogation and reply formats. He suggested that the existing 1030 MHZ and 1090 MHZ frequencies could be retained and existing ground interrogators and airborne transponders, with some modifications, could be used. Stevens[4] also proposed to protect against errors using a simple parity system. Monopoles would be used to determine the bearing of the aircraft thereby reducing to one corresponding to the number of interrogations/replies per aircraft on each scan of the antenna. Furthermore, each interrogation would be preceded by main beam pulses P_1 and P_2 separated by 2 μs so that transponders operating on modes A and C would take it as coming from the antenna sidelobe.

Secondary Surveillance Radar

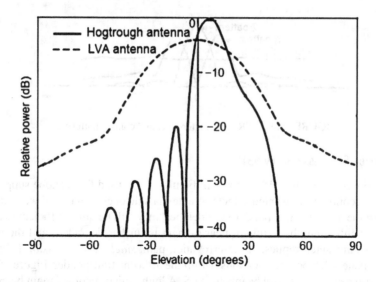

FIGURE 15.3 Vertical beam shapes of the old hogtrough and improved antennas.

15.4.1 Improved Antennas

The existing standard *hogtrough* SSR antennas have a relatively *large vertical aperture* (LVA) of 0.4 m with a beamwidth of 50°. This vertical beam has too much energy directed at high elevation angles and toward the ground. To circumvent this problem, it is necessary to properly design antennas with much larger vertical apertures to produce a much better match to the operational requirements. Typically, 10 or 12 rows of radiating elements are spaced across the vertical aperture. The implementation of an omnidirectional antenna for pulse P_2 requires that it be mounted above the main SSR antenna to overcome some problems associated with it. Multipath interference due to ground reflected energy, which can adversely affect the SSR performance, is much reduced. Figure 15.3 shows the vertical beam shapes of the old hogtrough and improved antennas.

15.4.2 Monopulse Techniques

The monopulse technique overcomes many of the performance limitations of conventional SSR. As its name implies, monopulse technique allows the target azimuth to be measured with high accuracy using a single transponder reply. In addition, the antenna used with a monopulse SSR has a relatively large vertical aperture and smaller vertical beamwidth, and can reduce the ground-reflected energy responsible for causing problems in SSR performance. Another advantage of monopulse SSR is that it is capable of decoding individual replies with much accuracy. It also does not need many replies from an aircraft on each scan of the antenna, thus providing a lower PRF leading to a reduction in co-channel interference.

In monopulse SSR the main beam is generally known as the *sum (Σ) pattern*. Another beam, known as the *difference (Δ) pattern*, is also introduced to obtain a more accurate estimation of the location of the transponder reply. The typical sum and difference patterns are illustrated in Figure 15.4. The sum pattern provides the normal reception and a reference for the monopulse, while the difference pattern provides monopulse location capability. The control pattern is used for sidelobe suppression. The sum and difference patterns can both be obtained from the same antenna, as summarized by Figure 15.5. A typical tapered amplitude distribution is applied across the horizontal aperture to reduce the sidelobe levels. The sum and difference patterns are produced by applying a uniform phase distribution across the aperture and by introducing phase reversal $\pm \pi / 2$ between horizontal elements, respectively, as shown at the lower part of the figure.

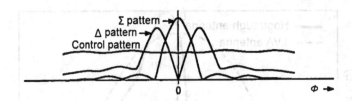

FIGURE 15.4 SSR sum, difference, and control patterns.

15.4.3 SIDELOBE SUPPRESSIONS IN SSR

As already mentioned, that pulse P_2 as shown in Figure 15.4 is used for sidelobe suppression. SSR is much more susceptible to unwanted effects of antenna sidelobes. Relatively weak interrogating signals, reaching the airborne transponder via sidelobes, may still be quite sufficient to trigger transponder replies at full strength. Azimuth measurement is further degraded, and the transponder, therefore, supplies unwanted replies. An interrogation-path, *sidelobe suppression* (SLS) system is designed to eliminate sidelobe interrogations within the airborne transponder. Figure 15.6 indicates the use of the main beam and control beams for SLS. A high-gain directional main beam is used for transmitting pulses P_1 and P_3, and a very broad control beam in the azimuth plane for transmitting P_2. The gain of the control beam exceeds that of the main beam in all directions except that of the main beam. Two cases of interrogations using main lobe and sidelobe interrogations are illustrated in Figure 15.7. When the transponder is interrogated from the main lobe, P_1 is of higher amplitude than P_2 indicating the transponder to reply. When P_2 pulse, transmitted on a collocated omnidirectional antenna 2 µs after P_1 pulse, is greater than P_1 indicating a sidelobe interrogation, the interrogation is suppressed and no reply is elicited.

Interrogation in sidelobe blanking is the primary defense in the SSR system to antenna sidelobe problems. In cases where many interrogations are present within an area, it may also be warranted to test replies to check if they were from the high-gain antenna main beam or sidelobes.

Multipath can cause problems as serious as sidelobes. Reflections from the main beam can cause the transponder to reply. Fortunately, the SLS eliminates most of the reflection interrogations in the sidelobe region since they arrive at the transponder delayed in time from the direct pulses. It is also possible to synthetically set up a false sidelobe by radiating the P_1 pulse on the omnidirectional antenna and cause the transponder to actively suppress almost all reflection ambiguities.

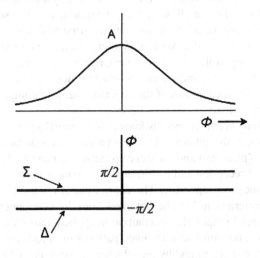

FIGURE 15.5 Techniques of producing sum and difference SSR patterns.

Secondary Surveillance Radar

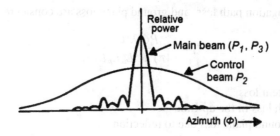

FIGURE 15.6 Main and control beams for SLS.

15.5 RANGE PERFORMANCE IN SSR

Secondary surveillance radars use a transmitter to interrogate airborne transponders. Using replies from the transponder as an enhanced signal return, and interpreting its reply code, specific information about the aircraft can be extracted. The range performance of an SSR system is determined by considering two separate range equations, corresponding to the up-link interrogation and down-link reply.

15.5.1 The Up-Link Range

The interrogation link, known as the up-link, is considered first. The interrogation power density at the airborne transponder located at a range R_i is

$$\hat{p}_i = \frac{P_i G_i}{4\pi R^2} \qquad (15.1)$$

where
$\hat{p}_i$ = interrogation power density at the transponder, W/m²
P_i = interrogation transmitted power, W
G_i = gain of the of the interrogator transmit antenna
R = interrogation range from interrogator transmit antenna to transponder, m.

The power received by the airborne transponder is

$$P_t = \left(\frac{P_i G_i}{4\pi R^2}\right)\left(\frac{\lambda_i^2 G_t}{4\pi}\right) \qquad (15.2)$$

where
P_t = power received by the transponder from the interrogator (W)
λ_i = wavelength of the interrogator signal (m)
G_t = transponder antenna gain

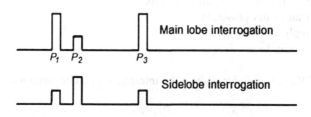

FIGURE 15.7 Main lobe and sidelobe interrogations.

If the system loss, propagation path loss, and ground plane loss are considered, (15.2) becomes

$$P_t = \frac{P_i G_i \lambda_i^2 G_t}{(4\pi)^2 R^2 L_{si} L_p L_g} \tag{15.3}$$

where
L_{si} = interrogator system loss
L_p = propagation path loss in the up-link
L_g = interrogation ground plane loss due to reflection

The SNR of the interrogation in the transponder usually limits the performance of SSR, and is given by

$$\frac{S_t}{N_t} = \frac{P_i G_i G_t \lambda_i^2}{(4\pi)^2 R^2 K T_0 B_i F_t L_{si} L_p L_g} \tag{15.4}$$

where
S_t / N_t = SNR at the transponder
K = Boltzmann's constant (1.38×10^{-23} J/K)
$T_0 = 290°$ K
B_i = noise bandwidth of the interrogation signal
F_t = noise factor of the transponder receiver.

Solving (15.4) for the maximum range $R_{i(\max)}$ corresponding to the minimum SNR $(S_t / N_t)_{\min}$ at which the transponder can be successfully interrogated yields

$$R_{i(\max)} = \left[\frac{P_i G_i G_t \lambda_i^2}{(4\pi)^2 R^2 K T_0 B_i F_t L_{si} L_p L_g (S_t / N_t)_{\min}} \right]^{\frac{1}{2}}. \tag{15.5}$$

It should be noted that the product $P_i G_i$, called the *effective radiated power* (ERP), must never be greater than 52.5 dBW conforming to the ICAO specification for SSR. In systems with a significantly large antenna height at the interrogator, the mismatch in space between lobe and gaps at two different wavelengths of the interrogator and transponder can affect performance of the SSR system.

15.5.2 THE DOWN-LINK RANGE

Another one-way communication corresponding to the transponder reply, known as the down-link, can be analyzed using the same reasoning as with the up-link. Assuming the range from transponder to interrogator to be the same as that from interrogator to transponder, the power P_i received by the interrogator receiver from the transponder is

$$P_i = \frac{P_t G_t \lambda_t^2 G_i}{(4\pi)^2 R^2 L_{st} L_p L_g} \tag{15.6}$$

where the symbols have their usual meanings except that,
P_t = transponder transmitter power, W
λ_t = transponder reply wavelength, m
L_{st} = transponder system loss.

The SNR (S_i / N_i) of the transponder reply at the interrogator can be written as

$$\frac{S_i}{N_i} = \frac{P_t G_i G_t \lambda_t^2}{(4\pi)^2 R^2 K T_0 B_i F_i L_{st} L_p L_g}. \tag{15.7}$$

Secondary Surveillance Radar

Thus, the maximum range $R_{t(max)}$ corresponding to the minimum SNR $(S_i / N_i)_{min}$ at which the transponder reply can be successfully detected by the interrogator is

$$R_{t(max)} = \left[\frac{P_t G_i G_t \lambda_t^2}{(4\pi)^2 KT_0 B_t F_i L_{st} L_p L_g (S_i / N_i)_{min}} \right]^{\frac{1}{2}}. \tag{15.8}$$

The up-link and down-link designs are normally balanced by making comparable transmit powers of the interrogator and the transponder. The antennas are both shared, with the result that the maximum ranges expressed by (15.5) and (15.8) are both about the same. The range performances in the up-link and down-link are essentially similar.

REFERENCES

1. M. C. Stevens, *Secondary Surveillance Radar* (Norwood, MA: Artech House, 1988).
2. H. W. Cole, "The Future for SSR," *ICAO Bulletin* (September 1980).
3. C. Ullyatt, "Sensors for the ATC Environment with Special Reference to SSR," *UK Symposium on Electronics for Civil Aviatiaon,* 3, C1–C3 (1969).
4. M. C. Stevens, "Secondary Surveillance Radar—Today and Tomorrow," *SERT Avionics Symposium,* Swansea (1974).

Appendixes

"His faith was like the radar of a bat, it took him through the darkness that surrounded him. It would guide him through everything that evil would throw at him. When his body gave up, God's powers lifted him up and kept his heart thumping."

—**Mark A. Cooper**

APPENDIX A: REVIEW OF DETERMINISTIC SIGNALS

In this appendix we review several important aspects of deterministic signals useful in the study of radar systems. A deterministic signal is a signal whose physical description is known completely, either in a mathematical form or a graphical form. We discuss deterministic periodic and nonperiodic signals through their respective Fourier series and Fourier transforms.

A.1 FOURIER SERIES

The Fourier series applies to *periodic signal*. Let $s(t)$ denote the periodic signal with the fundamental period T_0 such that $s(t) = s(t + nT_0)$ for any integer n. By using a Fourier series expansion of this periodic signal, it may be expressed in the trigonometric form:

$$s(t) = a_0 + 2\sum_{n=1}^{\infty}[a_n\cos(n\omega_0 t) + b_n\sin(n\omega_0 t)] \tag{A.1}$$

where

$$\omega_0 = \frac{2\pi}{T_0}. \tag{A.2}$$

The Fourier series coefficients a_n and b_n represent the amplitudes of the cosine and sine terms, respectively, and are given by

$$a_n = \frac{1}{T_0}\int_{-T_0/2}^{T_0/2} s(t)\cos(n\omega_0 t)\, dt, \quad n = 1,2,\ldots \tag{A.3}$$

$$b_n = \frac{1}{T_0}\int_{-T_0/2}^{T_0/2} s(t)\sin(n\omega_0 t)\, dt, \quad n = 1,2,\ldots \tag{A.4}$$

The periodic signal $s(t)$ can be expanded in a Fourier series if the signal is bounded by period T_0, and it has a finite number of discontinuities and at most a finite number of maxima and minima in the interval T_0. These conditions are called *Dirichlet conditions*. The Dirichlet conditions are satisfied by the periodic signals usually encountered in radar systems.

Complex Exponential Series

The Fourier series of (A.1) can be converted into a much simpler and elegant form by using complex exponentials as

$$s(t) = \sum_{n=-\infty}^{n=\infty} c_n \exp(jn\omega_0 t) \tag{A.5}$$

where

$$c_n = \frac{1}{T_0} \int_{-T_0/2}^{T_0/2} s(t) \exp(-jn\omega_0 t)\, dt, \quad n = 0,\ \pm 1,\ \pm 2,\ldots \tag{A.6}$$

This series expansion is referred to as the *complex exponential Fourier series*. The c_n are called the *complex Fourier series coefficients*, and are related to a_n and b_n as

$$c_n = \begin{cases} a_n - jb_n, & n > 0 \\ a_n, & n = 0 \\ a_n + jb_n, & n < 0 \end{cases} \tag{A.7}$$

A.2 FOURIER TRANSFORMS

We now express a similar representation for a nonperiodic signal in terms of complex exponential signals. Its Fourier transform, denoted by $S(\omega)$, is given by

$$S(\omega) = \int_{-\infty}^{\infty} s(t) \exp(-j\omega t)\, dt. \tag{A.8}$$

The inverse Fourier transform of $S(\omega)$ can be written as

$$s(t) = \frac{1}{2\pi} \int_{-\infty}^{\infty} S(\omega) \exp(j\omega t)\, d\omega. \tag{A.9}$$

Equations (A.8) and (A.9) are referred to as the Fourier transform pair. $S(\omega)$ is sometimes called the spectrum of $s(t)$. The magnitude of and phase of $S(\omega)$ are called the amplitude spectrum and phase spectrum of $s(t)$, respectively. Sufficient conditions for $S(\omega)$ to exist for a given $s(t)$ are that it must satisfy the Dirichlet conditions and

$$\int_{-\infty}^{\infty} |s(t)|\, dt < \infty. \tag{A.10}$$

Appendixes 293

APPENDIX B: REVIEW OF RANDOM SIGNALS

In Appendix A, we reviewed the Fourier transform as the mathematical tool to describe the *deterministic signals* that are completely specified functions of time. In this Appendix B, we deal with the statistical characterization of *random signals*, whose precise value cannot be predicted in advance. It may be described in terms of statistical properties such as average power or the spectral distribution of this power on the average. This constitutes another avenue of considerable importance in the study of radar signals.

The mathematical discipline that deals with the statistical characterization of random signals is probability theory. An understanding of statistical averages and random variables depend on the basic topics in probability.

B.1 PROBABILITY THEORY

The theory of probability[1] is rooted in phenomena that, explicitly or implicitly, can be modeled by an experiment with an outcome that is subject to *chance*. However, if the experiment is repeated, the outcome can differ because of the influence of an underlying random phenomenon. The set of all possible outcomes in a given random experiment, denoted by S, is called *sample space*. It is called *discrete* if it has a finite number of outcomes and called *continuous* if a continuum of outcomes is possible. Any one possible outcome of a sample space is called an *element* of the space. An event is a subset of the sample space. Two events that have no common elements are *mutually exclusive*.

Axioms of Probability

A measure of probability is assigned to each event A on a sample space S. $P(A)$, which denotes the probability of an event A, satisfies the following three axioms:

Axiom I. The probability of an event is positive:

$$0 \le P(A) \le 1. \tag{B.1}$$

Axiom II. The probability of sure event is unity,

$$P(S) = 1. \tag{B.2}$$

Axiom III. If $A + B$ is the union of two mutually exclusive events on a sample space S, then the probability of two mutually exclusive events is the sum of the probabilities of the individual events:

$$P(A + B) = P(A) + P(B). \tag{B.3}$$

Joint and Conditional Probabilities

The probability that two events A and B occur together is called *joint probability*, denoted by $P(AB)$. If the events A and B are not mutually exclusive, then the probability of the union event A or B equals

$$P(A + B) = P(A) + P(B) - P(AB). \tag{B.4}$$

Conditional probability of an event A, given that event B has occurred, is denoted by $P(A \mid B)$ and defined by

$$P(A \mid B) = \frac{P(AB)}{P(B)} \tag{B.5}$$

294 Appendixes

where $P(B) > 0$ is assumed. Similarly, if $P(A) > 0$,

$$P(B \mid A) = \frac{P(AB)}{P(A)} \tag{B.6}$$

so that

$$P(A \mid B)P(B) = P(B \mid A)P(A). \tag{B.7}$$

This relation is referred to as a special form of Bayes' rule.

Suppose that the conditional probability $P(A \mid B)$ is simply equal to the elementary probability of occurrence of event A, that is,

$$P(A \mid B) = P(A) \tag{B.8}$$

In that case, the probability of occurrence of the joint event AB is equal to the product of the elementary probabilities of the events A and B:

$$P(AB) = P(A)P(B) \tag{B.9}$$

so that

$$P(B \mid A) = P(B). \tag{B.10}$$

Events A and B that satisfy this condition are said to be *statistically independent*.

B.2 RANDOM VARIABLES, DISTRIBUTIONS, AND DENSITIES

Consider the outcome of an experiment as a variable that can wander over the set of sample points and whose value is determined by the experiment. *A function whose domain is a sample space and whose range is some set of real numbers is called a random variable.*

The *cumulative distribution function* (CDF) or simply the distribution function of the random variable X, is denoted by $F_X(x)$, and is interpreted as the total probability that the random variable X is less or equal to the value x. More precisely,

$$F_X(x) = P(X \le x). \tag{B.11}$$

An alternative description of the probability of the random variable X is often practical, and is expressed as the *probability density function* (PDF). This is the derivative of the distribution function:

$$f_X(x) = \frac{d}{dx} F_X(x). \tag{B.12}$$

The PDF has the following properties:

$$0 \le F_X(x) \le 1 \tag{B.13}$$

$$\int_{-\infty}^{\infty} f_X(x)\,dx = 1 \tag{B.14}$$

$$F_X(x) = \int_{-\infty}^{x} f_X(\xi)\,d\xi \tag{B.15}$$

$$P(x_1 < X \le x_2) = \int_{x_1}^{x_2} f_X(x)\,dx. \tag{B.16}$$

Appendixes

B.3 **STATISTICAL AVERAGES**

The average behavior of the outcomes arising in the random experiment, called the *statistical average,* also called the *expected value*, or *mean value*, is defined as

$$\mu_X = E[X] = \int_{-\infty}^{\infty} x f_X(x)\,dx \tag{B.17}$$

where $E[\cdot]$ is the statistical expectation operator.

Let X denote a random variable, and let $g(X)$ denote a real valued-function define on the real line. Thus the statistical average is obtained by applying the standard formula:

$$E[g(X)] = \int_{-\infty}^{\infty} g(x) f_X(x)\,dx. \tag{B.18}$$

Moments

For the special case of functions $g(X) = X^n$, using (B.18), we obtain the nth moment of the probability distribution of the random variable X:

$$E[X^n] = \int_{-\infty}^{\infty} x^n f_X(x)\,dx. \tag{B.19}$$

We may define *central moments* as the moments of the difference between a random variable X and its mean μ_X. Thus the nth central moment is

$$\mu_n = E[(X - \mu_X)^n] = \int_{-\infty}^{\infty} (x - \mu_X)^n f_X(x)\,dx. \tag{B.20}$$

Moment μ_2 is called the *variance* of X and is usually given the special symbol σ_X^2, written as

$$\sigma_X^2 = E[(X - \mu_X)^2] = \int_{-\infty}^{\infty} (x - \mu_X)^2 f_X(x)\,dx. \tag{B.21}$$

The positive square root of variance is called *standard deviation*. The Chebyshev inequality states that for any positive number $\in$, we have

$$P(|X - \mu_X|) \geq \in \leq \frac{\sigma_X^2}{\in^2}. \tag{B.22}$$

It follows that the mean and variance of a random variable provide a partial description of its probability distribution.

For two random variables X and Y, joint moments about the origin, denoted m_{ik}, are expressed as

$$m_{ik} = E[X^i Y^k] = \int_{-\infty}^{\infty} \int_{-\infty}^{\infty} x^i y^k f_{X,Y}(x,y)\,dx\,dy. \tag{B.23}$$

The joint moment m_{11} is a special one, called *correlation*, denoted by R_{XY}:

$$R_{XY} = E[XY] = \int_{-\infty}^{\infty} \int_{-\infty}^{\infty} xy f_{X,Y}(x,y)\,dx\,dy. \tag{B.24}$$

For $R_{XY} = 0$, X and Y are said to be *orthogonal*. If $R_{XY} = m_{01}m_{01}$, X and Y are called *uncorrelated*. *Joint central moments* of random variables X and Y are denoted by μ_{ik} and defined by

$$\mu_{ik} = E\left[(X - m_{10})^i (Y - m_{01})^k\right] = \int_{-\infty}^{\infty} \int_{-\infty}^{\infty} (x - m_{10})^i (y - m_{01})^k f_{X,Y}(x,y) \, dx \, dy. \qquad (B.25)$$

Central moment μ_{11}, known as *covariance* and denoted by C_{XY}, is written as

$$C_{XY} = E\left[(X - m_{10})(Y - m_{01})\right] = \int_{-\infty}^{\infty} \int_{-\infty}^{\infty} (x - m_{10})(y - m_{01}) f_{X,Y}(x,y) \, dx \, dy. \qquad (B.26)$$

If $C_{XY} = 0$, X and Y are uncorrelated. If $C_{XY} = -m_{10}m_{01}$, X and Y are orthogonal.

Characteristic Function

Another important statistical average is the characteristic function $\Phi_X(\upsilon)$ of the probability distribution of the random variable X, which is defined as the expectation of the complex exponential function $\exp(j\upsilon X)$ as

$$\Phi_X(\upsilon) = E\left[\exp(j\upsilon x)\right] = \int_{-\infty}^{\infty} f_X(x) \exp(j\upsilon x) \, dx. \qquad (B.27)$$

The PDF $f_X(x)$ can be obtained by taking the inverse Fourier transform:

$$f_X(x) = \frac{1}{2\pi} \int_{-\infty}^{\infty} \Phi_X(\upsilon) \exp(-j\upsilon x) \, d\upsilon. \qquad (B.28)$$

For two random variables X and Y, the joint characteristic function of X and Y, denoted by $\Phi_{X,Y}(\upsilon_1, \upsilon_2)$, is written as

$$\Phi_{X,Y}(\upsilon_1, \upsilon_2) = \int_{-\infty}^{\infty} \int_{-\infty}^{\infty} f_{X,Y}(x,y) \exp(j\upsilon_1 x + j\upsilon_2 y) \, dx \, dy. \qquad (B.29)$$

Thus the joint PDF $f_{X,Y}(x,y)$ can be written, following Eq. (B.28) as:

$$f_{X,Y}(x,y) = \frac{1}{4\pi^2} \int_{-\infty}^{\infty} \int_{-\infty}^{\infty} \Phi_{X,Y}(\upsilon_1, \upsilon_2) \exp(-j\upsilon_1 x - j\upsilon_2 y) \, d\upsilon_1 \, d\upsilon_2 \qquad (B.30)$$

B.4 RANDOM PROCESSES

A random signal is any waveform having randomly varying amplitude as a function of time. The sample of space or ensemble, of all possible random signals comprised of functions of time is called a *random or stochastic process*. Thus, when the random variable becomes a function of both the outcomes of the experiment as well as time, it is called a random process. It can be viewed as a random process as an ensemble of time domain functions that is the outcome of a certain random experiment.

Appendixes 297

Following the analysis of random variables, we denote the CDF and PDF of a random process as $F_X(x,t)$ and $f_X(x,t)$, respectively. The nth moment for random process $X(t)$ is given by

$$E\left[X^n(t)\right] = \int_{-\infty}^{\infty} x^n f_X(x,t)\, dx. \tag{B.31}$$

Broadly stated, a random process is called *stationary* if its statistical properties do not change with time. A random process $X(t)$ is said to be wide-sense stationary if

$$E\left[X(t)\right] = \bar{X} = \text{constant} \tag{B.32}$$

$$E\left[X(t)X(t+\tau)\right] = R_{XX}(\tau) \tag{B.33}$$

where $R_{XX}(\tau)$ is the autocorrelation function. If the time average and time correlation functions are equal to the statistical average and statistical correlation functions, the random process is referred to as an *ergodic* random process. Consider the sample function $x(t)$ of a wide-sense stationary process $X(t)$, with the observation interval defines as $-T \le t \le T$. The *dc value* of $x(t)$ is defined by the time average

$$\mu_x(T) = E\left[X(t)\right] = \frac{1}{2T} \int_{-T}^{T} x(t)\, dt \tag{B.34}$$

Since the process $X(t)$ is assumed to be wide-sense stationary, the mean of the time average $\mu_x(T)$ is written as

$$E\left[\mu_x(T)\right] = \frac{1}{2T} \int_{-T}^{T} E\left[x(t)\right] dt = \frac{1}{2T} \int_{-T}^{T} \mu_X\, dt = \mu_X \tag{B.35}$$

where μ_X is the mean of the process $X(t)$. Accordingly, the time average $\mu_x(T)$ represents an unbiased estimate of the ensemble-average mean μ_X. Thus an ergodic process is the one that satisfies the previous two relations.

The covariance of two random process $X(t)$ and $Y(t)$ is defined by

$$Cov_{XY}(t,t+\tau) = E[\{X(t) - E[X(t)]\}\{Y(t+\tau) - E[Y(t+\tau)]\}]. \tag{B.36}$$

This is written as

$$Cov_{XY}(t,t+\tau) = R_{XY}(\tau) - \bar{X}\bar{Y} \tag{B.37}$$

where $R_{XY}(\tau)$ is the cross-correlation function of $X(t)$ and $Y(t)$.

APPENDIX C: Z-TRANSFORMS TABLE

The z-transform for discrete-time signals is the counterpart of the Laplace transforms for continuous-time signals, and they have a similar relationship to the c-corresponding Fourier transform. In mathematics and signal processing, the *z-transform* converts a discrete-time signal, which is a sequence of real or complex numbers, into a complex frequency domain representation. It can be considered as a discrete-time equivalent of the Laplace transform. The z-transform, therefore, maps samples of a discrete time domain sequence $x[n]$ into a new domain known as the z-domain:

$$X(e^{j\Omega}) \triangleq \sum_{n=-\infty}^{\infty} x[n] e^{j\Omega n} \tag{C.1}$$

where Ω is the frequency-domain real variable called the *discrete frequency*.

A more general transform, called the *bilateral or two-sided z-transform* of the sequence $x[n]$ is defined as

$$X(z) = \sum_{n=-\infty}^{\infty} x[n] z^{-n} \tag{C.2}$$

where z is a frequency-domain complex variable. In polar form, z can be written as

$$z = |z| e^{j\Omega}. \tag{C.3}$$

where the magnitude and phase of z are $|z|$ and Ω, respectively. In the z-domain, the region over which $X(z)$ is finite is the *region of convergence* (ROC). It follows from (C.1) and (C.2) that the z-transform is evaluated on the unit circle $|z| = 1$.

TABLE OF COMMON Z-TRANSFORMS

Here we define

$$u[n] = \begin{cases} 1, & n \geq 0 \\ 0, & n < 0 \end{cases} \tag{C.4}$$

where $u[n]$ is the unit (or Heaviside) step function and

$$\delta[n] = \begin{cases} 1, & n = 0 \\ 0, & n \neq 0 \end{cases} \tag{C.5}$$

where $\delta[n]$ is the discrete-time unit impulse function (Dirac delta function, which is a continuous-time version). The two functions are chosen together so that the unit step function is the accumulation (running total) of the unit impulse function.

The commonly encountered z-transform pairs are summarized in Table C-1.

TABLE C.1
z-Transform Table

Signal, $x[n]$; $n \geq 0$	z-Transform, $X(z)$	ROC
$\delta[n]$	1	all z
1	$\dfrac{1}{1 - z^{-1}}$	$|z| > 1$
n	$\dfrac{z^{-1}}{(1 - z^{-1})^2}$	$|z| > 1$
n^2	$\dfrac{z^{-1}(1 + z^{-1})}{(1 - z^{-1})^3}$	$|z| > 1$
n^3	$\dfrac{z^{-1}(1 + 4z^{-1} + z^{-2})}{(1 - z^{-1})^4}$	$|z| > 1$
a^n	$\dfrac{1}{1 - az^{-1}}$	$|z| > |a|$
na^n	$\dfrac{az^{-1}}{(1 - az^{-1})^2}$	$|z| > |a|$
$n^2 a^n$	$\dfrac{az^{-1}(1 + az^{-1})}{(1 - az^{-1})^3}$	$|z| > |a|$
$\cos(\omega_0 n)$	$\dfrac{1 - z^{-1}\cos(\omega_0)}{1 - 2z^{-1}\cos(\omega_0) + z^{-2}}$	$|z| > 1$
$\sin(\omega_0 n)$	$\dfrac{z^{-1}\sin(\omega_0)}{1 - 2z^{-1}\cos(\omega_0) + z^{-2}}$	$|z| > 1$
$a^n \cos(\omega_0 n)$	$\dfrac{1 - az^{-1}\cos(\omega_0)}{1 - 2az^{-1}\cos(\omega_0) + a^2 z^{-2}}$	$|z| > |a|$
$a^n \sin(\omega_0 n)$	$\dfrac{az^{-1}\sin(\omega_0)}{1 - 2az^{-1}\cos(\omega_0) + a^2 z^{-2}}$	$|z| > |a|$

APPENDIX D: FOURIER TRANSFORMS TABLE

TABLE D.1
Fourier Transform Table

List	Function $s(t); \ a > 0$	Fourier Transform $S(\omega)$
1	$e^{-at}u(t)$	$\dfrac{1}{1 - j\omega}$
2	$e^{-at}u(-t)$	$\dfrac{1}{1 + j\omega}$
3	$te^{-at}u(t)$	$\dfrac{1}{(1 - j\omega)^2}$
5	$\delta(t)$	1
6	1	$2\pi\delta(\omega)$
7	$e^{j\omega_0 t}$	$2\pi\delta(\omega - \omega_0)$
8	$\cos(\omega_0 t)$	$\pi[\delta(\omega - \omega_0) + \delta(\omega + \omega_0)]$
9	$\sin(\omega_0 t)$	$j\pi[\delta(\omega + \omega_0) - \delta(\omega - \omega_0)]$
10	$u(t)$	$\pi\delta(\omega) + \dfrac{1}{j\omega}$
11	$e^{-at}\cos(\omega_0 t)u(t)$	$\dfrac{a + j\omega}{(a + j\omega)^2 + \omega_0^2}$
12	$e^{-at}\sin(\omega_0 t)u(t)$	$\dfrac{\omega_0}{(a + j\omega)^2 + \omega_0^2}$
13	$rect\left(\dfrac{t}{\tau}\right)$	$\tau\,\mathrm{sinc}\left(\dfrac{\omega\tau}{2}\right)$
14	$\Delta\left(\dfrac{t}{\tau}\right)$	$\dfrac{\tau}{2}\,\mathrm{sinc}^2\left(\dfrac{\omega\tau}{4}\right)$
15	$sgn(t)$	$\dfrac{2}{j\omega}$

Appendixes

APPENDIX E: PROBABILITY DENSITY FUNCTIONS

Gaussian Distribution

$$f_X(x) = \frac{1}{\sqrt{2\pi}\sigma_X} \exp\left[-\frac{1}{2}\left(\frac{x-\bar{X}}{\sigma_X}\right)^2\right] \tag{E.1}$$

where σ_X^2 is the variance of X given by

$$\sigma_X = \int_{-\infty}^{\infty} (x-\bar{X})^2 f(x)\,dx \tag{E.2}$$

and $\bar{X}$ is the expected or mean of the random variable X given by

$$\bar{X} = \int_{-\infty}^{\infty} xf(x)\,dx. \tag{E.3}$$

For zero mean and unit variance, we use the notation $f(x)$:

$$f(x) = \frac{1}{\sqrt{2\pi}} \exp\left(\frac{-x^2}{2}\right). \tag{E.4}$$

Rayleigh Distribution

$$f_X(x) = \frac{x}{\sigma^2} \exp\left[\frac{-x^2}{2\sigma^2}\right] u(x) \tag{E.5}$$

where σ^2 is the variance of the narrow-band noise $n(t)$, and

$$\bar{X} = \sqrt{\frac{\pi}{2}}\sigma; \quad \sigma_X^2 = \frac{\sigma^2}{2}(4-\pi). \tag{E.6}$$

Erlang Distribution

$$f_X(x) = \left(\frac{N}{\bar{X}}\right)^N \frac{\sigma^{N-1}}{(N-1)!} \exp\left[\frac{-N\sigma}{\bar{X}}\right] u(x), \quad N = 1,2,3,\ldots \tag{E.7}$$

Chi-Square Distribution

$$f(x) = \frac{M}{\Gamma(M)\bar{X}} \left(\frac{Mx}{\bar{X}}\right)^{M-1} \exp\left(\frac{-Mx}{\bar{X}}\right) u(x), \quad M > 0 \tag{E.8}$$

where

$$\Gamma(z) = \int_0^{\infty} \xi^{z-1} \exp(-\xi)\,d\xi, \quad \mathrm{Re}\{z\} > 0. \tag{E.9}$$

302 Appendixes

Uniform Distribution

$$f_X(x) = \frac{1}{b-a}; \quad a < b \tag{E.10}$$

$$\overline{X} = \frac{a+b}{2}; \quad \sigma_X^2 = \frac{(b-a)^2}{12} \tag{E.11}$$

Weibull Distribution

$$f_X(x) = \frac{b\Gamma(1+1/b)}{\overline{X}}\left[\frac{x\Gamma(1+1/b)}{\overline{X}}\right]^{b-1} \exp\left[\frac{x\Gamma(1+1/b)}{\overline{X}}\right]^b u(x) \tag{E.12}$$

Log-Normal Distribution

$$f_X(x) = \frac{1}{\sqrt{2\pi b^2}\, x} \exp\left\{\frac{[\ln(x)-a]^2}{2b^2}\right\} u(x) \tag{E.13}$$

where

$$a = \ln(x_m); \quad b = \sqrt{2\ln\left(\frac{\overline{x}}{x_m}\right)} \tag{E.14}$$

where x_m is the median value of X.

Appendixes

303

APPENDIX F: MATHEMATICAL FORMULAS

A number of mathematical formulas are listed that are useful in solving problems and in the interpretation of many radar mathematical relationships. Most formulas are readily available from many sources.

F.1 TRIGONOMETRIC IDENTITIES

$$\sin(x \pm y) = \sin x \cos y \pm \cos x \sin y \tag{F.1}$$

$$\cos(x \pm y) = \cos x \cos y \mp \sin x \sin y \tag{F.2}$$

$$\cos(2x) = \cos^2 x - \sin^2 x \tag{F.3}$$

$$\sin(2x) = 2 \sin x \cos x \tag{F.4}$$

$$e^{\pm jx} = \cos x \pm j \sin x \tag{F.5}$$

$$\cos x = \frac{1}{2}(e^{jx} + e^{-jx}) \tag{F.6}$$

$$\sin x = \frac{1}{2j}(e^{jx} - e^{-jx}) \tag{F.7}$$

$$\sin x \sin y = \frac{1}{2}[\cos(x - y) - \cos(x + y)] \tag{F.8}$$

$$\cos x \cos y = \frac{1}{2}[\cos(x - y) + \cos(x + y)] \tag{F.9}$$

$$\sin x \cos y = \frac{1}{2}[\sin(x - y) + \sin(x + y)] \tag{F.10}$$

$$\cos^3 x = \frac{1}{4}[3 \cos x + \cos 3x] \tag{F.11}$$

$$\sin^3 x = \frac{1}{4}[3 \sin x - \sin 3x] \tag{F.12}$$

F.2 INTEGRALS

Indefinite Integrals

$$\int x \sin(ax)\, dx = \frac{1}{a^2}[\sin(ax) - ax \cos(ax)] \tag{F.13}$$

$$\int x \cos(ax)\, dx = \frac{1}{a^2}[\cos(ax) - ax \sin(ax)] \tag{F.14}$$

$$\int x^2 \sin(ax)\, dx = \frac{1}{a^3}[2ax \sin(ax) + 2 \cos(ax) - (ax)^2 \cos(ax)] \tag{F.15}$$

$$\int x^2 \cos(ax)\, dx = \frac{1}{a^3}[2ax \cos(ax) - 2 \sin(ax) + (ax)^2 \sin(ax)] \tag{F.16}$$

$$\int xe^{ax}\,dx = \frac{e^{ax}}{a^2}(ax-1) \qquad (F.17)$$

$$\int xe^{ax^2}\,dx = \frac{e^{ax^2}}{2a} \qquad (F.18)$$

$$\int e^{ax}\sin(bx)\,dx = \frac{e^{ax}}{a^2+b^2}[a\sin(bx)-b\cos(bx)] \qquad (F.19)$$

$$\int e^{ax}\cos(bx)\,dx = \frac{e^{ax}}{a^2+b^2}[a\cos(bx)+b\sin(bx)] \qquad (F.20)$$

$$\int \frac{dx}{a^2+x^2} = \frac{1}{a}\tan^{-1}\left(\frac{x}{a}\right) \qquad (F.21)$$

$$\int \frac{xdx}{a^2+x^2} = \frac{1}{2}\ln(a^2+x^2) \qquad (F.22)$$

$$\int \frac{x^2 dx}{a^2+x^2} = x - a\tan^{-1}\left(\frac{x}{a}\right) \qquad (F.23)$$

$$\int \frac{dx}{(a^2+x^2)^2} = \frac{x}{2a^2(a^2+x^2)} + \frac{1}{2a^3}\tan^{-1}\left(\frac{x}{a}\right) \qquad (F.24)$$

$$\int \frac{x}{(a^2+x^2)^2} = \frac{-1}{2(a^2+x^2)} \qquad (F.25)$$

$$\int \frac{x^2}{(a^2+x^2)^2} = \frac{-x}{2(a^2+x^2)} + \frac{1}{2a}\tan^{-1}\left(\frac{x}{a}\right) \qquad (F.26)$$

Definite Integrals

$$\int_0^\infty \frac{x\sin(ax)}{b^2+x^2}\,dx = \frac{\pi e^{(-ab)}}{2}, \quad a>0, b>0 \qquad (F.27)$$

$$\int_0^\infty \frac{\cos(ax)}{b^2+x^2}\,dx = \frac{\pi e^{(-ab)}}{2b}, \quad a>0, b>0 \qquad (F.28)$$

$$\int_0^\infty \frac{x\sin(ax)}{b^2-x^2}\,dx = \frac{\pi}{4b^3}[\sin(ab)-ab\cos(ab)], \quad a>0, b>0 \qquad (F.29)$$

$$\int_0^\infty \operatorname{sinc} x\,dx = \int_0^\infty \operatorname{sinc}^2 x\,dx = \frac{1}{2} \qquad (F.30)$$

$$\int_0^\infty x^2 e^{-ax^2}\,dx = \frac{1}{4a}\sqrt{\frac{\pi}{a}}, \quad a>0 \qquad (F.31)$$

Appendixes

Bessel Functions

$$\int x^n J_{n-1}(x)\,dx = x^n J_n(x) \tag{F.32}$$

$$\int x^{-n} J_{n+1}(x)\,dx = -x^{-n} J_n(x) \tag{F.33}$$

$$\int x^n I_{n-1}(x)\,dx = x^n I_n(x) \tag{F.34}$$

$$\int x^{-n} I_{n+1}(x)\,dx = x^{-n} J_n(x) \tag{F.35}$$

F.3 INFINITE SERIES EXPANSIONS

Binomial Series

$$(1+x)^n = 1 + nx + \frac{n(n-1)}{2!}x^2 + \cdots, \quad |nx| < 1 \tag{F.36}$$

Exponential Series

$$e^x = 1 + x + \frac{x^2}{2!} + \frac{x^3}{3!} + \cdots \tag{F.37}$$

Logarithmic Series

$$\ln(1+x) = \sum_{n=1}^{\infty} \frac{(-1)^{n-1} x^n}{n}, \quad |x| < 1 \tag{F.38}$$

Taylor Series

$$f(x) = f(a) + \frac{f'(a)}{1!}(x-a) + \frac{f''(a)}{2!}(x-a)^2 + \frac{f'''(a)}{3!}(x-a)^3 + \cdots$$

$$\text{where} \quad f^{(n)}(a) = \left. \frac{d^n f(x)}{dx^n} \right|_{x=a} \tag{F.39}$$

Maclaurin Series

$$f(x) = f(a) + \frac{f'(0)}{1!}x + \frac{f''(0)}{2!}x^2 + \frac{f'''(0)}{3!}x^3 + \cdots$$

$$\text{where} \quad f^{(n)}(0) = \left. \frac{d^n f(x)}{dx^n} \right|_{x=0} \tag{F.40}$$

Trigonometric Series

$$\sin x = x - \frac{1}{3!}x^3 + \frac{1}{5!}x^5 - \cdots \tag{F.41}$$

$$\cos x = 1 - \frac{1}{2!}x^2 + \frac{1}{4!}x^4 - \cdots \tag{F.42}$$

$$\tan x = x - \frac{1}{3!}x^3 + \frac{2}{15!}x^5 - \cdots \qquad (F.43)$$

$$\sin^{-1} x = x + \frac{1}{6}x^3 + \frac{3}{40}x^5 + \cdots \qquad (F.44)$$

$$\tan^{-1} x = x - \frac{1}{3}x^3 + \frac{1}{5}x^5 + \cdots, \quad |x| < 1 \qquad (F.45)$$

$$e^{j\beta\cos x} = \sum_{n=-\infty}^{\infty} (j)^n J_n(\beta) e^{jnx} \qquad (F.46)$$

$$\cos[\beta\sin x] = J_0(\beta) + 2\sum_{n=1}^{\infty} J_n(\beta)\cos(2nx) \qquad (F.47)$$

$$\cos[\beta\cos x] = J_0(\beta) + 2\sum_{n=1}^{\infty} (-1)^n J_n(\beta)\cos(2nx) \qquad (F.48)$$

$$\sin[\beta\sin x] = 2\sum_{n=0}^{\infty} J_{2n+1}(\beta)\sin[(2n+1)x] \qquad (F.49)$$

$$\sin[\beta\cos x] = 2\sum_{n=0}^{\infty} (-1)^n J_{2n+1}(\beta)\sin[(2n+1)x] \qquad (F.50)$$

$$\sum_{n=1}^{\infty} \frac{\cos(nx)}{n} = -\ln(2)\sin(x/2), \quad 0 < x < 2\pi \qquad (F.51)$$

$$\sum_{n=1}^{\infty} \frac{\sin(nx)}{n} = \frac{1}{2}(\pi - x), \quad 0 < x < 2\pi \qquad (F.52)$$

REFERENCE

1. R. W. Hamming, *The Art of Probability for Scientists and Engineers* (Reading, MA: Addison Wesley, 1991).

Answers to Selected Problems

CHAPTER 2

2.1 $t_d = 2.56$ s.

2.3 (a) $P_{av} = 200$ W; (b) $B = 1$ MHz, $\Delta R = 150$ m.

2.5 $f_d = 24000.024$ Hz; (b) $f_d = 24000$ Hz.

2.7 $f_d = 30$ kHz, $R_u = 18.75$ km

2.10 (a) $\tau = 0.667 \mu s$; (b) $P_{av} = 0.5$ W.

2.11 $DF = 0.99999974$ for $\theta = 45°$; $DF = 1$ for $\theta = 90°$.

CHAPTER 3

3.1 $G_{max} = 38.46$ dB for $D = 1$; $G_{max} = 41.98$ dB for $D = 1.5$ m; $G_{max} = 44.48$ dB for $D = 2$ m.

3.3 $D = 0.360$ m for $G = 30$ dB; $D = 1.14$ m for $G = 40$ dB; $D = 3.60$ m for $G = 50$ dB.

3.5 $S_{min} = 16 \times 10^{-12}$ W; $P_t = 1.03$ MW; $\tau = 0.25$ μs.

3.7 $R_{max} = 156.33$ km.

3.9 $P_t = 2.122$ kW.

3.11 $D = 1.014$ m.

3.13 (a) $(S / N)_0 = 31.63$ dB for $R = 32$ km, $(S / N)_0 = 3.67$ dB for $R = 160$ km; (b) $R_0 = 197.66$ km at unity SNR.

3.15 (a) $P_{av}A = 36.65$ dB; (b) $P_t = 4.30$ kW corresponding to duty factor $d_t = 0.03$.

3.17 (a) $R_1 = 2.345$ km; (b) $T_i = 14.2$ ms; (c) $n = 142$; (d) $R_n = 6.81$ km; (e) The maximum unambiguous range is 15 km that justifies R_n is below R_u.

3.18 (a) $(R_{co})_{ssj} = 3.86$ km; (b) $(R_d)_{ssj} = 1.22$ km.

3.19 (a) $(R_{co})_{soj} = 57.250$ km; (b) $(R_d)_{soj} = 32.194$ km.

3.21 $R_t = 9.752$ km; $R_r = 5.910$ km.

CHAPTER 4

4.2 (a) $\sigma = -15.02$ dBsm; (b) $f_{min} = 2.387$ GHz.

4.3 $\sigma_{sp} = -1.05$ dBsm; $\sigma_{fp} = 37$ dBsm.

4.5 $F_{op} = 10.58$ dB.

4.7 $T_s = 210.55$ K.

4.9 (a) $T_{S1} = 530$ K; (b) $T_{S2} = 265$ K.

4.11 (a) $F_0 = 10.004$ dB; (b) $F_{op} = 6.30$ dB.

CHAPTER 5

5.1 (a) $F^4 = 10.0145$; (b) $P_r = 2.468 \times 10^{-10}$ W.

5.3 $\theta_{l1} = 0.026°$; $\theta_{n1} = 0.052°$.

5.7 $d_1 = 7.233$ m; $d_2 = 2355.76$ m; $d = 2.363$ km; $R_d = 2.45$ km.

5.8 (a) $d_2 = 32.51$ km; (b) $d_1 = 0.832$ km $\rightarrow d = d_1 + d_2 = 33.342$ km; (c) $D = 0.9984$.

Answers to Selected Problems

CHAPTER 6

6.1 $f_b = 10$ kHz.

6.2 $R = 150$ km.

6.3 $R = 175$ km.

6.5 $f_b^+ = 5.6$ kHz, $f_b^- = 42.4$ kHz.

6.6 (a) $R = 350$ km, $\dot{R} = 250$ m/s; (b) $R_{max} = 375$ km; (c) $R_{un} = 7500$ km.

6.7 $R = 13.851$ km; $v_r = 128.55$ m/s.

6.9 $\langle f_b \rangle = 20$ kHz.

6.11 $R_{un} = 100$ km.

CHAPTER 7

7.1 $R_{un} = 10.068$ km.

7.3 $f_{r1} = 73.75$ kHz, $f_{r2} = 80.00$ kHz, $R_{u1} = 2.034$ km, $R_{u2} = 1.875$ km.

7.5 $R_{u1} = 89.13$ km, $R_{u2} = 86.58$ km, $R_{u3} = 84.03$ km.

7.6 (a) $f_{d1} = 8$ kHz for $n_1 = 2$, $f_{d2} = 2$ kHz for $n_2 = 2$, $f_{d3} = 16$ kHz for $n_3 = 2$.

    (b) $f_{dn} = 8,\ 22,\ 36,\ 50$ ....kHz corresponding to 8 kHz

       $f_{dn} = 16,\ 33,\ 50,\ 67$ ....kHz corresponding to 16 kHz

        $f_{dn} = 10,\ 30,\ 50,\ 70$ ....kHz corresponding to 10 kHz.

    (c) $v_r = 750$ m/s.

CHAPTER 8

8.5 $B = 500$ MH, $\Delta R_u = 30$ m, $\Delta R_c = 30$ cm.

8.7 (a) $f_1 = 1.00$ GHz, $f_2 = 1.06$ GHz; (b) $N_{FFT} = 4{,}800$.

8.9 The possible Barker codes are $B_{5,11}$ or $B_{11,5,}$ which are

$$B_{5,11} = \{11100010010, 11100010010, 11100010010, 00011101101, 11100010010\}$$

$$B_{11,5} = \{11101, 11101, 11101, 00010, 00010, 00010, 11101, 00010, 00010, 11101, 00010\}$$

8.10 $F_9 = \{000; 024; 042\}$.

CHAPTER 9

9.1 $\Delta X = 3$ m, $L = 10$ m.

9.2 $d_x = 1.0$ m, $\ell_{max} = 2.0$ m.

9.4 $\delta R = 1.0$ m.

9.6 $SNR = 52.15$ dB.

9.7 $SNR = 34.19$ dB.

CHAPTER 10

10.1 $R_0 = 18$ km.

10.3 $\sigma_{id}^2 = 0.004 \times 10^{-12} s^2$.

10.4 $\sigma_{\hat{R}} = 15$ m.

10.5 $R_v = 36$ m.

10.7 $\sigma_{\hat{R}} = 2.4$ m.

10.9 $X_s = \begin{bmatrix} x_s & v_s \end{bmatrix}^T = \begin{bmatrix} 12 \times 10^3 & 340 \end{bmatrix}^T$.

CHAPTER 11

11.1 $P_{rad} = 10\pi^2$.

11.3 $G_D = 1.27$.

11.5 $\varepsilon_{ap} = 64\%$.

11.7 $D = 41.62$ dB.

11.8 $\varepsilon_{ap} = 80\%$, $A_e = 62.83$ m^2.

11.12 $\psi = 0°$, $FNBW = 38.85°$, $HPBW = 16.98°$, $\theta_s = 60°$ or $120°$.

11.13 $\psi = 90°$, $FNBW = 96.38°$, $HPBW = 63.02°$, $\theta_s = 60°$.

11.14 $FNBW = 36.8°$, $HPBW = 15.8°$, $\theta_s = 45°$ or $101°$.

Bibliography

This bibliography includes all the numerous books, articles, and reports that are referenced in the body and end-of-chapter notes and references of the text as well as some additional excellent books of general interest in the study of radar systems.

Abramowitz, M., and I. A. Stegun (eds.). *Handbook of Mathematical Functions with Formulas, Graphs, and Mathematical Tables, Applied Mathematical Series 55*. Washington, DC: National Burreau of Standards, 1964.

AN/FPS-118 Over-the-Horizon Radar. Jane's Radar and Electronic Warfare Systems, Land-Based Air Defense Radars, 2001.

Anderson, R. H., S. Kraut, and J. L. Krolik. "Robust Altitude Estimation for Over-the Horizon Radar Using a State-Space Multipath Fading Model." *IEEE Transaction on Aerospace and Electronic Systems 39, no. 1,* 2003: 192–201.

Austin, P. M. "Radar Measurement of the Distribution of Precipitation in New England Storm." *Proceedings of the 10th Weather Conference.* Boston, 1965: 247–254.

Balanis, C. A. *Antenna Theory Analysis and Design.* New York: Harper & Row Publishers, 1982.

Barker, R. H. *Group Synchronizing of Binary Digital Systems, Communication Theory*, edited by W. Jackson. New York: Academic Press, 1953.

Barnum, J. R. "Ship Detection with High-Resolution HF Skywave Radar." *IEEE Journal of Ocean Engineering (Invited Paper) OE-11,* no. 2, 1986: 196–209.

Barton, D. K. *Radar System Analysis.* Englewood Cliffs, NJ: Prentice-Hall, 1964.

Barton, D. K. *Modern Radar System Analysis.* Norwood, MA: Artech House, 1988.

Beard, C. I. "Coherent and Incoherent Scattering of Microwaves from the Ocean." *IEEE Transactions on Antennas and Propagation 9,* no. 5, 1961: 470–483.

Beckmann, P., and A. Spizzichino. *The Scattering of Electromagnetic Waves from Rough Surfaces.* Oxford, United Kingdom: Pergamon Press, 1963.

Benedict, R. T., and G. W. Brodner. "Synthesis of an Optimal Set of Radar Track-While-Scan Smoothing Equations." *IRE Transaction on Automatic Control, 7,* 1962: 27–32.

Berkowtz, R. S. *Modern Radar.* New York: John Wiley and Sons, 1965.

Blake, L. V. "Prediction of Radar Range." In *Radar Handbook*, 2nd Edition, edited by M. I. Skolnik. New York: McGraw-Hill Book Company, 1990.

_____. *Radar Range Performance Analysis.* Norwood, MA: Artech House, 1986.

_____. *Radar/Radio Tropospheric Absorption and Nosie Temperature.* Report 7461 and NTIS documents, AD753197, Washington, DC: US Naval Research Laboratory, 1972.

Boithias, L. *Radio Wave Propagation.* New York: McGraw-Hill Book Company, 1987.

Boothe, R. R. *The Weibull Distribution Applied to the Ground Clutter Backscatter Coefficient.* Report RE-TR-69-15, USAMC, 1969.

Bourgeosis, D., C. Morisseau, and M. Flecheux. "Over-the–Horizon Radar Target Tracking Using MQP Ionospheric Modeling." *IEEE 8th International Conference on Information Fusion (FUSION) 1,* 2005: 283–289.

Bradley, P. A. "Long-Term HF Predictions for Radio Circuit Planning." *Radio and Electronic Engineer, 45,* 1975: 31–41.

Brown, B. P. "Radar Height Finding," In *Radar Handbook*, edited by M. I. Skolnik. New York: McGraw-Hill Book Company, 1970.

Brown, W. M., and L. J. Porcello. "An Introduction to Synthetic Aperture Radar." *IEEE Spectrum 6,* 1969: 52–62.

Bryant, K. O. "Programmable 20-bit Pseudorandom (PRF) Generator." US Patent #3,662,386, 1972.

Burdic, W. S. *Radar Signal Analysis.* Englewood Cliffs, NJ: Perntice-Hall, 1968.

Burkowitz, R. W., ed. *Modern Radar.* New York: John Wiley, 1966.

Burrows, C. R., and S. S. Attwood. *Radio Wave Propagation.* New York: Academic Press, 1949.

Carlson, A. B. *Communication Systems: An Introduction to Signals and Noise in Electrical Communications.* New York: McGraw-Hill Book Company, 1986.

CCIR. *Atlas of Ionospheric Characteristics.* Geneva: ITU, 1983.

CCIR. *Characteristics and Applications of Atmospheric Radio Noise.* Data Report 322-2, Geneva: ITU, 1984.

Cole, H. W. "The Future for SSR." *ICAO Bull.,* September 1980.

311

Collins, R. E. *Antennas and Radio Wave Propagations*. New York: McGraw-Hill Book Company, 1985.

Crispin, J. W., and K. M. Siegel. *Method of Radar Cross-Section Analysis*. New York: Academic Press, 1968.

Curry, G. R. *Radar Essentials—A Concise Handbook for Radar Design and Performance Analysis*. Rayleigh, NC: SciTech Publishing, 2012.

Davenport, W. B. Jr., and W. L. Root. *An Introduction to the Theory of Random Signals and Noise*. New York: McGraw-Hill Book Company, 1958.

Davie, K. *Ionospheric Radio Propagation*. New York: Dover Publications, 1966.

Department of Defense, Joint Chief of Staff. *Dictionary of Military and Associated Terms*. JCS Pub-1, 1974.

Di Cenzo, A. "A New Look at Nonseparable Synthetic Aperture Radar Processing." *IEEE Trans. on Aerospace and Electronic Systems 24*, No. 3, 1988: 218–223.

Domb, C., and M. H. L. Pryce. "The Calculation of Field Strengths over a Spherical Earth." *Journal of the Institution of Electrical Engineers—Part III: Radio and Communication Engineering 94*, no. 31, 1947: 325–339.

Dunn, J. H., D. D. Howard, and A. M. King. "Phenomena of Scintillation Noise in Radar Tracking Systems." *Proceedings of the IRE 47*, no. 5, 1959: 855–863.

Dyer, R. D., and F. B. Hayes. *Computer Modeling for the Fire Control Radar Systems*. Technical Report 1, Contract DAA25-73-C-0256, Atlanta: Georgia Institute of Technology, 1974.

Dyson, P. L., and J. A. Bennett. "Exact Ray Path Calculations Using Realistic Ionospheres." *IEEE Proceedings H—Microwaves, Antenna and Propagation 139*, no. 5, 1992: 407–413.

Eave, R. *Principles of Modern Radar*. New York: Van Nostrand Reinhold Company, 1987.

Edde, B. *Radar Principle, Technology, Applications*. Englewood Cliffs, NJ: PTR Prentice-Hall, 1993.

Elliot, R. S. *Antenna Theory and Design*. Englewood Cliffs, NJ: Prentice-Hall, 1981.

Eustace, E. H., ed. *International Countermeasures Handbook*, 3rd Edition. Palo Alto, CA: EW Communications, 1977.

Evans, D. L., and J. J. Van Zyl. "Polarimetric Imaging Radar: Analysis Tools and Applications." *Progress in Electromagnetics Research 103*, 1990: 371–389.

Flock, W. L., and J. L. Green. "The Detection and Identification of Birds in Flight, Using Coherent and Noncoherent Radars." *Proceedings of the IEEE 62*, no. 6, 1974: 745–753.

Frank, R. L. "Polyphase Codes with Good Nonperiodic Correlation Properties." *IEEE Transaction on Information Theory 9*, 1963: 43–45.

Galelian, J. C. "Designing Radar Receivers to Overcome Jamming." *Electronics 36*, May 1963: 50–54.

Glover, K. M. et al. "Radar Observations of Insects in Free Flight." *Science 254*, 1966: 967–972.

Golay, M. J. E. "Complementary Series." *IRE Transaction on Information Theory 7*, no. 2, 1961: 82–87.

Goldstein, H. *A Primer of Sea Echo*. Report No. 157, San Diego: US Navy Electronics Laboratory, 1950.

Gunn, K. L. S., and T. W. R. East. "The Microwave Properties of Precipitation Particles." *Quarterly Journal of the Royal Meteorological Society 80*, no. 346, 1954: 522–545.

Hamming, R. W. *The Art of Probability for Scientists and Engineers*. Reading, MA: Addison-Wesley, 1991.

Hannan, P. W. "Microwave Antennas Derived from the Cassegrain Telescope." *IRE Transactions on Antennas and Propagation*, 1961: 140–153.

Harrington, R. F. *Time-Harmonic Electromagnetic Fields*. New York: McGraw-Hill Book Company, 1968.

Harrington, R. F., and J. R. Mautz. "Straight Wires with Arbitrary Excitation and Loading." *IEEE Transactions on Antennas and Propagation AP-15*, 1967.

Headrick, J. M., and M. I. Skolnik. "Over-the-Horizon Radar in the HF Band." *Proceedings of the IEEE 62*, no. 69, 1974.

Heimiller, R. C. "Theory and Evaluation of Gain Patterns of Synthetic Arrays." *IRE Transaction on Military Electronics 6*, 1962: 122–129.

Herman, E. E. *The Elevation Angle Computer for the AN/SPS-2 Radar*. Report 3896, Washington, DC: US Naval Research Laboratory, 1951.

Holt, O. "Technology Survey—A Sampling of Low Noise Amplifiers." *The Journal of Electronic Defense*, 2015.

Houghton, E. W., F. Blackwell, and T. A. Willmot. "Bird Strike and Radar Properties of Birds." *International Conference on Radar—Present and Future*. IEE Conference Pub. No. 105, 1973: 257–262.

Hovanessian, S. A. *Introduction to Synthetic Array and Imaging Radars*. Dedham, MA: Artech House, 1980.

————. *Radar Detection and Tracking Systems*. Dedham, MA: Artech House, 1978.

IEEE. "High Frequency Radar for Ocean and Ice Mapping and Ship Location." *IEEE Journal of Oceanic Engineering 11*, 1986: 145–332.

IEEE. "IEEE Standard Definitions of Terms for Antennas." *IEEE Transaction on Antenna and Propagation 22*, no. 1, 1974.

Bibliography

_____. *IEEE Standard Radar Definition, IEEE Standard 686-1982m*. New York: Electrical and Electronics Engineers Institute, 1982.

Johnston, S. L., ed. *Radar Electronic Counter-Countermeasures*. Norwood, MA: Artech House, 1979.

Johnston, S. L. "Radar Electronic Counter-Countermeasures against Chaff." *Proceedings of the International Conference on Radar*. Paris, France, 1984: 517–522.

Johnston, S. L. *Radio Defense System Electronic Countermeasures*. Course Notes, Course ENG 216., Continuing Education Institute, n.d.

Josefsson, L., and P. Persson. *Conformal Array Antenna Theory and Design*. Hoboken, NJ: Wiley Inter-Science Publication, IEEE Press, 2006.

Barton, D. K. *Radar System Analysis*. Englewood Cliffs, NJ: Prentice-Hall, n.d.

Kalman, R. E. "A New Approach to Linear Filtering and Prediction Problems." *Transaction of the ASME-Journal of Basic Engineering*, 1960: 35–45.

Kalmus, H. P. "Direction Sensitive Doppler Device." *Proceedings of the IRE 43,* 1955: 690–700.

Kelso, J. M. *Radio Ray Propagation in the Atmosphere*. New York: McGraw-Hill Book Company, 1964.

Kerr, D. E., ed. *Propagation of Short Wave*. Lexington, MA: Boston Technical Publishers, 1964.

Khan, R. H., and D. K. Mitchell. "Waveform Analysis for High-Frequency FMICW Radar." *IEE Proceedings on Radar and Signal Processing 138,* no. 59, 1991: 411–419.

Kluder, J. R., A. C. Price, S. Darlington, and W. J. Albersheim. "The Theory and Design of Chirp Radar." *Bell System Technical Journal 39,* no. 4, 1960: 745–808.

Knot, E. F., J. F. Shaeffer, and M. T. Tuley. *Radar Cross Section*. Norwood, MA: Artech House, 1985.

Konrad, T. G., J. J. Hicks, and E.B. Dobson. "Radar Characteristics of Birds in Flight." *Science 159,* 1968: 274–280.

Kosolov, A. A. et al. *Over-the-Horizon Radar*. Dedham, MA: Artech House, 1987.

Krestschmer F. F., Jr., and B. L. Lewis. *Doppler Properties of Polyphase Coded Pulse Compression Waveforms*. NRL Report 8635, Washington DC: Naval Research Laboratory, 1982.

Lake, J. S. "Observable Countermeasures." *Proceedings of the Military Microwaves Conference MM-84,* 1984: 391–396.

Leonov, K. I., and A. I. Fomichev. *Monopulse Radar*. Norwood, MA: Artech House, 1986.

Levanon, N. *Radar Princples*. New York: John Wiley & Sons, 1968.

Locke, A. S. *Guidance*. Princeton: NJ: D. Van Nostarnd Company, 1955.

Long, R. M. *Radar Reflectivity of Land and Sea*. Dedham, MA: Artech House, 1983.

Mahafza, B. R. *Introduction to Radar Analysis*. Boca Raton, FL: CRC Press, 1998.

_____. *Radar System Analysis and Design Using MatLab*, Second Edition. Boca Raton, FL: Chapman & Hall/CRC, 2005.

Maisel, L. "Performance of Sidelobe Blanking Systems." *IEEE Transactions on Aerospace and Electronics Systems AES-4,* no. 2, 1968: 174–180.

Marcum, J. I. "A Statistical Theory of Target Detection by Pulsed Radar: Mathematical Appendix." *IRE Transaction on Information Theory IT-6,* no. 2, 1960.

Maxwell, J. C. *A Treatise in Electricity and Magnetism*. Oxford: Clarendon Press, 1892.

McCandless, S. W. "SAR in Space: The Theory, Design, Engineering and Application of a Space-based SAR System." In *Space-Based Radar Handbook*, edited by L. J. Cantafio. Norwood, MA: Artech House, n.d.

Meeks, M. L. *Radar Propagation at Low Altitudes*. Dedham, MA: Artech House, 1982.

Melsa, J. L., and D. L. Cohn. *Decision and Estimation Theory*. New York: McGraw-Hill Book Company, 1978.

Mensa, D. L. *High Resolution Radar Imaging*. New York: McGraw-Hill Book Company, 1984.

Miller, A. R. et al. "New Derivation for the Rough Surface Reflection Coefficient and for the Distribution of Sea Wave Elevation." *Proceedings of the IEEE International Conference on Microwaves, Optics and Antennas 131,* no. 2, 1984: 114–116.

Milne, K. "The Combination of Pulse Compression with Frequency Scanning for Three-Dimensional Radars." *Radio and Electronic Engineer 28,* 1964: 89–106.

Nathanson, F. E. *Radar Design Principles*, 2nd Edition. New York: McGraw-Hill Book Company, 1991.

Naval/Coastal Surveillance and Navigation Radars. *CONDO-R Naval Radar*. Jane's Radar and Electronic Warfare Systems, 2006.

North, D. O. "An Analysis of the Factors Which Determine Signal/Noise Discrimination of Pulsed Carrier Systems." *Proceedings of the IEEE 51,* no. 7, 1963: 1015–1027.

Nyquist, H. "Thermal Agitation of Electric Charge in Conductors." *Physical Review 32,* no. 1, 1928: 110–113.

Oppenheim, A. V., A. S. Wilsky, and I. T. Young. *Signals and Systems*. Englewood Cliffs, NJ: Prentice-Hall, 1983.

314 Bibliography

Page, R. M. "Monopulse Radar." *IRE National Convention Record 3*, no. 8, 1955: 132–134.
Papazoglou, M., and J. Krolik. "Electromagnetic Matched-Field Processing for Target Height Finding with Over-the-Horizon Radar." *IEEE International Conference on Acoustic, Speech, and Signal Processing (ICASSP)*, 1997: 559–562.
Papoulis, A. *Probability, Random Variables, and Stochastic Processes*, 2nd Edition. New York: McGraw-Hill Book Company, 1984.
———. *Signal Analysis*. New York: McGraw-Hill Book Company, 1977.
Peebles, P. Z. Jr. *Radar Principles*. New York: John Wiley and Sons, 1998.
Pielou, J. M. "Sky-Wave Radar Propagation Predictions for HF Radar System Planning." *International Conference on Antennas and Propagation*, 1985: 510–514.
Price, A. *History of US Electronic Warfare*, vol. 1. Alexandria, VA: Association of Old Crows, 1984.
Pulliainen, J. T., K. J. Hyyppa, and M. T. Hallikainen. "Backscattering Properties of Boreal Forests at the C- and X-Bands." *IEEE Transactions on Geoscience and Remote Sensing 32*, no. 5, 1994: 1041–1050.
Raney, R. K. "The Delay/Doppler Radar Altimeter." *IEEE Transactions Geoscience and Remote Sensing 36*, 1998: 1578–1588.
Reed, H. R., and C. M. Russell. *Ultra High Frequency Propagation*. London: Chapman and Hall Ltd., 1966.
Rhodes, D. R. *Introduction to Monopulse*. New York: McGraw-Hill Book Company, 1959.
Ridenour, L. N. *Radar System Engineering, MIT Radiation Laboratory Series,* vol. 1. New York: McGraw-Hill Book Company, 1947.
Rihaczek, A. W. *Principle of High Resolution Radar*. New York: McGraw-Hill Book Company, 1959.
Roddy, D. *Satellite Communications*. New York: McGraw-Hill Book Company, 2001.
Ruck, G. T. et al. *Radar Cross-Section Handbook*. New York: Plenum Press, 1970.
Saillant, S. "NOSTRADAMUS Over the Horizon Radar—First Subway Experimentation." *IEEE International Conference on Radar*, 1992: 78–81.
Sales, G. S. *OTH-B Radar System: System Summary*. Philips Laboratory, Air Force Systems Command, Hanscom Air Force Base, MA, 1992.
Sandwell, D. T., and W. H. F. Smith. "Retracking ERS-1 Altimeter Waveforms for Optimal Gravity Field Recovery." *Geophysical Journal International 163*, 2005: 79–89.
Sandwell, D. T. *Radar Altimetry*. Lecture Notes, Scripps Institution of Oceanography, University of California, San Diego, CA, 2011.
Scanlon, M. J., ed. *Modern Radar Techniques*. New York: Macmillan Publishing Company, 1987.
Scheer, J. A., and J. L. Kutz, eds. *Coherent Radar Performance Estimation*. Dedham, MA: Artech House, 1993.
Schleher, D. C., ed. *MTI Performance Analysis in MTL Radar*. Dedham, MA: Artech House, 1978.
Shearman, E. D. R. "The Technique of Ionospheric Investigation Using Ground Backscatter." *Proceedings of the IEE 103b*, 1956: 210–223.
Sherman, S. M. *Monopulse Principles and Techniques*. Dedham, MA: Aretch House, 1984.
Shrader, W. W. "MTI Radar," In Radar *Handbook*, edited by M. I. Skolnik. New York: McGraw-Hill Book Company, 1970.
Six, M., J. Parent, A. Bourdillon, and J. Dellue. "A New Multibeam Receiving Equipment for the Valensole Skywave HF Radar: Description and Applications." *IEEE Transactions on Geoscience and Remote Sensing 34*, no. 3, 1996: 708–719.
Sklansky, J. K. *Optimizing the Dynamic Parameter of a Track-While Scan System*. Princeton, NJ: RCA Laboratories, 1957.
Skolnik, M. I. *A Perspective of Synthetic Aperture Radar for Remote Sensing*. Memorandum Report 3783, Washington, DC: Naval Research Laboratory, n.d.
———. *Introduction to Radar Systems*. New York: McGraw-Hill Book Company, 1980.
———. "Radar Horizon and Propagation Loss." *Proceedings of the IRE*, 1957: 697–698.
Skolnik, M. I., ed. *Radar Handbook*, 2nd Edition. New York: McGraw-Hill Book Company, 1990.
———. *Radar Handbook*. New York: McGraw-Hill Book Company, 1970.
Smith, E. K., and S. Weintraub. "The Constraints in the Equation for Atmospheric Refractive Index at Radio Frequencies." *Proceedings of the IRE 41*, 1953: 1032–1937.
Stark, L. "Microwave Theory of Phased Array Antennas—A Review." *Proceedings of the IEEE 62*, 1974: 1661–1701.
Stevens, M. C. *Secondary Surveillance Radar*. Norwood, MA: Artech House, 1988.
———. "Secondary Surveillance Radar—Today and Tomorrow." *SERT Avionics Symposium*. Swansea, 1974.
Stimson, G. W. *Introduction to Airborne Radar*. El Segundo, CA: Hughes Aircraft Company, 1983.

Bibliography

Straiton, A. W., and C. W. Tolbert. "Anomalies in the Absorption of Radio Waves by Atmospheric Gases." *Proceedings of the IRE 48,* 1960: 898–903.

Stutzman, W. L., and G. A. Thiele. *Antenna Theory and Design.* New York: John Wiley and Sons, 1981.

Swerling, Peter. "Detection of Fluctuating Pulsed Signal in the Presence of Noise." *IRE Transaction on Information Theory,* IT-3, no. 3, 1957.

Thomason, J. F. "Development of Over-the-Horizon Radar in the United States." *Proceedings of the IEEE International Conference on Radar,* 2003: 599–601.

Tzannes, N. S. *Communication and Radar Systems.* Englewood Cliffs, NJ: Prentice-Hall, 1985.

US Department of Defense (DOD). "Electronic Warfare Definitions." *Electronic Warfare 2, 4,* 29, 1970.

Ulaby, F. T., R. K. Moore, and A. K. Fung. *Microwave Remote Sensing: Active and Passive,* vol. 1. Norwood, MA: Artech House, 1981.

Ullyatt, C. "Sensors for the ATC Environment with Special Reference to SSR." *UK Symposium on Electronics for Civil Aviation 1,* cat.1, sect. C, 1969.

Urkowitz, H. "Analysis and Synthesis of Delay Line Periodic Filters." *IRE Transactions on Circuit Theory 4,* 1957: 41–53.

Vakman, D. E. *Sophistaceted Signals and the Uncertainty Principles in Radar.* New York: Springer-Verlag, 1968.

Van Vleek, J. H. "The Absorption of Microwaves by Oxygen." *Physical Review 71,* 1947: 413–424.

Van Vleek, J. H. "The Absorption of Microwaves by Uncondensed Water Vapor." *Physical Review 71,* 1947: 425–433.

Ward, K. D., and S. Watts. "Radar Sea Clutter." *Microwave Journal 29,* no., 6, 1985: 109–121.

Watanabe, M., T. Tamama, and N. Yamauchi. "A Japanese 3-D Radar for Air Traffic Control." *Electronics 44,* 1971: 68–72.

Welti, G. R. "Quaternary Codes for Pulsed Radar." *IRE Transaction on Information Theory 6,* no. 3, 1960: 82–87.

White, D. M. *Synthesis of Pulse Compression Waveform with Weighted Finite Frequency Combs,* TG-934, Applied Physics Laboratory Report, 1967.

White, W. D., and A. E. Ruvin. "Recent Advances in Synthesis of Comb Filters." *IRE National Convention Record 5,* no. 2, 1957: 186–199.

Wiley, R. G. *Electronic Intelligence: Interception of Radar Signals.* Norwood, MA: Artech House, 1985.

Wise, J. C. "Summary of Recent Australian Radar Developments." *IEEE Aerospace and Electronic System Magazine,* December 2004.

Wise, John C. *PLA Air Defense Radars.* Technical Report APA-TR-2009-0103, 2009.

Woodward, P. M. *Probability and Information Theory with Applications to Radar.* New York: McGraw-Hill Book Company, 1963.

Wyatt, L. R., G. D. Burrows, and M. D. Moorhead. "An Assessment of an FMICW Ground-Wave Radar System for Ocean Wave Studies." *International Journal of Remote Sensing 6,* 1985: 275–282.

Zhang, Y., M. G. Amin, and G. J. Frazer. "High-Resolution Time-Frequency Distributions for Maneuvering Target Detection in Over-the-Horizon Radars." *IEE Proceedings Radar, Sonar and Navigation 150,* no. 4, 2003: 299–304.

Index

A

α-β tracker, 189, 190, 192
α-β-γ tracker, 191
Absorption, deviating, 93
Absorption, nondeviating, 93
Absorptive network, 68
Ambiguity function, radar, 135
Ambiguity functions, properties of, 136
Ambiguity, Doppler, 127
Ambiguity, range, 12, 125
Ampere's law, 3
Angle of incidence, 158
Antenna, AN/FPS-95, 276
Antennas
    antenna efficiency, 22, 201
    aperture type, 197, 202
    array, 211
    array factor, 211
    bandwidth, 202
    beam solid angle, 199, 200
    broadside array, 218
    Cassegrain reflector, 210
    circular array, 223
    conformal array, 198, 225
    directive gain, 200
    directivity, 208, 209
    effective aperture, 201
    end-fire array, 218
    main lobe, 198
    minor lobe, 198
    parabolic reflector, 197, 202, 206
    pattern multiplication in array
        factor, 211, 212
    phased array, 197, 219
    planar array, 223
    polarization, 165, 202
    power gain, 201
    radiation integral, 203
    radiation pattern, 81, 198, 199, 206, 211, 234
    sidelobe, 198
Anti-transmit-receive (ATR), 5
A-scope, 115
Attenuation, ionospheric, 92
Attenuation, resonance peaks in, 90
Automatic gain control (AGC), 177

B

B-26, RCS of, 51
Beam splitting, 235

Beamwidth, half-power, 158, 159,
    198, 204
Beaufort wind scale, 57
Bessel functions, 305
Binary phase codes, 108
B-scope, 71

C

Chain Home, 1, 234, 263
Chirp. *See* Linear frequency modulation
Clutter spectrum, 64
Clutter, scattering parameter of, 56
Coherent integration, 272
Conjugate filter. *See* Matched filter
Coulomb's law, 3
Cross-section density. *See* Radar clutter
CW radars
    beat frequency in, 102
    linear frequency modulation, 102
    multiple-frequency, 107
    phase-modulatatin in, 108
    sinusoidal FM in, 105

D

Definite integrals, 304
Digital beamforming, 239
Dilation factor, 15
Dirichlet conditions, 291, 292
Doppler aliasing, 275
Doppler effect, 14, 97, 124
Doppler frequency in CW
    radar, 98
Doppler resolution, 275
Doppler shift, 14
Doppler spectral density, 19
Doppler tolerance, 110, 153
Dot angels. *See* Radar clutter
Duty factor, 12

E

ECCM. *See* Electronic counter-
    countermeasures
ECCM design, 255
    operational doctrines, 260
    opration philosophy, 260
    radar parameters in, 255
    signal processing in, 260

ECCM, antenna-related, 256
   sidelobe blanker (SLB) in, 256
   sidelobe canceler (SLC) in, 257
ECCM, receiver-related, 259
   CFAR, 259
   Dicke fix in, 259
   log receiver in, 259
ECCM, transmitter-related, 258
   burn-through mode, 258
   frequency in, 258
   power in, 258
   spotlighting, 258
   waveform coding in, 258
Echolocation, 3
ECM. *See* Electronic countermeasures
Electroamgnetic compatibilty, 247
Electronic counter-countermeasures, 247
Electronic countermeasures, 251
Electronic warfare, 245–61
   active ECM, 251
   anti-jamming (AJ) fixes, 260
   defined, 246
   denial ECM, 252
   electromagnetic compatibility, 247
   electronic counter-countermeasures,
      255–61
   electronic countermeasures,
      251–55
   electronic intelligence, 248
   electronic support measures,
      248–50
   ESM receivers, 248–49
   radar warning receiver, 248
EMC. *See* Electromagnetic compatibility
ESM. *See* Electronic support measures

F

False alarm, 11, 255, 259
Faraday's law, 3
Fourier series, 291, 292
Fourier transform, 292
Fourier trnsform table, 300
Four-third earth model, 87
Frank polyphase codes, 110
Frequency scanners, 238

G

Golay codes, 110

H

Hertz, H., 1
Hulsmeyer, C., 1

I

Indefinite integrals, 303
Infinite series, 305
Interferometry height-finder, 239
Ionmosphere, thin layer model of, 269
Ionosphere, D layer in, 266
Ionosphere, E layer in, 267
Ionosphere, electron density in, 267
Ionosphere, F layer in, 267
Ionosphere, F_1 layer in, 267
Ionosphere, multiple layers model of, 270
Ionosphere, refractive index *in*, 267
Ionosphere, thick layer model of, 270
Ionsphere, F_2 layer in, 267

J

Jamming in ECM, 252
   azimuth deception, 253
   barrage jamming, 252
   chaff in, 254
   chemical jamming, 254
   deception jamming, 252
   decoys, 255
   flares, 255
   imitative, 253
   inverse gain repeater, 253
   manipulatve deception, 253
   mechanical jamming, 254–55
   range deception, 253
   range gate stealer, 253
   spoofers, 252
   spot jammimg, 252
   sweep jamming, 252
   velocity deception, 253
JORN. *See* Over-the-horizon radars

L

LFM. *See* Linear frequency modulation
Light detection and ranging (LIDAR), 1
Low angle grazing region, 54
Low noise amplifier (LNA), 68
Low probability of intercept (LPI). *See also*
      Electronic warfare

M

MADRE. *See* Over-the-horizon radars
Marconi, M. G., 1
Matched filter, 131
Maximum unambiguous range, 13
Maxwell, J., 1
Mie region, 44
Minimum detectable signal, 11, 24

Index

MLS codes. *See* Pseudorandom codes
Monopulse, amplitude comparison, 179
Monopulse, phase comparison, 181
Moving target indication (MTI), 113–23
  blind speeds, 118
  delay-line canceler, double, 119
  delay-line canceler, single, 117
  nonrecursive filter in, 121
  recursive filter in, 120
  staggered PRF in, 122
Multipath propagation, 75, 80, 233,
      234, 286
  anomalous propagation in, 86, 269
  diffractiion in, 76, 88, 265, 278
  divergence factor in, 81, 82
  propagation factor in, 77
  reflection coefficients in, 80
  roughness factor in, 83
MUSIC. *See* Over-the-horizon radars

N

Noise, radar receiver, 64
  effective noise temperature, 66
  noise factor, 66
  thermal noise, 65
  white noise, 65
North filter. *See* Matched flter
Nyquist sampling rate, 146

O

Optical region, 44
OTHR. *See* Over-the-horizon radars
Over-the-horizon radars, 263–79
  backscatter radar (OTH-B), 277
  Cobra Mist, 276
  FMCW in, 274
  frequency, least usable, 265
  frequency, maximum usable, 265
  ionospheric effects on, 266
  Jindalee Operational Radar Network, 264
  Podsolnuh, 279
  pulse Doppler waveform in, 274
  range eqiation of, 271
  simple pulsed waveform in, 274
  skywave, 265
  Steel Yard, 278
  surface-wave, 265
  Woodpecker Blankers, 279

P

Plateau region, 54
Point jamming. *See* Jamming in ECM

Polyphase codes, 108, 110
Probability theory, 293–97
  covariance, 296
  cumulative distribution function, 294
  expected value, 295
  PDF defined, *294*
  PDF properties, *294*
  probability, axioms of, *293*
  probability, conditional, *294*
  probability, joint, *293*
  probabilty density functions, 301–2
  random process, ergodic, 297
  random process, stationary, 297
  random variable, 295
  stanadrd deviation, 295
  variance, 295
Pseudorandom codes, 109
Pulse compression, 131
  Barker codes for, 150
  correlation processing, 145
  Frank code for, 151
  frequency stepping in, 143
  linear frequency modulation in, 138
  phase-coded modulation in, 149
  stretch processing in, 146
Pulse Doppler radar, 2, 113, 124
Pulse repetition frequency (PRF), 12
Pulse repetition interval (PRI)., 12

R

Radar altimeter
  beam-limited altimeter, 240
  defined, 240
  pulse-limited altimeter, 240
  SAR altimeter, 242
Radar clutter
  chaff as, 60
  chaffs, rain, or weather as, 57
  cllutter reflectivity of, 54
  defined, 53
  ghosts or angels as, 61
  land as, 56
  rain as, 58
  sea as, 57
  surface reflectivity of, 55
  volume clutters, 57
Radar cross section (RCS)
  defined, 43
  glint in, 51
  of birds, 61
  of complex targets, 50
  of corner reflectors, 49
  of cylinder, 47
  of dipole, 49

320   Index

of smooth surface, 47
of sphere, 44
of surface clutter, 54
scintillation in, 51
Swerling models of, 51
Radar elements
antenna, 6
display, 6–7
duplexer, 4, 5, 25, 71, 98
transmitter, 5
Radar equations
high PRF SNR, 27
low PRF SNR, 25
of beacon radar, 37
of bistatic radar, 30
of SAR system, 163
of search radar, 29
of tracking radars, 30
simple form of, 21
with pulse compression, 32
with self-screening jamming (SSJ), 34
with stand-off jamming (SOJ), 36
Radar signatures, 18
Radar system losses, 289
attenuation loss, 72
collapsing loss, 71
field degradation, 72
integration loss, 71
limiting loss, 71
operator loss, 71
plumbing loss, 70
random loss, 71
reflection loss, 72
refraction loss, 72
scanning loss, 71
Radar types
air traffic control, 9, 281
bistatic, 7
continuous wave (CW), 8
GOME, 10
MIMO, 7
monostatic, 7
primary, 7, 281
search, 8, 28
secondary, 7
tracking, 8
wind scatterometer (WSC), 10
Random signals, 15, 293, 296
Range measurement, 11
Range resolution, 13
Range-height-indicator, 235
Rayleigh region, 44
Receiver, radar
homodyne, 98
superheterodyne, 100
Region of convergence (ROC), 298

S

SAR. *See* Synthetic aperture radar
Secondary surveillance radar
flight levels in, 283
framing pulse in, 283
FRUIT, 284
Garbling in, 284
Gilham code pattern in, 283
hogtrough SSR antennas, 285
Identification Friend or Foe, 281
interrogation modes in, 282
interrogation pulses in, 282
interrogation, down-link, 288
interrogation, up-link, 287
monopulse technique in, 285
replies, 283
sidelobe blanking, 286
sidelobe suppressions in, 286
sliding window, 283
spoofing, 281
transponder, 281
Second-time-around echoes, 12
Sequential lobing, 174, 179, 186, 235, *See also*
        Taracking radars
Squint angle, 142, 160, 164, 174, 176
Stacked-beam radar, 235, 237, 238
Statistical distribution, clutter, 62
K-distribution, 63
log-normal distribution, 63
Rayleigh distributiion, 63
Weibull distribution, 63
Sunflower. *See* Over-the-horizon radars
Synthetic aperture radar, 155
airborne system, 166
cross-range resolution in, 157
depression angle, 158
endeavor, 156
polarimetric SAR, 165
range resolution along LOS in, 157
range-Doppler algorithm, 162
real aperture radar, 155
side-looking airborne radar (SLAR), 159
signal processing in, 162
space-borne system, 166
spotlight-mode SAR, 165
squint-mode SAR, 165
swath width, 166
two-dimension algorithm, 162

T

Tapped delay-line filter, 121
Threshold detection, 11, 260
Tracking radars
angle accuracy in, 186

Index

conical-scan in, 174
cossover loss in, 177
early gate, 171
late gate, 171
lobe switching in, 173
range accuracy in, 172
split-gate range in, 171
track-while-scan (TWS) in, 187
Transmit-receive (TR), 5
Trigonometric series, 305
Trogonometric identities, 303

U

Uncertainty function, radar, 136

V

V-beam radar, 235, 236
Vertical incidence region, 54

W

Watson-Watt, R., 1
Welti codes, 110
Woodpecker, 264, 279

Z

z-transform, 15, 195, 298
z-transform table, 298